Energy Systems

Series editor

Panos M. Pardalos, Gainesville, USA

More information about this series at http://www.springer.com/series/8368

Stephan Dempe · Vyacheslav Kalashnikov
Gerardo A. Pérez-Valdés
Nataliya Kalashnykova

Bilevel Programming Problems

Theory, Algorithms and Applications to Energy Networks

Springer

Stephan Dempe
Fak. für Mathematik und Informatik
TU Bergakademie Freiberg
Freiberg, Sachsen
Germany

Vyacheslav Kalashnikov
Department of Systems and Industrial Engineering
Tecnologico de Monterrey (ITESM)
Monterrey, NL
Mexico

and

Department of Social Modeling
Central Economics and Mathematics Institute (CEMI)
Russian Academy of Sciences
Moscow
Russian Federation

and

Deparment of Computing
Sumy State University
Sumy
Ukraine

Gerardo A. Pérez-Valdés
Department of Industrial Economics
Norwegian University of Science and Technology (NTNU)
Trondheim
Norway

Nataliya Kalashnykova
Facultad Ciencias Físico-Mathemáticas (FCFM)
Universidad Autónoma de Nuevo León (UANL)
San Nicolás de los Garza
Mexico

ISSN 1867-8998 ISSN 1867-9005 (electronic)
Energy Systems
ISBN 978-3-662-45826-6 ISBN 978-3-662-45827-3 (eBook)
DOI 10.1007/978-3-662-45827-3

Library of Congress Control Number: 2014958578

Springer Heidelberg New York Dordrecht London

Printed on acid-free paper

Springer-Verlag GmbH Berlin Heidelberg is part of Springer Science+Business Media
(www.springer.com)

Preface

Bilevel optimization is a vital field of active research. Depending on its formulation it is part of nonsmooth or nondifferentiable optimization, conic programming, optimization with constraints formulated as generalized equations, or set-valued optimization. The investigation of many practical problems as decision making in hierarchical structures, or situations where the reaction of nature on selected actions needs to be respected, initiated modeling them as bilevel optimization problems. In this way, new theories have been developed with new results obtained.

A first attempt was the use of the Karush-Kuhn-Tucker conditions in situations when they are necessary and sufficient optimality conditions for the lower level problem, or dual problems in case strong duality holds to model the bilevel optimization problem. The result is a special case of the mathematical program with equilibrium constraints (MPEC), or complementarity constraints (MPCC). The latter has motivated the investigation of optimality conditions and the development of algorithms solving such problems. Unfortunately, it has been shown very recently that stationary points of an MPEC need not be related to stationary solutions of the bilevel optimization problem. Because of that, the solution algorithms must select the Lagrange multipliers associated with the lower level problem very carefully. Another option is to avoid the explicit use of Lagrange multipliers resulting in the so-called primal KKT transformation, which is an optimization problem with a generalized equation as the constraint. Violation of the constraint qualifications, often used to verify the optimality conditions and convergence of the solution algorithms, at every feasible point are other challenges for research.

The idea of using the optimal value function of the lower level problem to model the bilevel optimization problem is perhaps self-explanatory. The result yet is a nondifferentiable equality constraint. One promising approach here is based on variational analysis, which is also exploited to verify the optimality conditions for the MPCC. So, bilevel optimization initiated some advances in variational analysis, too.

Applications often force the use of integer variables in the respective models. Besides suitable formulations, mixed-integer bilevel optimization problems renew the question of existence of an optimal solution, leading to the notion of a weak solution. Surprisingly, adding some constraints that are inactive at a global optimum of the continuous bilevel problem, as well as replacing a discrete bilevel problem with its continuous relaxation can destroy the global optimality of a feasible point.

These and other questions are the topic of the first part of the monograph. In the second part, certain applications are carefully investigated, especially a natural gas cash-out problem, an equilibrium problem in a mixed oligopoly, and a toll assignment problem. For these problems, besides the formulation of solution algorithms, results of the first numerical experiments with them are also reported.

Bilevel optimization is a quickly developing field of research with challenging and promising contributions from different topics of mathematics like optimization, as well as from other sciences like economics, engineering, or chemistry. It was not a possible aim of the authors to provide an overview of all the results available in this area. Rather than that, we intended to show some interactions with other topics of research, and to formulate our opinion about some directions for explorations in the future.

Stephan Dempe
Vyacheslav Kalashnikov
Gerardo A. Pérez-Valdés
Nataliya Kalashnykova

Acknowledgment

Kalashnikov V. and Kalashnykova N., was supported by the SEP-CONACYT grant CB-2013-01-221676 (Mexico).

Contents

Chapter 1
Introduction

1.1 The Bilevel Optimization Problem

Since its first formulation by Stackelberg in his monograph on market economy in 1934 [294] and the first mathematical model by Bracken and McGill in 1972 [27] there has been a steady growth in investigations and applications of bilevel optimization. Formulated as a hierarchical game (the Stackelberg game), two decision makers act in this problem. The so-called leader minimizes his objective function subject to conditions composed (in part) by optimal decisions of the so-called follower. The selection of the leader influences the feasible set and the objective function of the follower's problem, who's reaction has strong impact on the leader's payoff (and feasibility of the leader's initial selection). Neither player can dominate the other one completely. The bilevel optimization problem is the leader's problem, formulated mathematically using the graph of the solution set of the follower's problem.

The bilevel optimization has been shown to be $\mathscr{N}\mathscr{P}$-hard, even verification of local optimality for a feasible solution is in general $\mathscr{N}\mathscr{P}$-hard, often used constraint qualifications are violated in every feasible point. This makes the computation of an optimal solution a challenging task.

Bilevel optimization problems are nonconvex optimization problems, tools of variational analysis have successfully been used to investigate them. The results are a larger number of necessary optimality conditions, some of them are presented in Chap. 3 of this monograph.

A first approach to investigate bilevel optimization problems is to replace the lower level problem by its (under certain assumptions necessary and sufficient) optimality conditions, the Karush-Kuhn-Tucker conditions. This replaces the bilevel optimization problem by a so-called mathematical program with complementarity conditions (MPCC). MPCCs are nonconvex optimization problems, too. Algorithms solving them compute local optimal solutions or stationary points. Recently it has been shown that local optimal solutions of an MPCC need not to be related to local optimal solutions of the corresponding bilevel optimization problem, new attempts

S. Dempe et al., *Bilevel Programming Problems*, Energy Systems,
DOI 10.1007/978-3-662-45827-3_1

for the development of solution approaches for the bilevel problem are necessary. Some results can be found in different chapters of this monograph.

The existence of an optimal solution, verification of necessary optimality conditions, and convergence of solution algorithms are strongly related to continuity of certain set-valued mappings. These properties can often not be guaranteed for mixed-discrete bilevel optimization problems. This is perhaps one reason for the small number of references on those class of problems. But, applied problems do often lead to mixed-integer bilevel problems. One such problem is investigated in Chap. 6. Focus on mixed-discrete bilevel optimization including the notion of a weak optimal solution and some ideas for solving these problems is in Chap. 5.

The solution set of an optimization problem does in general not reduce to a singleton, leading to the task of selecting a "good" optimal solution. If the quality of an optimal solution is measured by a certain function, this function needs to be minimized on the solution set of a second optimization problem. This is the so-called simple bilevel optimization problem, investigated in Chap. 4.

Interest in bilevel optimization is largely driven by applications. Two of them are investigated in details in Chaps. 6 and 7. The gas cash-out problem is a bilevel optimization problem with one Boolean variable, formulated using nondifferentiable functions. Applying results of the previous chapters, after some transformations and the formulation of an approximate problem, a model is obtained which can efficiently be solved. The obtained solutions have successfully be used in practice.

Due to its complexity, the dimension of bilevel optimization models is of primarily importance for solving them. Large-scale problems can perhaps not be solved in reasonable time. But, e.g. the investigation of stochastic bilevel optimization problems using methods to approximate the probability distributions leads to large-scale problems and not all data are deterministic ones on many applications. This makes ideas to reduce the number of variables important. Such ideas are the topic of Chap. 8.

1.2 Possible Transformations into a One-Level Problem

Bilevel optimization problems are optimization problems where the feasible set is determined (in part) using the graph of the solution set mapping of a second parametric optimization problem. This problem is given as

$$\min_y \{f(x, y) : g(x, y) \le 0,\ y \in T\}, \tag{1.1}$$

where $f : \mathbb{R}^n \times \mathbb{R}^m \to \mathbb{R}$, $g : \mathbb{R}^n \times \mathbb{R}^m \to \mathbb{R}^p$, $T \subseteq \mathbb{R}^m$ is a (closed) set.

Let $Y : \mathbb{R}^n \rightrightarrows \mathbb{R}^m$ denote the *feasible set mapping*:

$$Y(x) := \{y : g(x, y) \le 0\},$$

$$\varphi(x) := \min_y \{f(x, y) : g(x, y) \le 0,\ y \in T\}$$

the *optimal value function*, and $\Psi : \mathbb{R}^n \rightrightarrows \mathbb{R}^m$ the *solution set mapping* of the problem (1.1) for a fixed value of x:

$$\Psi(x) := \{y \in Y(x) \cap T : f(x, y) \leq \varphi(x)\}.$$

Let

$$\mathbf{gph}\,\Psi := \{(x, y) \in \mathbb{R}^n \times \mathbb{R}^m : y \in \Psi(x)\}$$

be the graph of the mapping Ψ. Then, the bilevel optimization problem is given as

$$\text{“}\min_x\text{”}\{F(x, y) : G(x) \leq 0,\ (x, y) \in \mathbf{gph}\,\Psi,\ x \in X\}, \tag{1.2}$$

where $F : \mathbb{R}^n \times \mathbb{R}^m \to \mathbb{R}$, $G : \mathbb{R}^n \to \mathbb{R}^q$, $X \subseteq \mathbb{R}^n$ is a closed set.

Problem (1.1), (1.2) can be interpreted as an hierarchical game of two decision makers (or players) which make their decisions according to an hierarchical order. The first player (which is called the *leader*) makes his selection first and communicates it to the second player (the so-called *follower*). Then, knowing the choice of the leader, the follower selects his *response* as an optimal solution of problem (1.1) and gives this back to the leader. Thus, the leader's task is to determine a best decision, i.e. a point $\widehat{x}$ which is feasible for the problem (1.2): $G(\widehat{x}) \leq 0$, $\widehat{x} \in X$, minimizing together with the response $\widehat{y} \in \Psi(\widehat{x})$ the function $F(x, y)$. Therefore, problem (1.1) is called the follower's problem and (1.2) the leader's problem. Problem (1.2) is the *bilevel optimization problem*.

Example 1.1 In case of a linear bilevel optimization problem with only one upper and one lower level variables, where all functions F, f, g_i are (affine) linear functions, the bilevel optimization problem is illustrated in Fig. 1.1. Here, $G(x) \equiv 0$ and the set $\{(x, y) : g(x, y) \leq 0\}$ of feasible points for all values of x is the hatched area. If x is fixed to $x0$ the feasible set of the lower level problem (1.1) is the set of points $(x0, y)$ above $x0$. Now, if the lower level objective function $f(x, y) = -y$ is minimized on this set, the optimal solution of the lower level problem on the thick line is obtained. Then, if x is moved along the x-axis, the thick line as the set of feasible solutions of the upper level problem arises. In other words, the thick line equals the $\mathbf{gph}\,\Psi$ of the solution set mapping of the lower level problem. This is the feasible set of the upper level (or bilevel) optimization problem. Then, minimizing the upper level objective function on this set, the (in this case unique) optimal solution of the bilevel optimization problem is obtained as indicated in Fig. 1.1. □

It can be seen in Fig. 1.1 that the bilevel optimization problem is a nonconvex (since $\mathbf{gph}\,\Psi$ is nonconvex) optimization problem. Hence, local optimal solutions and also stationary points can appear.

Example 1.2 Consider the problem

$$\text{“}\min_x\text{”}\{x^2 + y : y \in \Psi(x)\},$$

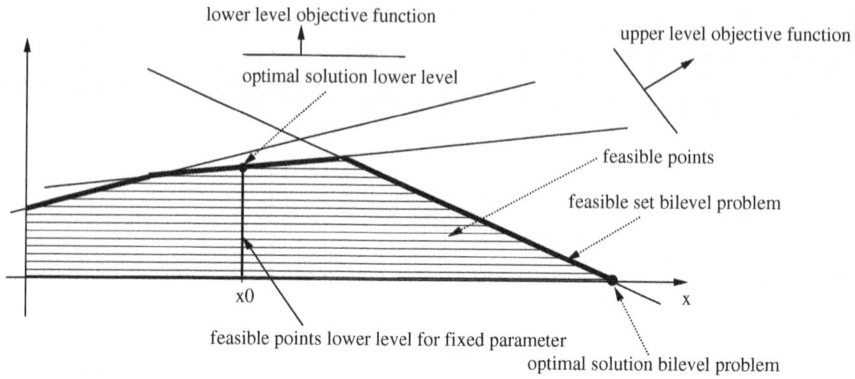

Fig. 1.1 Illustration of the linear bilevel optimization problem

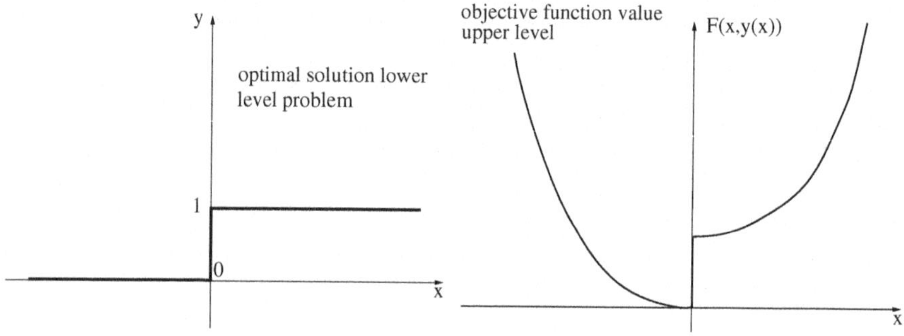

Fig. 1.2 Mapping to be "minimized" in Example 1.2

where

$$\Psi(x) := \operatorname*{Argmin}_y\{-xy : 0 \le y \le 1\}.$$

Then, the graph of the mapping Ψ is given in the figure on the left hand side of Fig. 1.2 and the graph of the mapping $x \mapsto F(x, \Psi(x))$ of the upper level objective function is plotted in the figure on the right-hand side. Note, that this is not a function and that its minimum is unclear since its existence depends on the response $y \in \Psi(x)$ of the follower on the leader's selection at $x = 0$. If the solution $y = 0$ is taken for $x = 0$, an optimal solution of the bilevel optimization problem exists. This is the *optimistic bilevel optimization problem* introduced below. In all other cases, the minimum does not exist, the infimum function value of the upper level objective function is again zero but it is not attained. If $y = 1$ is taken, the so-called *pessimistic bilevel optimization problem* arises. □

Hence, strictly speaking, the problem (1.2) is not well-posed in the case that the set $\Psi(x)$ is not a a singleton for some x, the mapping $x \mapsto F(x, y(x))$ is not a function. This is implied by an ambiguity in the computation of the upper level objective function value, which is rather an element in the set $\{F(x, y) : y \in \Psi(x)\}$. We have used quotation marks in (1.2) to indicate this ambiguity. To overcome such an unpleasant situation, the leader has a number of possibilities:

1. The leader can assume that the follower is willing (and able) to cooperate. In this case, the leader can take that solution within the set $\Psi(x)$ which is a best one with respect to the upper level objective function. This leads then to the function

$$\varphi_o(x) := \min\{F(x, y) : y \in \Psi(x)\} \tag{1.3}$$

to be minimized on the set $\{x : G(x) \leq 0, \ x \in X\}$. This is the optimistic approach leading to the *optimistic bilevel optimization problem*. The function $\varphi_o(x)$ is called *optimistic solution function*. Roughly speaking, this problem is closely related to the problem

$$\min_{x,y}\{F(x, y) : G(x) \leq 0, \ (x, y) \in \mathbf{gph}\, \Psi, \ x \in X\}. \tag{1.4}$$

If the point $\overline{x}$ is a local minimum of the function $\varphi_o(\cdot)$ on the set

$$\{x : G(x) \leq 0, \ x \in X\}$$

and $\overline{y} \in \Psi(\overline{x})$, then the point $(\overline{x}, \overline{y})$ is also a local minimum of problem (1.4). The opposite implication is in general not correct, as the following example shows:

Example 1.3 Consider the problem of minimizing the function $F(x, y) = x$ subject to $x \in [-1, 1]$ and $y \in \Psi(x) := \operatorname*{Argmin}_y\{xy : y \in [0, 1]\}$. Then,

$$y(x) \in \begin{cases} [0, 1] & \text{for } x = 0, \\ \{1\} & \text{for } x < 0, \\ \{0\} & \text{for } x > 0. \end{cases}$$

Hence, the point $(\overline{x}, \overline{y}) = (0, 0)$ is a local minimum of the problem

$$\min_{x,y}\{x : x \in [-1, 1], \ y \in \Psi(x)\}$$

since, for each feasible point (x, y) with $\|(x, y) - (\overline{x}, \overline{y})\| \leq 0.5$ we have $x \geq 0$. But, the point $\overline{x}$ does not minimize the function $\varphi_o(x) = x$ on $[-1, 1]$ locally. □

For more information about the relation between both problems, the interested reader is referred to Dempe [52].

2. The leader has no possibility to influence the follower's selection neither he/she has an intuition about the follower's choice. In this case, the leader has to accept

the follower's opportunity to take a worst solution with respect to the leader's objective function and he/she has to bound the damage resulting from such an unpleasant selection. This leads to the function

$$\varphi_p(x) := \max\{F(x, y) : y \in \Psi(x)\} \tag{1.5}$$

to be minimized on the set $\{x : G(x) \leq 0,\ x \in X\}$:

$$\min\{\varphi_p(x) : G(x) \leq 0,\ x \in X\} \tag{1.6}$$

This is the pessimistic approach resulting in the *pessimistic bilevel optimization problem*. The function $\varphi_p(x)$ is the *pessimistic solution function*. This problem is often much more complicated than the optimistic bilevel optimization problem, see Dempe [52].
In the literature there is also another pessimistic bilevel optimization problem. To describe this problem consider the bilevel optimization problem with connecting upper level constraints and an upper level objective function depending only on the upper level variable x:

$$\text{“}\min_x\text{”}\{F(x) : G(x, y) \leq 0,\ y \in \Psi(x)\}. \tag{1.7}$$

In this case, a point x is feasible if there exists $y \in \Psi(x)$ such that $G(x, y) \leq 0$, which can be written as

$$\min_x\{F(x) : G(x, y) \leq 0 \text{ for some } y \in \Psi(x)\}.$$

Now, if the quantifier $\exists$ is replaced by $\forall$ we derive a second pessimistic bilevel optimization problem

$$\min_x\{F(x) : G(x, y) \leq 0 \text{ for all } y \in \Psi(x)\}. \tag{1.8}$$

This problem has been investigated in Wiesemann et al. [316]. The relations between (1.8) and (1.6) should to be investigated in future.

3. The leader is able to predict a selection of the follower: $y(x) \in \Psi(x)$ for all x. If this function is inserted into the upper level objective function, this leads to the problem

$$\min_x\{F(x, y(x)) : G(x) \leq 0,\ x \in X\}.$$

Such a function $y(\cdot)$ is called a selection function of the point-to-set mapping $\Psi(\cdot)$. Hence, we call this approach the *selection function approach*. One special case of this approach arises if the optimal solution of the lower level problem is unique for all values of x. It is obvious that the optimistic and the pessimistic problems are special cases of the selection function approach.

Even under restrictive assumptions (as in the case of linear bilevel optimization or if the follower's problem has a unique optimal solution for all x), the function $y(\cdot)$ is in general not differentiable. Hence, the bilevel optimization problem is a nonsmooth optimization problem.

Definition 1.1 A point $\overline{z} \in Z$ is a local optimal solution of the optimization problem

$$\min\{w(z) : z \in Z\}$$

provided that there is a positive number $\varepsilon > 0$ such that

$$w(z) \geq w(\overline{z}) \; \forall \, z \in Z \text{ satisfying } \|z - \overline{z}\| \leq \varepsilon.$$

$\overline{z}$ is a global optimal solution of this problem if ε can be taken arbitrarily large.

This well-known notion of a (local) optimal solution can be applied to the bilevel optimization problems and, using e.g. Weierstraß Theorem we obtain that problem (1.4) has a global optimal solution if the function F is continuous and the set $Z := \{(x, y) : G(x) \leq 0, \ (x, y) \in \mathbf{gph}\, \Psi, \ x \in X\}$ is not empty and compact. If this set is not bounded but only a nonempty and closed set and the function F is continuous and coercive (i.e. $F(x, y)$ tends to infinity for $\|(x, y)\|$ tending to infinity) problem (1.4) has a global optimal solution, too. For closedness of the set Z we need closedness of the graph of the solution set mapping of the lower level problem. We will come back to this issue in Chap. 3, Theorem 3.3.

With respect to problem

$$\min\{\varphi_0(x) : G(x) \leq 0, \ x \in X\} \tag{1.9}$$

existence of an optimal solution is guaranteed if the function $\varphi_0(\cdot)$ is lower semicontinuous (which means that $\liminf_{x \to x^0} \varphi_0(x) \geq \varphi_0(x^0)$ for all x^0) and the set Z is not empty and compact by an obvious generalization of the Weierstraß Theorem. Again boundedness of this set can be replaced by coercivity. Lower semicontinuity of the function is again an implication of upper semicontinuity of the mapping $x \mapsto \Psi(x)$, see for instance Bank et al. [8] in combination with Theorem 3.3. It is easy to see that a function $w(\cdot)$ is lower semicontinuous if and only if its epigraph $\operatorname{epi} w := \{(z, \alpha) : w(z) \leq \alpha\}$ is a closed set.

Example 1.2 showed already that an optimal solution of the problem (1.6) does often not exist. Its existence is guaranteed e.g. if the function $\varphi_p(\cdot)$ is lower semicontinuous and the set Z is not empty and compact (Lucchetti et al. [207]). But, for lower semicontinuity of the function $\varphi_p(\cdot)$ lower semicontinuity of the solution set mapping $x \mapsto \Psi(x)$ is needed which can often only be shown if the optimal solution of the lower level problem is unique (see Bank et al. [8]).

If an optimal solution of problem (1.6) does not exist we can aim to find a weak (global) optimum by replacing the epigraph of the objective function by its closure: Let $\overline{\varphi}_p$ be defined such that

$$\text{epi } \overline{\varphi}_p = \text{cl epi } \varphi_p.$$

Then, a local or global optimal solution of the problem

$$\min\{\overline{\varphi}_p(x) : G(x) \leq 0,\ x \in X\} \tag{1.10}$$

is called a (local or global) weak solution of the pessimistic bilevel optimization problem (1.6). Note that $\overline{\varphi}_p(x^0) = \liminf_{x \to x^0} \varphi_p(x)$. The function $\overline{\varphi}_p(\cdot)$ is the largest lower semicontinuous function which is not larger than $\varphi_p(\cdot)$, see Fanghänel [105]. Hence, a weak global solution of problem (1.6) exists provided that $Z \neq \emptyset$ is compact.

1.3 An Easy Bilevel Optimization Problem: Continuous Knapsack Problem in the Lower Level

To illustrate the optimistic/pessimistic approaches to the bilevel optimization problem consider

$$\text{“}\min_b\text{”}\{d^\top y + fb : b_u \leq b \leq b_o, \ldots, y \in \Psi(b)\}, \tag{1.11}$$

where

$$\Psi(b) := \operatorname*{Argmin}_y\{c^\top y : a^\top y \geq b,\ 0 \leq y_i \leq 1\ \forall\, i = 1, \ldots, n\}$$

and $a,\ c,\ d \in \mathbb{R}^n_+$. Note that the upper level variable is called b in this problem. Assume that the indices are ordered such that

$$\frac{c_i}{a_i} \leq \frac{c_{i+1}}{a_{i+1}},\ i = 1, 2, \ldots, n-1.$$

Then, for fixed $b \in \left\{b : \sum_{i=1}^{k-1} a_i \leq b \leq \sum_{i=1}^{k} a_i\right\}$, the point

$$y_i = \begin{cases} 1, & i = 1, \ldots, k-1 \\ \frac{b - \sum_{j=1}^{k-1} a_j}{a_k}, & i = k \\ 0, & i = k+1, \ldots, n \end{cases} \tag{1.12}$$

is an optimal solution of the lower level problem. Its optimal function value in the lower level is

$$\varphi(b) = \sum_{i=1}^{k-1} c_i + \frac{c_k}{a_k}(b - \sum_{j=1}^{k-1} a_j),$$

which is an affine linear function of b. The function $b \mapsto \varphi(b)$ is convex. The optimal solution of the lower level problem is unique provided that $\frac{c_k}{a_k}$ is unique in the set $\left\{\frac{c_i}{a_i} : i = 1, \ldots, n\right\}$. Otherwise, the indices $i \in \left\{j : \frac{c_i}{a_i} = \frac{c_k}{a_k}\right\}$ need to be ordered such that

$$d_t \leq d_{t+1} : t, t+1 \in \left\{j : \frac{c_i}{a_i} = \frac{c_k}{a_k}\right\}$$

for the optimistic and

$$d_t \geq d_{t+1} : t, t+1 \in \left\{j : \frac{c_i}{a_i} = \frac{c_k}{a_k}\right\}$$

for the pessimistic approaches. As illustration consider the following example:

Example 1.4 The lower level problem is

$$\begin{array}{l} 10x_1 + 30x_2 + 8x_3 + 60x_4 + 4x_5 + 16x_6 + 32x_7 + 30x_8 + 120x_9 + 6x_{10} \rightarrow \min \\ 5x_1 + 3x_2 + 2x_3 + 5x_4 + x_5 + 8x_6 + 4x_7 + 3x_8 + 6x_9 + 3x_{10} \geq b \\ \forall i : \quad 0 \leq \quad y_i \quad \leq \quad 1, \end{array}$$

and the upper level objective function is

$$\begin{aligned} F(x, f) = {} & 20x_1 + 15x_2 - 24x_3 + 20x_4 - 40x_5 \\ & + 80x_6 - 32x_7 - 60x_8 - 12x_9 - 60x_{10} + fb. \end{aligned}$$

This function is to be minimized subject to $y \in \Psi(b)$ and b is in some closed interval $[b_u, b_o]$. Note that the upper level variable is b and the lower level one is x in this example.

Using the above rules we obtain the the following sequence of the indices in the optimistic approach:

$i =$	10	1	6	5	3	7	8	2	4	9
$\frac{c_i}{a_i} =$	2	2	2	4	4	8	10	10	12	20
$\frac{d_i}{a_i} =$	−20	4	10	−40	−12	−8	−20	5	4	−2

Using the pessimistic approach we get

$i =$	6	1	10	3	5	7	2	8	4	9
$\frac{c_i}{a_i} =$	2	2	2	4	4	8	10	10	12	20
$\frac{d_i}{a_i} =$	10	4	−20	−40	−12	−8	−20	5	4	−2

Fig. 1.3 The optimistic and pessimistic objective value functions in Example 1.4, see Winter [317]

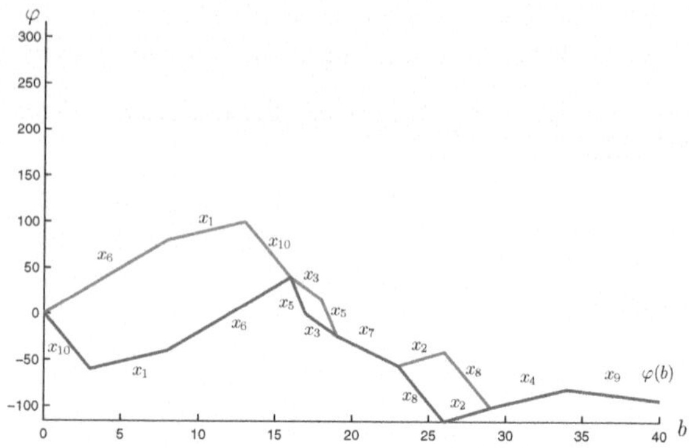

Now, the functions $\varphi_o(b)$ and $\varphi_p(b)$ are plotted in Fig. 1.3. The upper function is the pessimistic value function, computed according to (1.5), the lower one the optimistic value function, cf. (1.3).

Both functions are continuous, but not convex. Local optima of both functions can either be found at $b \in (b_u, b_o)$ or at points

$$b \in \left\{ \sum_{i \in I} a_i : I \subseteq \{1, 2, \ldots, n\} \right\}.$$

Note that local optima can be found at vertices of the set

$$\left\{ (x, b) : b_u \leq b \leq b_0,\ 0 \leq x_i \leq 1,\ i = 1, \ldots, n,\ \sum_{i=1}^{n} a_i x_i = b \right\}. \qquad \square$$

1.4 Short History of Bilevel Optimization

The history of bilevel optimization dates back to H.v. Stackelberg who in 1934 formulated in the monograph [294] an hierarchical game of two players now called Stackelberg game. The formulation of the bilevel optimization problem goes back to Bracken and McGill [27], the notion "Bilevel Programming" has been coined probably by Candler and Norton [39], see also Vicente [305]. With the beginning of the 80s of the last century a very intensive investigation of bilevel optimization started. A number of monographs, see e.g. Bard [10], Shimizu et al. [288] and Dempe [52], edited volumes, see Dempe and Kalashnikov [57], Talbi [297] and Migdalas et al. [231] and (annotated) bibliographies, see e.g. Vicente and Calamai [306], Dempe [53] have been published in that field.

One possibility to investigate bilevel optimization problems is to transform them into one-level (or ordinary) optimization problems. This will be the topic of Chap. 3.

In the first years linear bilevel optimization problems (where all the problem functions are affine linear and the sets X and T equals the whole spaces) have been transformed using linear optimization duality or, equivalently, the Karush-Kuhn-Tucker conditions for linear optimization. Applying this approach, solution algorithms have been suggested, see e.g. Candler and Townsley [40]. The transformed problem is a special case of a mathematical program with equilibrium constraints MPEC (now sometimes called mathematical program with complementarity constraints MPCC). We can call this the KKT transformation of the bilevel optimization problem. This approach is also possible for convex parametric lower level problems satisfying some regularity assumption.

General MPCC's have been the topic of some monographs, see e.g. Luo et al. [208] and Outrata et al. [259]. Solution algorithms for MPCC's (see for instance Outrata et al. [259], Demiguel et al. [48], Leyffer et al. [201], and many others) have been suggested also for solving bilevel optimization problems.

Since MPCC's are nonconvex optimization problems, solution algorithms will hopefully compute local optimal solutions of the MPCCs. Thus, it is interesting if a local optimal solution of an the KKT transformation of a bilevel optimization problem is related to a local optimal solution of the latter problem. This has been the topic of the paper [55] by Dempe and Dutta. We will come back to this in Chap. 3.

Later on, the selection function approach to bilevel optimization has been investigated in the case when the optimal solution of the lower level problem is uniquely determined and strongly stable in the sense of Kojima [191]. Then, under some assumptions, the optimal solution of the lower level problem is a PC^1-function, see Ralph and Dempe [265] and Scholtes [283] for the definition and properties of PC^1-functions. This can then be used to determine necessary and sufficient optimality conditions for bilevel optimization (see Dempe [50]).

Using the optimal value function $\varphi(x)$ of the lower level problem (1.1), the bilevel optimization problem (1.4) can be replaced with

$$\min_{x,y}\{F(x, y) : G(x) \leq 0,\ g(x, y) \leq 0,\ f(x, y) \leq \varphi(x),\ x \in X\}.$$

This is the *optimal value transformation*. Since the optimal value function is nonsmooth even under restrictive assumptions, this is a nonsmooth, nonconvex optimization problem. Using nonsmooth analysis (see e.g. Mordukhovich [241, 242], Rockafellar and Wets [274]), optimality conditions for the optimal value transformation can be obtained (see e.g. Outrata [260], Ye and Zhu [324], Dempe et al. [56]).

Nowadays, a large number of PhD thesis have been written on bilevel optimization problems, very different types of (necessary and sufficient) optimality conditions can be found in the literature, the number of applications is huge and both exact and heuristic solution algorithms have been suggested.

1.5 Applications of Bilevel Optimization

1.5.1 Optimal Chemical Equilibria

In the monograph Dempe [52] the following application of bilevel optimization in the chemical industry is formulated:

In producing substances by chemical reactions we have often to answer the question of how to compose a mixture of chemical substances such that

- the substance we like to produce really arises as a result of the chemical reactions in the reactor and
- the amount of this substance should clearly be as large as possible or some other (poisonous or etching) substance is desired to be vacuous or at least of a small amount.

It is possible to model this problem as a bilevel optimization problem where the first aim describes the lower level problem and the second one is used to motivate the upper level objective function.

Let us start with the lower level problem. Although the chemists are technically not able to observe in situ the single chemical reactions at higher temperatures, they described the final point of the system by a convex optimization problem. In this problem, the entropy functional $f(y, p, T)$ is minimized subject to the conditions that the mass conservation principle is satisfied and masses are not negative. Thus, the obtained equilibrium state depends on the pressure p and the temperature T in the reactor as well as on the masses x of the substances which have been put into the reactor:

$$\sum_{i=1}^{N} c_i(p, T) y_i + \sum_{i=1}^{G} y_i \ln \frac{y_i}{z} \to \min_y$$
$$z = \sum_{j=1}^{G} y_j, \; Ay = \overline{A}x, \; y \geq 0,$$

where $G \leq N$ denotes the number of gaseous and N the total number of reacting substances. Each row of the matrix A corresponds to a chemical element, each column to a substance. Hence, a column gives the amount of the different elements in the substances; y is the vector of the masses of the substances in the resulting chemical equilibrium whereas x denotes the initial masses of substances put into the reactor; $\overline{A}$ is a submatrix of A consisting of the columns corresponding to the initial substances. The value of $c_i(p, T)$ gives the chemical potential of a substance which depends on the pressure p and the temperature T (Smith and Missen [291]). Let $y(p, T, x)$ denote the unique optimal solution of this problem. The variables p, T, x can thus be considered as parameters for the chemical reaction. The problem is now that there exists some desire about the result of the chemical reactions which should be reached as best as possible, as e.g. the goal that the mass of one substance should be as large or as small as possible in the resulting equilibrium. To reach this goal the parameters

p, T, x are to be selected such that the resulting chemical equilibrium satisfies the overall goal as best as possible (Oeder [258]):

$$\langle c, y\rangle \to \min_{p,T,x}$$
$$(p, T, x) \in Y,\ y = y(p, T, x).$$

1.5.2 Optimal Traffic Tolls

In more and more regions of the world, traffic on the streets is due to tolls. To model such a problem, use a directed graph $G = (V, E)$ where the nodes $v \in V := \{1, 2, \ldots, n\}$ stand for the junctions in some region and the directed edges (or arcs) $(i, j) \in E \subset V \times V$ are used to implement the streets leading from junction i to junction j. Then, the graph is used to model the map of the streets in a certain region. The streets are modeled as one-way roads here. If a street can be passed in both directions, there are opposite directed edges in the graph. The streets are assumed to have certain capacities which are modeled as a function $u : E \to \mathbb{R}$ and the cost (or time) to pass one street by a driver is given by a second function $c : E \to \mathbb{R}$. We assume here for simplicity that the costs are independent of the flow on the street. Assume further that there is a set T of pairs of nodes $(q, s) \in V \times V$ for which there is a certain demand d_{qs} of traffic running from the origin q to the destination nodes s, $(q, s) \in T$. Then, if x_e^{qs} is used to denote the part of the traffic with respect to the origin-destination pair (O-D pair in short) $(q, s) \in T$ using the street $e = (i, j) \in E$, the problem of computing the system optimum for the traffic can be modeled as a multicommodity flow problem (Ahuja et al. [1]):

$$\sum_{(q,s)\in T}\sum_{e\in E} c_e x_e^{qs} \longrightarrow \min \tag{1.13}$$

$$x_e + \sum_{(q,s)\in T} x_e^{qs} = u_e \quad \forall e \in E \tag{1.14}$$

$$\sum_{e\in O(j)} x_e^{qs} - \sum_{e\in I(j)} x_e^{qs} = \begin{cases} d_{qs}, & j = q \\ 0, & j \in V \setminus \{q, s\} \\ -d_{qs}, & j = s \end{cases} \quad \forall (q, s) \in T \tag{1.15}$$

$$x_e, x_e^{qs} \geq 0 \quad \forall (q, s) \in T, \forall e \in E. \tag{1.16}$$

Here $O(j)$ and $I(j)$ denote the set of arcs e having the node j as tail or as head, respectively, and x_e is a slack variable for arc e, x is used to abbreviate all the lower level variables (including slack variables).

Now, assume that the cost for passing a street does also depend on toll costs c_e^t which are added to the cost c_e for passing a street. Then, the objective function (1.13) is changed to

$$\sum_{(q,s)\in T}\sum_{e\in E}(c_e + c_e^t)x_e^{qs} \longrightarrow \min. \tag{1.17}$$

Let $\Psi(c^t)$ denote the set of optimal solutions of the problem of minimizing the function (1.17) subject to (1.14)–(1.16), then the problem of computing best toll costs is

$$f(c^t, x) \to \min \text{ s.t. } x \in \Psi(c^t),\ \ c^t \in C. \tag{1.18}$$

Here, C is a set of admissible toll costs and the objective function $f(c^t, x)$ can be used to express different aims, as e.g.:

1. Maximizing the revenue. In this case it makes sense to assume that, for each origin-destination pair $(q, s) \in T$ there is one (directed) path from q to s in the graph which is free of tolls (Didi-Biha et al. [88] and other references),
2. Reducing traffic in ecologically exposed areas (Dempe et al. [58]) or
3. Forcing truck drivers to use trains from one loading station to another one (Wagner [310]).

1.5.3 Optimal Operation Control of a Virtual Power Plant

Müller and Rehkopf investigated in the paper [247] the optimal control of a virtual power plant. This power plant consists of a number of decentralized micro-cogeneration units located in the residential houses of their owners and use natural gas to produce heat and electricity. This is a very efficient possibility for heat and energy supply. Moreover, the micro-cogeneration units can produce much more electricity than used in the houses and the superfluous electricity is injected into the local electricity grid. For that, the residents get a compensation helping them to cover the costs of the micro-cogeneration units. To realize this, the decentralized micro-cogeneration units are joined into a virtual power plant (VP) which collects the superfluous electricity from the decentralized suppliers and sells it on the electricity market. For the VP, which is a profit maximizing unit, it is sensible to sell the electricity to the market in time periods when the revenue on the market is high. Hence, the owner of the VP wants to ask the decentralized suppliers to inject power into the system when the national demand for electricity is large. For doing this he can apply ideas from principal-agent theory establishing an incentive system to motivate the suppliers to produce and inject power into the grid in the desired time periods. In this sense, the owners of the decentralized units are the followers (agents) and the owner of the VP is the leader (principal).

To derive a mathematical model for the VP consider the owners of the micro-cogeneration units first. It is costly to switch the units on implying that it makes sense to restrict the number of time units when the system is switched on. This and failure probability imply that a producing unit should keep working for a minimum time length and the time the system is switched off is also bounded from below after turning it off. To abbreviate these and perhaps other restrictions for the decentralized systems (which are in fact linear inequalities), we use the system $Ay \leq b$.

Under these conditions, the owner of the decentralized systems has to minimize the costs for power and heat generation depending on the costs of the used natural gas, the expenses for switching on the unit and the prices for buying and selling power. Let this function be abbreviated as $f(y)$.

Now, assume that the owners of the decentralized micro-cogeneration units sell their superfluous electricity to the VP which establishes an incentive system to control time and amount of the injected power. Let z denote the premium payed for the power supply. This value is, of course, bounded from below by some values, depending on the expenses of the decentralized units resulting from switching them on and from additional costs of natural gas. Moreover, since the costs for power and heat generation do also depend on the premium payed, the owner of the decentralized units now minimizes a function $\widetilde{f}(y, z)$ subject to the constraints $Ay \leq b$ and some (linear) conditions relating the received bonuses to the working times of the power units. Let $\Psi(z)$ denote the set of solutions of the owners of the decentralized units (production periods of the units, delivered amount of power) depending on the premium z.

Then, the upper level problem of the VP consists of maximizing the revenue from the electricity market for the power supply minus the bonuses payed to the subunits. This function is maximized subject to restrictions from the above conditions that the bonus payed is bounded by some unit costs in the lower level.

1.5.4 Spot Electricity Market with Transmission Losses

In the paper Aussel et al. [5] deregulated spot electricity markets are investigated. This problem is modeled as a generalized Nash equilibrium problem, where each player solves a bilevel optimization problem. To formulate the problem assume that a graph $G = G(V, E)$ is given where each agent (or player) is located at one of the nodes $i \in V$. The arcs E are the electricity lines. The demand D_i at each node is supposed to be known and also that the real cost for generating q_i units of electricity at node i equals $A_i q_i + B_i q_i^2$.

Now, assume that there is an independent system operator (ISO) in the electricity network who is responsible for the trade of electricity. Moreover, each agent bids his cost $b_i q_i^2 + a_i q_i$ of producing q_i units of electricity and his demand to the ISO, who distributes the electricity between the agents. The goal of the ISO is to minimize the total bid costs subject to satisfaction of the demand of the agents. Assume that $L_{ij} t_{ij}^2$ are the thermal losses along $(i, j) \in E$ which are covered equally between agents at nodes i and j if t_{ij} is the amount of electricity delivered along $(i, j) \in E$. Then, the problem of the ISO reads as

$$\sum_{i=1}^{|V|} \left(b_i q_i^2 + a_i q_i\right) \to \min_{q,t}$$

$$q_i \geq 0, \quad i \in V$$

$$q_i - \sum_{k:(i,k)\in E} \left(t_{ik} + 0.5 L_{ik} t_{ik}^2\right) + \sum_{k:(k,i)\in E} \left(t_{ki} - 0.5 L_{ki} t_{ki}^2\right) \geq D_i, \quad i \in V$$

$$t_{ij} \geq 0, \quad (i,j) \in E.$$

Let $Q(a, b)$ denote the set of optimal solutions of this problem depending on the bid vectors announced by the producers. Then, the agents intend to maximize their profit which equals the difference between the real and the bid function for the production subject to the decision of the ISO. This leads to the following problem:

$$\left(b_i q_i^2 + a_i q_i\right) - \left(B_i q_i^2 + A_i q_i\right) \to \max_{a_i, b_i, q, t}$$

$$\underline{A}_i \leq a_i \leq \overline{A}_i$$

$$\underline{B}_i \leq b_i \leq \overline{B}_i$$

$$(q, t) \in Q(a, b).$$

This is a bilevel optimization problem with multiple leaders where the leaders act according to a Nash equilibrium.

1.5.5 Discrimination Between Sets

In many situations as e.g. in robot control, character and speech recognition, in certain finance problems as bank failure prediction and credit evaluation, in oil drilling, in medical problems as for instance breast cancer diagnosis, methods for discriminating between different sets are used for being able to find the correct decisions implied by samples having certain characteristics (cf. DeSilets et al. [86], Hertz et al. [144], Mangasarian [215, 216], Shavlik et al. [286], Simpson [289]). In doing so, a mapping $\mathscr{T}_0$ is used representing these samples according to their characteristics as points in the input space (usually the n-dimensional Euclidean space), see Mangasarian [215]. Assume that this leads to a finite number of different points. Now, these points are classified according to the correct decisions implied by their originals. This classification can be considered as a second mapping $\mathscr{T}_1$ from the input space into the output space given by the set of all possible decisions. This second mapping introduces a partition of the input space into a certain number of disjoint subsets such that all points in one and the same subset are mapped to the same decision (via its inverse mapping). For being able to determine the correct decision implied by a new sample we have to find that partition of the input space without knowing the mapping $\mathscr{T}_1$.

Consider the typical case of discriminating between two disjoint subsets $\mathscr{A}$ and $\mathscr{B}$ of the input space $\mathbb{R}^n$ [215]. Then, for approximating this partition, piecewise

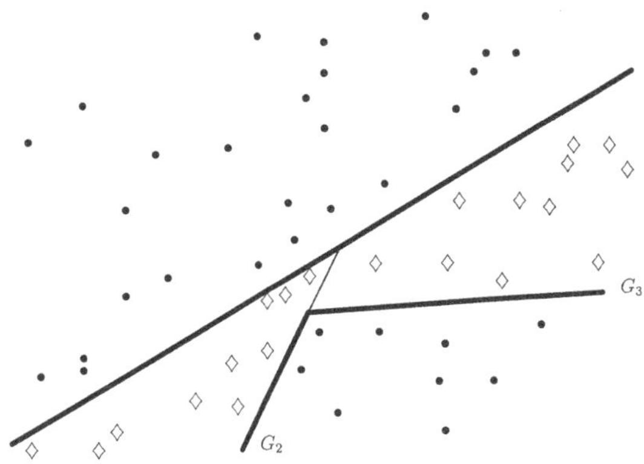

Fig. 1.4 Splitting of R^2 into three subsets each containing points of one of the sets $\mathscr{A}$ respectively $\mathscr{B}$ only

affine surfaces can be determined separating the sets $\mathscr{A}$ and $\mathscr{B}$ (cf. Fig. 1.4 where the piecewise affine surfaces are given by the bold lines). For the computation of these surfaces an algorithm is given by Mangasarian [215] which starts with the computation of one hyperplane (say G_1) separating the sets $\mathscr{A}$ and $\mathscr{B}$ as best as possible. Clearly, if both sets are separable, then a separating hyperplane is constructed. In the other case, there are some misclassified points. Now, discarding all subsets containing only points from one of the sets, the remaining subsets are partitioned in the same way again, and so on. In Fig. 1.4 this means that after constructing the hyperplane G_1 the upper-left half-space is discarded and the lower-right half-space is partitioned again (say by G_2). At last, the lower-right corner is subdivided by G_3.

This algorithm reduces this problem of discriminating between two sets to that of finding a hyperplane separating two finite sets $\mathscr{A}$ and $\mathscr{B}$ of points as best as possible. Mangasarian [216] formulated an optimization problem which selects the desired hyperplane such that the number of misclassified points is minimized. For describing that problem, let A and B be two matrices the rows of which are given by the coordinates of the s and t points in the sets $\mathscr{A}$ and $\mathscr{B}$, respectively. Then, a separating hyperplane is determined by an n-dimensional vector w and a scalar γ as $H = \{x \in \mathbb{R}^n : \langle w, x\rangle = \gamma\}$ with the property that

$$Aw > \gamma e^s, \;\; Bw < \gamma e^t$$

provided that the convex hulls of the points in the sets $\mathscr{A}$ and $\mathscr{B}$ are disjoint. Up to normalization, the above system is equivalent to

$$Aw - \gamma e^s - e^s \geq 0, \;\; -Bw + \gamma e^t - e^t \geq 0. \tag{1.19}$$

Then, a point in $\mathscr{A}$ belongs to the correct half-space if and only if the given inequality in the corresponding line of the last system is satisfied. Hence, using the step function a_* and the plus function a_+ which are component-wise given as

$$(a_*)_i = \begin{cases} 1 & \text{if } a_i > 0 \\ 0 & \text{if } a_i \leq 0 \end{cases}, \quad (a_+)_i = \begin{cases} a_i & \text{if } a_i > 0 \\ 0 & \text{if } a_i \leq 0 \end{cases}$$

we obtain that the system (1.19) is equivalent to the equation

$$e^{s\top}(-Aw + \gamma e^s + e^s)_* + e^{t\top}(Bw - \gamma e^t + e^t)_* = 0. \qquad (1.20)$$

It is easy to see that the number of misclassified points is counted by the left-hand side of (1.20). For $a, c, d, r, u \in \mathbb{R}^l$, Mangasarian [216] characterized the step function as follows:

$$r = a_*, u = a_+ \iff \begin{cases} \begin{pmatrix} r \\ u \end{pmatrix} = \begin{pmatrix} r - u + a \\ r + u - e^l \end{pmatrix}_+ \\ \text{and } r \text{ is minimal in case of uncertainty} \end{cases}.$$

Hence,

$$c = d_+ \iff c - d \geq 0,\ c \geq 0,\ c(c - d) = 0.$$

Using both relations, we can transform the problem of minimizing the number of misclassified points or, equivalently, the minimization of the left-hand side function in (1.20) into the following optimization problem, see Mangasarian [216]

$$e^{s\top} r + e^{t\top} s \to \min_{w,\gamma,r,u,p,v}$$

$$\begin{aligned} u + Aw - \gamma e^s - e^s &\geq 0 & v - Bw + \gamma e^t - e^t &\geq 0 \\ r &\geq 0 & p &\geq 0 \\ r^\top(u + Aw - \gamma e^s - e^s) &= 0 & p^\top(v - Bw + \gamma e^t - e^t) &\geq 0 \\ -r + e^s &\geq 0 & -p + e^t &\geq 0 \\ u &\geq 0 & v &\geq 0 \\ u^\top(-r + e^s) &= 0 & v^\top(-p + e^t) &= 0. \end{aligned}$$

This problem is an optimization problem with linear complementarity constraints, a generalized bilevel optimization problem. Mangasarian has shown in [215] that the task of training neural networks can be modeled by a similar problem.

1.5.6 Support Vector Machines

Closely related to the topic of Sect. 1.5.5 are support vector machines (SVM) (Cortes and Vapnik [45], Vapnik [302]), kernel methods (Shawe-Taylor and Christianini

[287]). Especially in kernel methods, the learning task is to construct a linear function that minimizes a convex loss function. The resulting optimization problem is a convex one, but typically it depends on hyper-parameters which needs to be selected by the users. To select good values for the hyper-parameters, cross validation is usually applied. Thus, an estimate of the out-of-sample generalization error is minimized. Bennett et al. [19, 20, 195] formulated an optimization problem for more efficiently selecting encouraging values for the hyper-parameters.

For that, assume that l points $\{(x_1, y_1), (x_2, y_2), \ldots, (x_l, y_l)\} \subset \mathbb{R}^{n+1}$ $(n, l \in \mathbb{N})$ are given and investigate the regression problem of finding a function $f : \mathbb{R}^n \mapsto \mathbb{R}$ among a given class of functions that minimizes the regularized risk functional

$$R(f) = P(f) + \frac{C}{l}\sum_{i=1}^{l} L(y_i, f(x_i)).$$

Here, L is a loss function of the observed data and the model output, P is a regularization operator and C a regularization parameter. One possibility is to use the ε-insensitive loss function

$$L(y, f(x)) = \max\{|y - f(x)| - \varepsilon, 0\}$$

for some $\varepsilon > 0$. For functions f in the class of linear functions $f(x) = w^\top x$ the regularization operator $P(f) = \|w\|^2$ can be used. The result of the regression problem depends on the hyper-parameters C, ε which are not easy to be estimated since one does not know beforehand how accurately the data $/(x_i, y_i)$ are given.

To formulate an optimization problem for computing encouraging values for the hyper-parameters, Bennett et al. [19] partition the l data points into T distinct partitions $\Omega_t, t = 1, \ldots, T$ such that

$$\bigcup_{t=1}^{T} \Omega_t = \{1, \ldots, l\}, \ \overline{\Omega}_l = \{1, \ldots, l\} \setminus \Omega_t.$$

Then, the following bilevel optimization problem can be used to find values for the hyper-parameters C, ε:

$$\begin{aligned} \frac{1}{T}\sum_{t=1}^{T}\frac{1}{|\Omega_t|}\sum_{i\in\Omega_i} |x_i^\top w^t - y_i| \to & \min_{C,\varepsilon,\lambda,w^t,w^0,\underline{w},\overline{w}} \\ \varepsilon, C, \lambda & \geq 0 \\ \underline{w} & \leq \overline{w} \\ w^t & \in \Psi_t(C, \varepsilon, \lambda, w^0, \underline{w}, \overline{w}), \end{aligned} \tag{1.21}$$

where

$$\Psi_t(C, \varepsilon, \lambda, w^0, \underline{w}, \overline{w}) = \operatorname*{Argmin}_{\underline{w} \le w \le \overline{w}} \left\{ C \sum_{j \in \overline{\Omega}_i} \max\{|x_j^\top w - y_j| - \varepsilon, 0\} \right.$$
$$\left. + \frac{1}{2}\|w\|^2 + \frac{\lambda}{2}\|w - w^0\|^2 \right\}.$$

Chapter 2
Linear Bilevel Optimization Problem

2.1 The Model and First Properties

The linear bilevel optimization problem illustrated in Example 1.1 is the problem of the following structure

$$\min_{x,y}\{a^\top x + b^\top y : Ax + By \le c,\ (x, y) \in \mathbf{gph}\,\Psi\}, \tag{2.1}$$

where $\Psi(\cdot)$ is the solution set mapping of the lower level problem

$$\Psi(x) := \operatorname*{Argmin}_y \{d^\top y : Cy \le x\}. \tag{2.2}$$

Here, A is a (p, n)-, B a (p, m)- and C a (n, m)-matrix and all variables and vectors used are of appropriate dimensions. Note that we have used here the so-called optimistic bilevel optimization problem, which is related to problem (1.4).

We find so-called connecting constraints $Ax+By \le c$ in the upper level problem. Validity of such constraints is beyond the choice of the leader and can be verified only after the follower has selected his/her possibly not unique optimal solution. Especially in the case when $\Psi(x)$ does not reduce to a singleton this can be difficult. For investigating the bilevel programming problem in the case that $\Psi(x)$ does not reduce to a singleton, Ishizuka and Aiyoshi [153] introduced their double penalty method. In general, *connecting constraints* may imply that the feasible set of the bilevel programming problem is disconnected. This situation is illustrated by the following example:

Example 2.1 (Mersha and Dempe [227]). Consider the problem

$$\min_{x,y} -x - 2y$$

$$\text{subject to} \quad \begin{aligned} 2x - 3y &\ge -12 \\ x + y &\le 14 \end{aligned}$$

$$\text{and } y \in \operatorname*{Argmin}_y \{-y : -3x + y \le -3,\ 3x + y \le 30\}.$$

S. Dempe et al., *Bilevel Programming Problems*, Energy Systems,
DOI 10.1007/978-3-662-45827-3_2

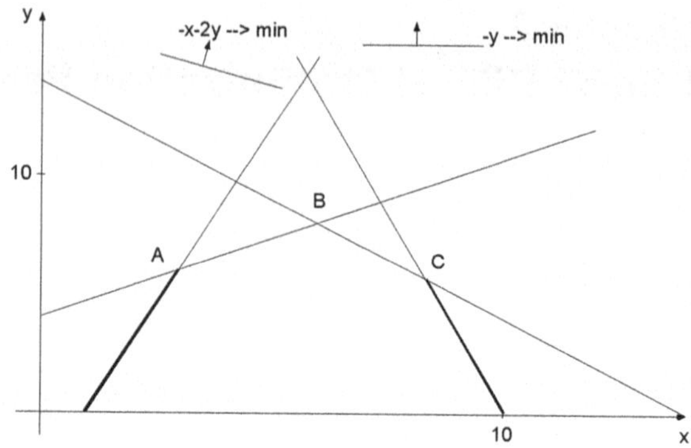

Fig. 2.1 The problem with *upper level* connecting constraints. The feasible set is depicted with *bold lines*. The point C is global optimal solution, point A is a local optimal solution

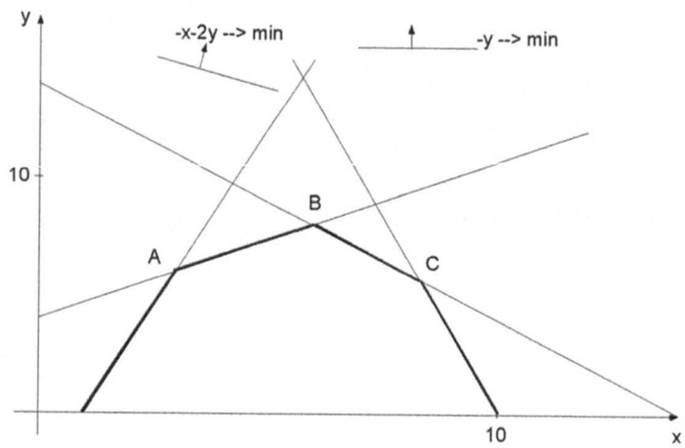

Fig. 2.2 The problem when the *upper level* connecting constraints are shifted into the *lower level* problem. The feasible set is depicted with *bold lines*. The global optimal solution is point B

The optimal solution for this problem is point C at $(\overline{x}, \overline{y}) = (8, 6)$ (see Fig. 2.1). But if we shift the two upper level constraints to the lower level we get point B at $(\widetilde{x}, \widetilde{y}) = (6, 8)$ as an optimal solution (see Fig. 2.2). From this example it can easily be noticed that if we shift constraints from the upper level to the lower one, the optimal solution obtained prior to shifting is not optimal any more in general. Hence ideas based on shifting constraints from one level to another will lead to a solution which may not be a solution prior to shifting constraints. □

In Example 2.1 the optimal solution of the lower level problem was unique for all x. If this is not the case, feasibility of a selection of the upper level decision maker possibly depends on the selection of the follower. In the optimistic case this means that the leader selects within the set of optimal solutions of the follower's problem one point which is at the same time feasible for the upper level connecting constraints and gives the best objective function value for the upper level objective function.

As we can see in Example 2.1 the existence of connecting upper level constraints will lead in general to disconnected feasible sets in the bilevel programming problem. Therefore, solution algorithms will live in one of the connected components of the feasible set (i.e. a sequence of feasible points which all belong to one of the connected parts is computed) or they need to jump from one of the connected parts of the feasible set to another one. This would use then ideas of discrete optimization.

In the following we will avoid this additional difficulty in assuming that the upper level constraints will depend on the upper level variables only. Hence, we consider the *linear bilevel optimization problem*

$$\min_{x,y}\{a^\top x + b^\top y : Ax \le c,\ (x, y) \in \mathbf{gph}\, \Psi\}, \tag{2.3}$$

where $\Psi(\cdot)$ is the solution set mapping of the lower level problem

$$\Psi(x) := \underset{y}{\text{Argmin}}\, \{d^\top y : Cy \le x\}. \tag{2.4}$$

In this problem, parametric linear optimization (see e.g. Nožička et al. [257]) can be used to show that the graph of the mapping $\Psi(\cdot)$ equals the connected union of faces of the set $\{(x, y)^\top : Cy \le x\}$.

Here, a set M is connected if it is not contained in the union of two disjoint open sets $M \subset M_1 \cup M_2$, M_1, M_2 are open and not empty, $M_1 \cap M_2 = \emptyset$, having nonempty intersection with both of these sets: $M \cap M_i \neq \emptyset,\ i = 1, 2$.

Hence, the convex hull of this set is a convex polyhedron implying that problem (2.3) is a linear optimization problem. Thus, its optimal solution can be found at a vertex of the set

$$\{(x, y)^\top : Cy \le x,\ Ax \le c\}.$$

Theorem 2.1 *If problem* (2.3) *has an optimal solution, at least one global optimal solution occurs at a vertex of the set*

$$\{(x, y)^\top : Cy \le x,\ Ax \le c\}.$$

This theorem can be found in the article [40] by Candler and Townsley, it is the basis of many algorithms using (implicit or not complete) enumeration to compute a global optimum of problem (2.3) (see e.g. Bard [10]).

This property is lost if problem (2.1) with upper level connecting constraints is considered.

As it can be seen in Fig. 2.2, the bilevel optimization problem is a nonconvex optimization problem, it has a feasible set which is not given by explicit constraints. As a result, besides a global optimal solution bilevel optimization problems can have local extrema and stationary solutions which are not local optimal solutions.

In Sect. 1.2, the bilevel optimization problem has been interpreted as an hierarchical game of two players, the leader and the follower where the leader is the first to make a choice and the follower reacts optimally on the leader's selection. It has been shown in the article [11] by Bard and Falk that the solution strongly depends on the order of play: the leader may take advantage from having the first selection.

The following theorem shows that the (linear) bilevel optimization problem is $\mathcal{NP}$-hard in the strong sense which implies that it is probably not possible to find a polynomial algorithm for computing a global optimal solution of it. For more results on complexity theory the interested reader is referred to the monograph [126] by Garey and Johnson.

Theorem 2.2 (Deng [85]) *For any $\varepsilon > 1$ it is $\mathcal{NP}$-hard to find a solution of the linear bilevel optimization problem* (2.3) *with not more than ε times the global optimal function value of this problem.*

In the next example we will see that the bilevel programming problem depends on constraints being not active in the lower level problem. Hence, a global optimal solution of the bilevel problem can loose its optimality if an inequality is added which is not active at the global minimum. This behavior may be a bit surprising since it is not possible in problems on continuous (nonsmooth) optimization.

Example 2.2 (*Macal and Hurter* [210]) Consider the unconstrained bilevel optimization problem

$$\begin{aligned} &(x-1)^2 + (y-1)^2 \to \min_{x,y}, \\ &\text{where } y \text{ solves} \\ &0.5y^2 + 500y - 50xy \to \min_{y}. \end{aligned} \tag{2.5}$$

Since the lower level problem is unconstrained and convex we can replace it by its necessary optimality conditions. Then, problem (2.5) becomes

$$\min_{x,y}\{(x-1)^2 + (y-1)^2 : y - 50x + 500 = 0\}.$$

The unique optimal solution of this problem is $(x^*, y^*) = (50102/5002, 4100/5002)$ with an optimal objective function value of $z^* = 81, 33$.

Now, add the constraint $y \geq 0$ to the lower level problem and consider the problem

$$\begin{aligned} &(x-1)^2 + (y-1)^2 \to \min_{x,y}, \\ &\text{where } y \text{ solves} \\ &y \in \operatorname*{Argmin}_y \{0.5y^2 + 500y - 50xy : y \geq 0\}. \end{aligned} \tag{2.6}$$

The unique global optimal solution of problem (2.6) is $(\overline{x}, \overline{y}) = (1, 0)$. This point is not feasible for (2.5). Its objective function value in problem (2.6) is 1 showing that (x^*, y^*) is a local optimum but not the global optimal solution of problem (2.6). □

In the next theorem we need the notion of an inner semicontinuous mapping.

Definition 2.1 (*Mordukhovich* [241]) A point-to-set mapping $\Gamma : \mathbb{R}^n \rightrightarrows \mathbb{R}^m$ is said to be *inner semicontinuous* at $(\overline{z}, \overline{\alpha}) \in \mathbf{gph}\,\Gamma$ provided that, for each sequence $\{z^k\}_{k=1}^\infty$ converging to $\overline{z}$ there is a sequence $\{\alpha^k\}_{k=1}^\infty$, $\alpha^k \in \Gamma(z^k)$ converging to $\overline{\alpha}$.

Theorem 2.3 (Dempe and Lohse [68]) *Let $(\overline{x}, \overline{y})$ be a global optimal solution of the problem* (1.4). *Let Ψ be inner semicontinuous at $(\overline{x}, \overline{y})$. Then, $(\overline{x}, \overline{y})$ is a local optimal solution of the problem*

$$\min_{x,y}\{F(x, y) : x \in X,\ (x, y) \in \mathbf{gph}\,\Psi^1\} \tag{2.7}$$

with

$$\Psi^1(x) := \operatorname*{Argmin}_y \{f(x,y) : g(x,y) \le 0,\ h(x,y) \le 0\}$$

with $h : \mathbb{R}^n \times \mathbb{R}^m \to \mathbb{R}$ *provided that* $h(\overline{x}, \overline{y}) < 0$ *and that the function* h *is continuous.*

Proof A point $y \in \Psi(x)$ is an optimal solution of the lower level problem of (2.7) if it is feasible for this problem, $\overline{y} \in \Psi^1(\overline{x})$. Hence, the point $(\overline{x}, \overline{y})$ is feasible for (2.7).

Assume that $(\overline{x}, \overline{y})$ is not a local optimum of problem (2.7). Then, there exists a sequence $\{(x^k, y^k)\}_{k=1}^\infty$ converging to $(\overline{x}, \overline{y})$ such that $x^k \in X$, $y^k \in \Psi^1(x^k)$ and $F(x^k, y^k) < F(\overline{x}, \overline{y})$. Note that (x^k, y^k) is feasible for problem (1.4) for large k.

Since Ψ is inner semicontinuous at $(\overline{x}, \overline{y})$ there exists a sequence $\widehat{y}^k \in \Psi(x^k)$ converging to $\overline{y}$. By continuity of the function h, $h(x^k, \widehat{y}^k) < 0$ and $\widehat{y}^k \in \Psi^1(x^k)$. Hence, $f(x^k, \widehat{y}^k) = f(x^k, y^k)$,

$$\begin{aligned}\Psi^1(x^k) &= \{y : g(x,y) \le 0,\ h(x,y) \le 0,\ f(x^k, y) = f(x^k, \widehat{y}^k)\}\\ &\subseteq \{y : g(x,y) \le 0,\ f(x^k, y) = f(x^k, \widehat{y}^k)\} = \Psi(x^k)\end{aligned}$$

and, hence,

$$\min_y\{F(x^k, y) : y \in \Psi(x^k)\} \le \min_y\{F(x^k, y) : y \in \Psi^1(x^k)\} \le F(x^k, y^k) < F(\overline{x}, \overline{y})$$

for sufficiently large k. This contradicts global optimality of $(\overline{x}, \overline{y})$. □

In the article Dempe and Lohse [68] an example is given which shows that the restrictive assumption of inner semicontinuity of the solution set mapping of the lower level problem is essential.

A similar result to Example 2.2 can be shown if one variable is added in the lower level problem: a global optimal solution can loose global optimality.

Consider the bilevel programming problem

$$\min_{x,y}\{F(x,y) : x \in X,\ (x,y) \in \mathbf{gph}\, \Psi_L\}, \tag{2.8}$$

with a linear lower level problem parameterized in the objective function

$$\Psi_L(x) := \operatorname*{Argmin}_y \{x^\top y : Ay = b,\ y \ge 0\}, \tag{2.9}$$

where $X \subseteq \mathbb{R}^n$ is a closed set,

Let $(\overline{x}, \overline{y})$ be a global optimal solution of problem (2.8). Now, add one new variable y_{n+1} to the lower level problem with objective function coefficient x_{n+1} and a new column A_{n+1} in the coefficient matrix of the lower level problem, i.e. replace the lower level problem with

$$\Psi_{NL}(x) := \operatorname*{Argmin}_y \{x^\top y + x_{n+1}y_{n+1} : Ay + A_{n+1}y_{n+1} = b,\ y, y_{n+1} \ge 0\} \tag{2.10}$$

and investigate the problem

$$\min_{x,y}\{\widetilde{F}(x, x_{n+1}, y, y_{n+1}) : (x, x_{n+1}) \in \widetilde{X},\ (x, x_{n+1}, y, y_{n+1}) \in \mathbf{gph}\, \Psi_{NL}\}. \tag{2.11}$$

Here $\widetilde{X} \subseteq \mathbb{R}^{n+1}$ and $\widetilde{X} \cap \mathbb{R}^n \times \{0\} = X$.

Example 2.3 (*Dempe and Lohse* [68]) Consider the following bilevel programming problem with the lower level problem

$$\Psi_L(x) := \operatorname*{Argmin}_y \{x_1 y_1 + x_2 y_2 : y_1 + y_2 \leq 2,\ -y_1 + y_2 \leq 0,\ y \geq 0\} \tag{2.12}$$

and the upper level problem

$$\min\{(x_1 - 0.5)^2 + (x_2 - 0.5)^2 - 3y_1 - 3y_2 : (x, y) \in \mathbf{gph}\, \Psi_L\}. \tag{2.13}$$

Then, the unique global optimum is $\overline{x} = (0.5; 0.5)$, $\overline{y} = (1; 1)$ with optimal objective function value -6. Now, adding one variable to the lower level problem

$$\Psi_{NL}(x) := \operatorname*{Argmin}_y \{x_1 y_1 + x_2 y_2 + x_3 y_3 : y_1 + y_2 + y_3 \leq 2,\ -y_1 + y_2 \leq 0,\ y \geq 0\} \tag{2.14}$$

and investigating the bilevel optimization problem

$$\min\{(x_1 - 0.5)^2 + (x_2 - 0.5)^2 + x_3^2 - 3y_1 - 3y_2 - 6y_3 : (x, y) \in \mathbf{gph}\, \Psi_{NL}\} \tag{2.15}$$

the point $x = (0.5; 0.5; 0.5)$, $y = (0; 0; 2)$ has objective function value -11.75. Hence, global optimality of $(\overline{x}, \overline{y})$ is destroyed. But, the point $((\overline{x}, 0), (\overline{y}, 0))$ remains feasible and it is a strict local minimum. □

Theorem 2.4 (Dempe and Lohse [68]) *Let* $(\overline{x}, \overline{y})$ *be a global optimal solution for problem* (2.8) *and assume that the functions* F, $\widetilde{F}$ *are concave,* X, $\widetilde{X}$ *are polyhedra. Let*

$$\overline{x}_B^\top B^{-1} A_{n+1} < 0 \text{ for each basic matrix } B \text{ for } \overline{y} \text{ and } \overline{x} \tag{2.16}$$

and $(\overline{x}, 0)$ *be a local minimum of the problem*

$$\min\{\widetilde{F}((x, x_{n+1}), (y, 0)) : (x, x_{n+1}) \in \widetilde{X},\ y \in \Psi_L(\overline{x})\}.$$

Then, the point $((\overline{x}, 0), (\overline{y}, 0))$ *is a local optimal solution of problem* (2.11).

Proof Assume that $((\overline{x}, 0), (\overline{y}, 0))$ is not a local optimum. Then, there exists a sequence $((x^k, x_{n+1}^k), (y^k, y_{n+1}^k))$ converging to $((\overline{x}, 0), (\overline{y}, 0))$ with

$$F((x^k, x_{n+1}^k), (y^k, y_{n+1}^k)) < F((\overline{x}, 0), (\overline{y}, 0)) \text{ for all } k.$$

Since $((x^k, x^k_{n+1}), (y^k, y^k_{n+1}))$ is feasible for (2.11) and $\mathbf{gph}\, \Psi_{NL}$ equals the union of faces of the set (see e.g. Dempe [52])

$$\{(x, y) : x \in \widetilde{X},\ Ay + A_{n+1}y_{n+1} = b,\ y, y_{n+1} \geq 0\},$$

then, since $((x^k, x^k_{n+1}), (y^k, y^k_{n+1}))$ converges to $((\overline{x}, 0), (\overline{y}, 0))$ there exists, without loss of generality, one facet M of this set with $((x^k, x^k_{n+1}), (y^k, y^k_{n+1})) \in M$ for all k. Moreover, by upper semicontinuity of $\Psi_{NL}(\cdot)$, $((\overline{x}, 0), (\overline{y}, 0)) \in M$. By Schrijver [285] there exists $c \in \mathbb{R}^{n+1}$ such that M equals the set of optimal solutions of the problem

$$\min\{c^\top (y, y_{n+1})^\top : Ay + A_{n+1}y_{n+1} = b,\ y, y_{n+1} \geq 0\}.$$

Since $(\overline{y}, 0) \in M$ there exists a basic matrix for $(\overline{y}, 0)$ and c. Then, the assumptions of the theorem imply that $(\overline{x}, 0) \neq c$ if x_{n+1} is a basic variable in (y^k, y^k_{n+1}) (since this implies that $c_B^\top B^{-1} A_{n+1} - c_{n+1} = 0$ by linear optimization). This implies that there is an open neighborhood V of $(\overline{x}, 0)$ such that $\Psi_{NL}(x, x_{n+1}) \subseteq \{(y, y_{n+1}) : y_{n+1} = 0\}$ for $(x, x_{n+1}) \in V$.

Hence, $y^k_{n+1} = 0$ for sufficiently large k.

By parametric linear optimization, $\Psi_L(x) \subseteq \Psi_L(\overline{x})$ for x sufficiently close to $\overline{x}$. Hence, the assertion follows. □

Similar results are shown in the paper Dempe and Lohse [68] in the case when the lower level problem is a right-hand side perturbed linear optimization problem.

2.2 Optimality Conditions

Consider the bilevel optimization problem

$$\min_{y,b,c}\{F(y) : b \in \mathscr{B},\ c \in \mathscr{C},\ y \in \Psi(b, c)\}, \tag{2.17}$$

where

$$\mathscr{B} = \{b : Bb = \widetilde{b}\},\ \mathscr{C} = \{Cc = \widetilde{c}\}$$

for some matrices $B,\ C$ of appropriate dimension, $c \in \mathbb{R}^n$, $\widetilde{c} \in \mathbb{R}^q$ and $b \in \mathbb{R}^m$, $\widetilde{b} \in \mathbb{R}^p$. Here, the function $F : \mathbb{R}^n \to \mathbb{R}$ depends only on the optimal solution of the lower level problem. This makes the formulation of optimality conditions, which can be verified in polynomial time, possible.

The mapping $(b, c) \mapsto \Psi(b, c)$ is again the set of optimal solutions of a linear optimization problem:

$$\Psi(b, c) = \operatorname*{Argmin}_y \{c^\top y : Ay = b,\ y \geq 0\}.$$

We have $\widehat{y} \in \Psi(b, c)$ if and only if there is a vector $\widehat{z}$ such that $(\widehat{y}, \widehat{z})$ satisfies the following system of equations and inequalities:

$$\begin{aligned} Ay = b, \; y \geq 0, \\ A^\top z \leq c, \\ y^\top (A^\top z - c) = 0. \end{aligned}$$

Thus, the graph **gph** Ψ of the mapping Ψ equals the projection of the union of faces of a certain polyhedron in $\mathbb{R}^n \times \mathbb{R}^m \times \mathbb{R}^m \times \mathbb{R}^n$ into the space $\mathbb{R}^n \times \mathbb{R}^m \times \mathbb{R}^m$. Hence, the tangent (Bouligand) cone

$$C_{\mathscr{M}}(\widehat{u}) := \left\{ \begin{array}{l} d : \exists \{u^k\}_{k=1}^\infty \subset \mathscr{M}, \; \exists \; \{t_k\}_{k=1}^\infty \subset \mathbb{R}_+ \\ \quad \text{with} \lim_{k\to\infty} t_k = 0, \; \lim_{k\to\infty} u^k = \widehat{u}, \; d = \lim_{k\to\infty} \frac{u^k - \widehat{u}}{t_k} \end{array} \right\}$$

at a point $(\overline{y}, \overline{b}, \overline{c})$ to the feasible set

$$\mathscr{M} := \{(y, b, c) : b \in \mathscr{B}, \; c \in \mathscr{C}, \; y \in \Psi(b, c)\}$$

equals the union of convex polyhedra, too. Thus, to check local optimality of some feasible point $(\overline{y}, \overline{b}, \overline{c}) \in \mathscr{M}$ for problem (2.17) it is necessary to verify that there is no direction of descent in any one of these convex polyhedra. Unfortunately, the number of these polyhedra cannot be bounded by a polynomial in the number of variables. This can be seen as a reason for $\mathscr{N}\mathscr{P}$-hardness of proving local optimality in general bilevel optimization (see Hansen et al. [136] where an exact proof for this result is given).

The following result can be found in the paper [67] by Dempe and Lohse. Let for a moment $\mathscr{B} = \{\overline{b}\}$ reduce to a singleton. Take an arbitrary vertex $\overline{y}$ of the set $\{y : Ay = b, \; y \geq 0\}$. Then, by parametric linear optimization, there exists $\widehat{c}$ such that $\Psi(\overline{b}, c) = \{\overline{y}\}$ for all c sufficiently close to $\widehat{c}$, formally $\forall c \in U(\widehat{c})$ for some open neighborhood $U(\widehat{c})$ of $\widehat{c}$. Hence, if $U(\widehat{c}) \cap \mathscr{C} \neq \emptyset$, there exists $\overline{z}$ satisfying $A^\top \overline{z} \leq \overline{c}$, $\overline{y}^\top (A^\top \overline{z} - \overline{c}) = 0$ for some $\overline{c} \in U(\widehat{c}) \cap \mathscr{C}$ such that $(\overline{y}, \overline{z}, \overline{b}, \overline{c})$ is a local optimal solution of the problem

$$\begin{aligned} F(y) &\to \min_{y,z,b,c} \\ Ay &= b, \; y \geq 0, \\ A^\top z &\leq c, \\ y^\top (A^\top z - c) &= 0 \\ Bb &= \widetilde{b}, \\ Cc &= \widetilde{c}. \end{aligned} \tag{2.18}$$

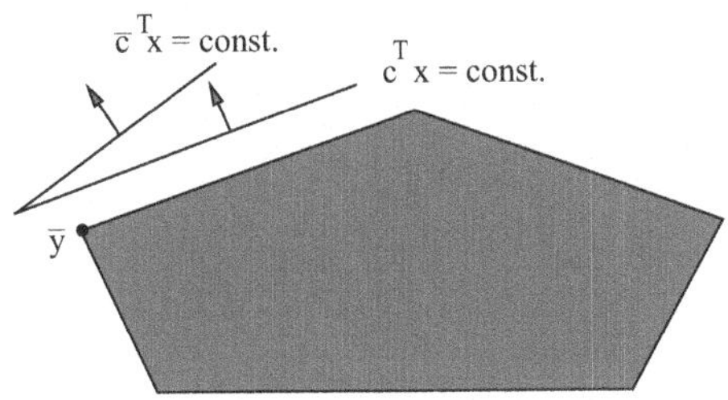

Fig. 2.3 Definition of local optimality

Theorem 2.5 (Dempe and Lohse [67]) *Let $\mathscr{B} = \{\overline{b}\}$, $\{\overline{y}\} = \Psi(\overline{b}, c)$ for all c in an open neighborhood $U(\widehat{c})$ of $\widehat{c}$ with $U(\widehat{c}) \cap \mathscr{C} \neq \emptyset$. Then, $(\overline{y}, \overline{b}, \overline{c}, \overline{z})$ is a locally optimal solution of* (2.18) *for some dual variables $\overline{z}$ and a certain $\overline{c} \in U(\widehat{c}) \cap \mathscr{C}$.*

Figure 2.3 can be used to illustrate this fact. The points $\overline{y}$ satisfying the assumptions of Theorem 2.5 are the vertices of the feasible set of the lower level problem given by the dashed area in this figure. Theorem 2.5 implies that each vertex of the set $\{y : Ay = b,\ y \geq 0\}$ is a local optimal solution of problem (2.17) which is not desired. To circumvent this difficulty the definition of a local optimal solution is restricted to variable y only:

Definition 2.2 (*Dempe and Lohse* [67]) A point $\overline{y}$ is a local optimal solution of problem (2.17) if there exists an open neighborhood $U(\overline{y})$ of $\overline{y}$ such that $F(y) \geq F(\overline{y})$ for all (y, b, c) with $b \in \mathscr{B}$, $c \in \mathscr{C}$ and $y \in U(\overline{y}) \cap \Psi(b, c)$.

To derive a necessary optimality condition for problem (2.17) according to this definition, a formula for a tangent cone to its feasible set depending only on y is needed. Let $(\overline{y}, \overline{z}, \overline{b}, \overline{c})$ be a feasible solution for problem (2.18) and define the index sets

$$\begin{aligned}
I(\overline{y}) &= \{i : \overline{y}_i = 0\},\\
I(z, c) &= \{i :\ (A^\top z - c)_i > 0\},\\
\mathscr{I}(\overline{y}) &= \{I(z, c) :\ A^\top z \geq c,\ (A^\top z - c)_i = 0\ \ \forall i \notin I(\overline{y}),\ c \in \mathscr{C}\}\\
I^0(\overline{y}) &= \bigcap_{I \in \mathscr{I}(\overline{y})} I.
\end{aligned}$$

Remark 2.1 If an index set I belongs to the family $\mathscr{I}(\overline{y})$ then $I^0(\overline{y}) \subseteq I \subseteq I(\overline{y})$.

This remark and also the following one are obvious consequences of the definitions of the above sets.

Remark 2.2 We have $j \in I(\overline{y}) \setminus I^0(\overline{y})$, if and only if the system

$$\begin{aligned}
(A^\top z - c)_i &= 0\,\forall i \notin I(\overline{y})\\
(A^\top z - c)_j &= 0\\
(A^\top z - c)_i &\geq 0\,\forall i \in I(\overline{y}) \setminus \{j\}\\
Cc &= \widetilde{c}
\end{aligned}$$

has a solution. Furthermore $I^0(\overline{y})$ is an element of $\mathscr{I}(\overline{y})$ if and only if the system

$$\begin{aligned}(A^\top z - c)_i &= 0\,\forall i \notin I^0(\overline{y})\\ (A^\top z - c)_i &\geq 0\,\forall i \in I^0(\overline{y})\\ Cc &= \widetilde{c}\end{aligned}$$

has a solution.

This result makes an efficient computation of the set $I^0(\overline{y})$ possible.

Now, it turns out that the dual feasible solution $\overline{z}$ for the lower level problem as well as the objective function coefficients $\overline{c}$ are not necessary for solving problem (2.17), it is only necessary to consider possible index sets $I \in \mathscr{I}(\overline{y})$.

Theorem 2.6 (Dempe and Lohse [67]) $\overline{y}$ *is a local optimum for* (2.17) *if and only if* $\overline{y}$ *is a (global) optimal solution for all problems* (A_I)*:*

$$\begin{aligned}F(y) &\to \min_{y,b}\\ Ay &= b\\ y &\geq 0\\ y_i &= 0 \quad \forall i \in I\\ Bb &= \widetilde{b}\end{aligned}$$

with $I \in \mathscr{I}(\overline{y})$.

Proof Let $\overline{y}$ be a local optimal solution of (2.17) and assume that there is a set $I \in \mathscr{I}(\overline{y})$ with $\overline{y}$ being not optimal for (A_I). Then there exists a sequence $\{y^k\}_{k=1}^{\infty}$ of feasible solutions of (A_I) with $\lim_{k\to\infty} y^k = \overline{y}$ and $F(y^k) < F(\overline{y})$ for all k. Consequently $\overline{y}$ can not be local optimal for (2.17) since $I \in \mathscr{I}(\overline{y})$ implies that all y^k are also feasible for (2.18).

Conversely, let $\overline{y}$ be an optimal solution for all problems (A_I) and assume that there is a sequence $\{y^k\}_{k=1}^{\infty}$ of feasible points of (2.17) with $\lim_{k\to\infty} y^k = \overline{y}$ and $F(y^k) < F(\overline{y})$ for all k. For k sufficiently large the elements of this sequence satisfy the condition $y_i^k > 0$ for all $i \notin I(\overline{y})$ and due to the feasibility of y^k for (2.17) there are sets $I \in \mathscr{I}(\overline{y})$ such that y^k is feasible for problem (A_I). Because $\mathscr{I}(\overline{y})$ consists only of a finite number of sets, there is a subsequence $\{y^{k_j}\}_{j\in\mathbb{N}}$ where y^{k_j} are all feasible for a fixed problem (A_I). So we get a contradiction to the optimality of $\overline{y}$ for this problem (A_I). □

Using the set I as a new variable in problem (A_I), the following problem is obtained which is equivalent to problem (2.18) by Theorem 2.6:

$$\begin{aligned}F(y) &\to \min_{y,b,I}\\ Ay &= b\end{aligned}$$

$$\begin{aligned} y &\geq 0 \\ y_i &= 0 \quad \forall i \in I \\ Bb &= \widetilde{b} \\ I &\in \mathscr{I}(\overline{y}) \end{aligned} \tag{2.19}$$

The following tangent cone can be used to express the feasible set of problem (A_I) near a feasible point $\overline{y}$ for a fixed set $I \in \mathscr{I}(\overline{y})$:

$$T_I(\overline{y}) = \{d|\ \exists r:\ Ad = r,\ Br = 0,\ d_i \geq 0,\ \forall i \in I(\overline{y}) \setminus I,\ d_i = 0,\ \forall i \in I\}$$

Using Theorem 2.6 a necessary optimality condition is derived:

Corollary 2.1 *If $\overline{y}$ is a local optimal solution of problem* (2.17)*, and F is directionally differentiable then $F'(\overline{y}; d) \geq 0$ for all $d \in T(\overline{y}) := \bigcup_{I \in \mathscr{I}(\overline{y})} T_I(\overline{y})$.*

Since $d \in \operatorname{conv} T(\overline{y})$ is equal to a convex linear combination of elements in $T(\overline{y})$, $\nabla F(\overline{y})d < 0$ for some $d \in \operatorname{conv} T(\overline{y})$ only if $\nabla F(\overline{y})\overline{d} < 0$ for a certain $\overline{d} \in T(\overline{y})$. This leads to the necessary optimality condition

$$\nabla F(\overline{y})d \geq 0\, \forall\, d \in \operatorname{conv} T(\overline{y})$$

provided that the objective function F is differentiable.

Consider the relaxed problem to (2.19):

$$\begin{aligned} F(y) &\to \min_{y,b} \\ Ay &= b \\ y_i &\geq 0 \quad \forall i \notin I^0(\overline{y}) \\ y_i &= 0 \quad \forall i \in I^0(\overline{y}) \\ b &\in \mathscr{B} \end{aligned} \tag{2.20}$$

and the tangent cone

$$T_R(\overline{y}) = \{d : Ad = r,\, Br = 0,\, d_i \geq 0,\ i \in I(\overline{y})\setminus I^0(\overline{y}),\, d_i = 0,\ i \in I^0(\overline{y})\}$$

to the feasible set of this problem at the point $\overline{y}$ again relative to y only.

Due to $I^0(\overline{y}) \subseteq I$ for all $I \in \mathscr{I}(y^0)$ we derive

$$\operatorname{conv} T(\overline{y}) = \operatorname{cone} T(\overline{y}) \subseteq T_R(\overline{y}), \tag{2.21}$$

where cone S denotes the conical hull of the set S, i.e. the set of all linear combinations of elements in S with nonnegative coefficients. Let span$S$ denote the set of all linear combinations of elements in S.

Definition 2.3 The point $\overline{y}$ is said to satisfy the *full rank condition* (FRC), if

$$\text{span}(\{A_i : i \notin I(\overline{y})\}) = \mathbb{R}^m, \tag{2.22}$$

where A_i denotes the ith column of the matrix A.

Example 2.4 All non degenerated vertices of $Ay = b,\ y \geq 0$ satisfy the full rank condition.

This condition allows us now to establish equality between the cones above.

Theorem 2.7 (Dempe and Lohse [67]) *Let (FRC) be satisfied at the point $\overline{y}$. Then equality holds in* (2.21).

Proof Let $\overline{d}$ be an arbitrary element of $T_R(\overline{y})$, that means there is a $\overline{r}$ with $A\overline{d} = \overline{r},\ B\overline{r} = 0,\ \overline{d}_i \geq 0,\ i \in I(\overline{y})\setminus I^0(\overline{y}),\ \overline{d}_i = 0,\ i \in I^0(\overline{y})$. Without loss of generality assume $I(\overline{y}) = \{1, 2, \ldots, l\}$.

We consider the following linear systems (S_1)

$$\begin{aligned} Ad &= \overline{r} \\ d_1 &= \overline{d}_1 \\ d_i &= 0,\ i \in I(\overline{y}) \setminus \{1\} \end{aligned}$$

and (S_j)

$$\begin{aligned} Ad &= 0 \\ d_j &= \overline{d}_j \\ d_i &= 0,\ i \in I(\overline{y}) \setminus \{j\} \end{aligned}$$

for $j = 2, \ldots, l$. These systems have all feasible solutions since $\overline{y}$ satisfies the full rank condition.

Let $d^1, \ldots, d^l$ be (arbitrary) solutions of the systems (S_j) and define the direction $d = \sum_{j=1}^{l} d^j$. Then, $d_i = \overline{d}_i$ for $i \in I(\overline{y})$ as well as $Ad = A\overline{d} = \overline{r}$.

If $d = \overline{d}$ we are done since $d \in \text{cone}\, T(\overline{y}) = \text{conv}\, T(\overline{y})$. Assume that $d \neq \overline{d}$. (Fig. 2.4).

Define $\widehat{d}^1 := d^1 + \overline{d} - d$. Since d^1 is feasible for (S_1) and $d_i = \overline{d}_i$ for $i = 1, \ldots, k$ as well as $Ad = A\overline{d} = \overline{r}$ we obtain $\widehat{d}_i^1 = 0$ for all $i = 2, \ldots, k$ and

$$A\widehat{d}^1 = A(d^1 + \overline{d} - d) = \overline{r} + \overline{r} - \overline{r} = \overline{r}.$$

Hence $\widehat{d}^1$ is also a solution of (S_1).

Thus, $\widehat{d}^1 + \sum\limits_{j=2}^{l} d^j = \overline{d} - d + \sum\limits_{j=1}^{l} d^j = \overline{d}$.

Due to the definition of I and of the tangent cones $T(\overline{y})$ and $T_R(\overline{y})$ the conclusion $T_R(\overline{y}) \subseteq T(\overline{y})$ follows. □

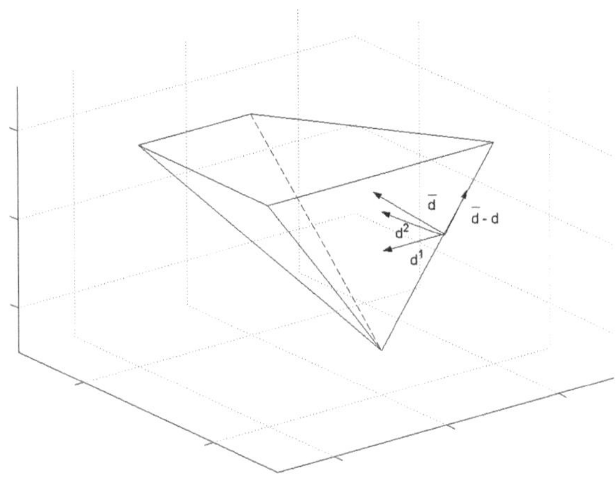

Fig. 2.4 Illustration (taken from Dempe and Lohse [67]) of the proof of Theorem 2.7

Due to Remark 2.2 at most n systems of linear (in-) equalities need to be investigated to compute the index set $I^0(\overline{y})$. Hence, by Theorem 2.7, verification of local optimality of a feasible point of problem (2.17) is possible in polynomial time.

2.3 Solution Algorithms

2.3.1 Computation of a Local Optimal Solution

We consider the linear bilevel optimization problem (2.3), (2.4). y^0 is an optimal solution of the lower level problem iff there exists u such that

$$C^\top u = d,\ u \leq 0,\ u^\top (Cy - x) = 0.$$

Let the rank of the matrix C be equal to m: $r(C) = m$. An optimal solution of problem (2.4) can be found at a vertex of the feasible set, which means that there are m linearly independent rows C_i, $i = 1, \ldots, m$ (without loss of generality, these are the first m rows) of the matrix C such that

$$C_i y = x_i,\ i = 1, \ldots, m$$

and

$$C_i y \geq x_i,\ i = m + 1, \ldots, q.$$

Then, if the first m rows of C compose a matrix D, N is build up of the last $q - m$ rows, $x = (x_D\ x_N)$ is accordingly decomposed, we obtain $C = (D\ N)^\top$ and

$y = D^{\top -1} x_D$ is a solution of problem (2.4). A solution u^0 of the dual problem is given by $u = (u_D\ u_N)^\top$ with $u_D = D^{\top\ -1} d,\ u_N = 0$. Then,

$$u_D \leq 0,\ Dy = x_D,\ Ny \leq x_N,\ u_N = 0,\ Ny \leq x.$$

For D with $D^{\top\ -1} d \leq 0$ the set

$$\mathscr{R}_D = \{x : Dy = x_D,\ Ny \leq x_N \text{ for some } y \in \mathbb{R}^m\}$$

is the so-called *region of stability* for the matrix D. It consists of all parameter vectors x for which an optimal solution of the primal problem (2.4) can be computed using the matrix D.

For other values of x, the basic matrix D consists of other rows of C. This, together with basic matrices for the upper level constraints $Ax \leq c$ can be used to describe an algorithm enumerating all these basic matrices to find a global optimum of the bilevel optimization problem. For this, Theorem 2.1 is of course needed. Many algorithms for solving the linear bilevel optimization problem suggested in the last century used this approach (see e.g. Bard [10]; Candler and Townsley [40]; Bard and Falk [11]).

The idea of the following algorithm can be found in Dempe [49]:

Descent algorithm for the linear bilevel problem:
Input: Linear bilevel optimization problem (2.3).
Output: A local optimal solution.

Algorithm: **Start** Select an optimal basic solution (x^1, y) of the problem

$$\min\{a^\top x + b^\top y : Ax \leq c,\ \ Cy \leq x\}.$$

Compute an optimal basic solution y^1 of the problem (2.4) for $x = x^1$. Set $k := 1$.

Step 1 Select a basic matrix D for y^k, compute the region of stability $\mathscr{R}_D$ and solve the problem

$$\min_x\{a^\top x + b^\top y : x = (x_D\ \ x_N)^\top,\ \ y = D^{\top -1} x_D,\ \ x \in \mathscr{R}_D\}.$$

Let $(\widehat{x}, D^{\top -1}\widehat{x}_D)$ be an optimal solution.

Step 2 Set $x^{k+1} = \widehat{x}$ and compute an optimal basic solution y^{k+1} of the problem

$$\min_y\{b^\top y : y \in \Psi(x^{k+1})\}$$

Stop if the optimal solution has not changed:

$$(x^{k+1}, y^{k+1}) = (x^k, y^k).$$

Otherwise goto Step 1.

This algorithm computes a local optimal solution since either one of the problems in Steps 1 or 2 of the algorithm would lead to a better solution. For a rigorous proof, the interested reader is referred to the original paper Dempe [49].

2.3.2 A Global Algorithm

Consider the linear bilevel optimization problem

$$\min_{x,y}\{a^\top x + b^\top y : Ax \le c,\ (x,y) \in \mathbf{gph}\,\Psi^1\} \tag{2.23}$$

with

$$\Psi^1(x) = \operatorname*{Argmin}_y \{x^\top y : By \le d\} \tag{2.24}$$

and the optimization problem

$$\min_{x,y}\{a^\top x + b^\top y : Ax \le c,\ By \le d,\ x^\top y \le \varphi^1(x)\}, \tag{2.25}$$

where

$$\varphi^1(x) = \min_y\{x^\top y : By \le d\}$$

is the optimal value function of problem (2.24). Both problems (2.23) and (2.25) are fully equivalent. It follows from parametric linear optimization (see e.g. Dempe and Schreier [77] and Beer [15]) that the function $\varphi(\cdot)$ is an in general nondifferentiable, concave, piecewise affine-linear and Lipschitz continuous function. It is equal to

$$\varphi^1(x) = \min\{x^\top y^1, x^\top y^2, \dots, x^\top y^p\},$$

where $\{y^1, y^2, \dots, y^p\}$ is the set of vertices of the convex polyhedron $\{y : By \le d\}$. Strictly speaking, formula (2.25) is correct only on the set of all x for which $|\varphi^1(x)| \le \infty$. If $\varphi^1(\widehat{x}) = \widehat{x}^\top y^k$, then $y^k \in \partial^{Cl}(\widehat{x})$ is an element of the generalized derivative in the sense of Clarke [see (3.10)]. Using the results from convex analysis (see Clarke [42] and Rockafellar [272]) we have

$$\varphi^1(x) \le \varphi^1(\widehat{x}) + \widehat{y}^\top (x - \widehat{x})\ \forall\, x,\ \forall\, \widehat{y} \in \partial^{cl}\varphi^1(\widehat{x}).$$

Hence,

$$\begin{aligned}&\{(x,y) :\ Ax \le c,\ By \le d,\ x^\top y \le \varphi^1(x)\}\\ &\quad \subseteq \{(x,y) : Ax \le c,\ By \le d,\ x^\top y \le \widehat{y}^\top x\}\end{aligned} \tag{2.26}$$

for $\widehat{y} \in \Psi^1(\widehat{x})$. This implies that the problem

$$\min_{x,y}\{a^\top x + b^\top y : Ax \le c,\ By \le d,\ x^\top y \le \widehat{y}^\top x\} \tag{2.27}$$

cannot have a worse objective function value than problem (2.25).

A solution algorithm for the linear bilevel optimization problem (2.23), (2.24)

Algorithm: Start `Select` x^0 `satisfying` $Ax^0 \le c$, `compute` $y^0 \in \Psi(x^0)$. `Set` $k := 1$ `and` $\mathscr{Y} := \{y^0\}$.

Step 1 `Solve problem`

$$\min_{x,y}\{a^\top x + b^\top y : Ax \le c,\ By \le d,\ x^\top y \le \widehat{y}^\top x\ \forall\, \widehat{y} \in \mathscr{Y}\} \tag{2.28}$$

`globally and let` (x^k, y^k) `be an optimal solution.`

Step 2 `If` $y^k \in \Psi(x^k)$, `stop:` (x^k, y^k) `is a global optimal solution of problem` (2.23), (2.24). `Otherwise, set`

$$\mathscr{Y} := \mathscr{Y} \cup \{\widehat{y}^k\}$$

`where` $\widehat{y}^k \in \Psi(x^k)$ `and go to Step 1.`

Theorem 2.8 (Dempe and Franke [60]) *Let* $\{(x, y) : Ax \le c,\ By \le d\}$ *be bounded. The above algorithm computes a global optimal solution of the linear bilevel optimization problem* (2.23), (2.24).

Proof If the algorithm stops in Step 2, the last point (x^k, y^k) is feasible for the linear bilevel optimization problem. Hence, due to (2.26) it is also globally optimal.

Let $\{(x^k, y^k)\}_{k=1}^\infty$ be an infinite sequence computed by the algorithm. Since the set $\mathscr{Y}$ is increasing, more and more constraints are added to problem (2.28) implying that the sequence of its optimal objective function values is nondecreasing. On the other hand, it is bounded from above by e.g. $a^\top x^1 + b^\top b^1$. Hence, this sequence converges to, say, v^*. Let, without loss of generality (x^*, y^*) be a limit point of the sequence $\{x^k, y^k\}_{k=1}^\infty$. Continuity of the function $\varphi(\cdot)$ leads to

$$\lim_{k\to\infty} \varphi(x^k) = \lim_{k\to\infty} x^{k\top} \widehat{y}^k = x^{*\top} \widehat{y}^*,$$

where $\widehat{y}^*$ is again without loss of generality a limit point of the sequence $\{\widehat{y}^k\}_{k=1}^\infty$. Then, we have

$$x^{k\top} y^k \le x^{k\top} \widehat{y}^{k-1}$$

by the formulae in the algorithm. Hence, by convergence of the sequences, we derive

$$x^{*\top} y^* \le \varphi(x^*) = x^{*\top} \widehat{y}^*.$$

Consequently, the point (x^*, y^*) is feasible and, thus, also globally optimal. □

It is clear that the algorithm can be implemented such that it stops after a finite number of iterations if the feasible set of the lower level problem is compact. The reason for this is that the feasible set has only a finite number of vertices which correspond to the vertices of the generalized derivative of the function $\varphi^1(\cdot)$.

One difficulty in realizing the above algorithm is that we need to solve the optimization problem (2.28) globally in each iteration. This is a nonconvex optimization problem and usually solution algorithms compute only stationary or local optimal solutions for such problems. Hence, it is perhaps more suitable to try to compute a local optimal solution of problem (2.23) respectively its equivalent problem (2.25). Often this is related to the use of a sufficient optimality condition. Since (2.25) is a nonsmooth optimization problem we can use a sufficient optimality condition of first order demanding that the directional derivative of the objective function is not negative on a suitable tangent cone to the feasible set.

Let M_R be the feasible set of problem (2.28) for some set $\mathcal{Y} \subseteq \{y : By \leq d\}$. Then, the Bouligand (or tangent) cone to M_R at some point (x^*, y^*) reads as

$$C_{M_R}(x^*, y^*) = \{d \in \mathbb{R}^{2n} : \exists\, \{(x^k, y^k)\}_{k=1}^{\infty} \subseteq M_R,\ \exists \{t_k\}_{k=1}^{\infty} \subseteq \mathbb{R}_+ \setminus \{0\}$$

$$\text{satisfying} \lim_{k\to\infty} (x^k, y^k) = (x^*, y^*),\ \lim_{k\to\infty} t_k = 0 \text{ and}$$

$$d = \lim_{k\to\infty} \frac{1}{t_k}((x^k, y^k) - (x^*, y^*))\}.$$

The local algorithm solving problem (2.23) is identical with the algorithm on the previous page with the only distinction that problem (2.28) in Step 1 of the algorithm is solved locally.

Theorem 2.9 (Dempe and Franke [60]) *Let the set* $\{(x, y) : Ax \leq c,\ By \leq d\}$ *be nonempty and compact. Assume that there is* $\gamma > 0$ *such that* $(a^\top\ b^\top)d \geq \gamma$ *for all* $d \in C_{M_R}(x^k, y^k)$ *and all sufficiently large k. Then, all accumulation points of the sequences computed using the local algorithm are locally optimal for problem* (2.23).

Proof The existence of accumulation points as well as their feasibility for problem (2.23) follow analogously to the proof of Theorem 2.8. Let

$$M_B := \{(x, y) : Ax \leq c,\ By \leq d,\ x^\top y \leq \varphi(x)\}$$

denote the feasible set of problem (2.23). Then, $M_B \subseteq M_R$ and C_{M_R} is a Bouligand cone to a convex set. Hence, for $(x, y) \in M_B$ sufficiently close to (x^k, y^k) we have $d^k := ((x, y) - (x^k, y^k))/\|(x, y) - (x^k, y^k)\| \in C_{M_R}(x^k, y^k)$ and $(a^\top\ b^\top)d^k \geq \gamma$ for sufficiently large k. The Bouligand cone to M_B is defined analogously to the Bouligand cone to M_R.

Let $(\overline{x}, \overline{y})$ be an arbitrary accumulation point of the sequence $\{(x^k, y^k)\}_{k=1}^{\infty}$ computed by the local algorithm. Assume that $(\overline{x}, \overline{y})$ is not a local optimal solution. Then there exists a sequence $\{\overline{x}^k, \overline{y}^k\}_{k=1}^\infty \subset M_B$ converging to $(\overline{x}, \overline{y})$ with $a^\top \overline{x}^k + b^\top \overline{y}^k < a^\top \overline{x} + b^\top \overline{y}$ for all k. Then, by definition, without loss of generality

$$\lim_{k\to\infty} \frac{(\overline{x}^k, \overline{y}^k) - (\overline{x}, \overline{y})}{\|(\overline{x}^k, \overline{y}^k) - (\overline{x}, \overline{y})\|} = \overline{d} \in C_{M_B}(\overline{x}, \overline{y}) \subseteq C_{M_R}(\overline{x}, \overline{y})$$

and $(a^\top\ b^\top)\overline{d} \le 0$. On the other hand, $\overline{d}$ is a limit point of a sequence $\{d^k\}_{k=1}^\infty$ with $d^k \in C_{M_R}(x^k, y^k)$ with $(a^\top\ b^\top)d^k \ge \gamma$ for all k by assumption.

This contradicts the assumption, thus proving the Theorem. □

The following example is presented in Dempe and Franke [60] to illustrate the algorithm.

Example 2.5 Consider the bilevel optimization problem

$$\begin{aligned} \min_{x,y}\ & 2x_1 + x_2 + 2y_1 - y_2 \\ \text{s.t.}\ & |x_1| \le 1 \\ & -1 \le x_2 \le -0.75 \\ & y \in \Psi(x) := \operatorname*{Argmin}_y \{x^\top y : -2y_1 + y_2 \le 0,\ y_1 \le 2,\ 0 \le y_2 \le 2\}. \end{aligned}$$

The concave optimal value function $\varphi(x)$ of the lower level problem reads

$$\varphi(x) = \begin{cases} 2x_1 + 2x_2 & \text{if } x_1 \in [-1, 0], x_2 \in [-1, -0.75] \\ x_1 + 2x_2 & \text{if } x_1 \in (\,0, 1\,], x_2 \in [-1, -0.75] \end{cases}$$

The values of the upper level objective function over the feasible set are

$$a^\top x + b^\top y = \begin{cases} -2 + 2x_1 + x_2 & \text{if } x_1 \in [-1, 0], x_2 \in [-1, -0.75] \\ 2x_1 + x_2 & \text{if } x_1 \in (\,0, 1\,], x_2 \in [-1, -0.75] \end{cases}$$

with the optimal solution at $x = (-1, -1)$ and the optimal function value 5. For $\mathscr{Y} \subseteq \{y : By \le d\}$, the problem (2.28) is

$$\begin{aligned} \min\nolimits_{x,y}\ & 2x_1 + x_2 + 2y_1 - y_2 \\ \text{s.t.}\ & |x_1| \le 1 \\ & -1 \le x_2 \le -0.75 \\ & x^\top y \le \min_{z\in\mathscr{Y}} x^\top z \\ & -2y_1 + y_2 \le 0 \\ & y_1 \le 2 \\ & 0 \le y_2 \le 2. \end{aligned}$$

Now, the above algorithm works as follows.
Start $\mathscr{Y} := \{(0,0)\}, k := 1$.
Step 1 The optimal solution of (2.28) is $(x_1^1, x_2^1, y_1^1, y_2^1) = (-1, -1, 2, 0)$.
Step 2 The lower level with $(x_1^1, x_2^1) = (-1, -1)$ leads to $(z_1^1, z_2^1) = (2, 2)$ which is added to $\mathscr{Y}$. Go to Step 1.
Step 1 The optimal solution of (2.28) is $(x_1^2, x_2^2, y_1^2, y_2^2) = (-1, -1, 2, 2)$.
Step 2 The lower level with $(x_1^2, x_2^2) = (-1, -1)$ leads to $(z_1^2, z_2^2) = (2, 2)$ which coincides with the solution of Step 1, hence the algorithm terminates with the optimal solution $(x_1^2, x_2^2, y_1^2, y_2^2) = (-1, -1, 2, 2)$. □

Chapter 3
Reduction of Bilevel Programming to a Single Level Problem

3.1 Different Approaches

The usually used approach to solve the bilevel optimization problem (1.1), (1.4) or to formulate (necessary or sufficient) optimality conditions, is to transform it into a one-level optimization problem. There are at least the following three possibilities to realize this:

Primal KKT transformation: The lower level problem can be replaced with its necessary (and sufficient) optimality conditions. This can only be done if the lower level problem is (for a fixed value of the parameter x) convex. Otherwise, the feasible set of the bilevel optimization problem is enlarged by local optimal solutions and stationary points of the lower level problem. In this case, the global optimal solution of the bilevel problem is in general not a stationary solution for the resulting problem (Mirrlees [232]). Let

$$Y(x) := \{y : g(x, y) \leq 0\}$$

denote the feasible set of the lower level problem and assume that $Y(x)$ is a convex set, $y \mapsto f(x, y)$ is a convex function, and $T \subseteq \mathbb{R}^m$ is a convex set. Then, $y \in \Psi(x)$ if and only if

$$0 \in \partial_y f(x, y) + N_{Y(x) \cap T}(y).$$

Here, $\partial_y f(x, y)$ is the subdifferential of the convex function $y \mapsto f(x, y)$ and $N_{Y(x) \cap T}(y)$ is the *normal cone* of convex analysis to the set $Y(x) \cap T$ at the point $y \in Y(x) \cap T$:

$$N_{Y(x) \cap T}(y) := \{d \in \mathbb{R}^m : d^\top (z - y) \leq 0 \quad \forall z \in Y(x) \cap T\}.$$

Note that $N_{Y(x) \cap T}(y) = \emptyset$ for $y \notin Y(x) \cap T$.

S. Dempe et al., *Bilevel Programming Problems*, Energy Systems,
DOI 10.1007/978-3-662-45827-3_3

The problem (1.4) is then equivalent to

$$\min\{F(x, y) : G(x) \le 0,\ 0 \in \partial_y f(x, y) + N_{Y(x)\cap T}(y),\ x \in X\}, \tag{3.1}$$

see Dempe and Zemkoho [80]. This is the *primal Karush-Kuhn-Tucker reformulation* (*primal KKT reformulation*).
Problem (3.1) is fully equivalent to the bilevel optimization problem (1.1), (1.4) both if global and local optimal solutions of the problems are considered.

Classical KKT transformation: If $T = \mathbb{R}^m$, and $Y(x) = \{y : g(x, y) \le 0\}$, the function $y \mapsto g(x, y)$ is convex for each fixed x, and a regularity condition, as e.g.

Slater's condition: There exists $\widehat{y}$ with $g(x, \widehat{y}) < 0$,

is fulfilled, then $y \in \Psi(x)$ if and only if the Karush-Kuhn-Tucker conditions (KKT conditions) are satisfied:

$$0 \in \partial_y f(x, y) + \lambda^\top \partial_y g(x, y),\ \lambda \ge 0,\ \lambda^\top g(x, y) = 0.$$

This leads to a second reformulation of the bilevel optimization problem:

$$\begin{aligned} F(x, y) &\to \min \\ G(x) &\le 0 \\ 0 &\in \partial_y f(x, y) + \lambda^\top \partial_y g(x, y) \\ \lambda &\ge 0,\ g(x, y) \le 0,\ \lambda^\top g(x, y) = 0 \\ x &\in X. \end{aligned} \tag{3.2}$$

This is the reformulation most often used. Problem (3.2) is a (nonsmooth) mathematical program with complementarity constraints (or better with a generalized equation constraint). It can be called the *classical KKT transformation*. Using the relation

$$N_{Y(x)}(y) = \{d \in \mathbb{R}^m : \exists \lambda \ge 0,\ \lambda^\top g(x, y) = 0,\ d = \lambda^\top \partial_y g(x, y)\}$$

which is valid if Slater's condition is satisfied at x (see e.g. Dhara and Dutta [87]), problem (3.1) reduces to (3.2).
The introduction of additional variables λ causes that problems (3.2) and (1.1), (1.4) are no longer fully equivalent. Clearly, each global optimal solution $(\overline{x}, \overline{y}, \overline{\lambda})$ of problem (3.2) is related to a global optimal solution $(\overline{x}, \overline{y})$ of (1.1), (1.4). Also, a local optimal solution $(\overline{x}, \overline{y})$ of (1.1), (1.4) is related to a local optimal solution $(\overline{x}, \overline{y}, \overline{\lambda})$ for each

$$\overline{\lambda} \in \Lambda(\overline{x}, \overline{y}) := \{\lambda \ge 0 : \lambda^\top g(\overline{x}, \overline{y}) = 0,\ 0 \in \partial_y f(\overline{x}, \overline{y}) + \lambda^\top \partial_y g(\overline{x}, \overline{y})\},$$

provided that Slater's condition is satisfied. With respect to the opposite direction we have

Theorem 3.1 (Dempe and Dutta [55]) *Let the lower level problem* (1.1) *be a convex optimization problem and assume that Slater's condition is satisfied for all $x \in X$ with $\Psi(x) \neq \emptyset$. A feasible point $(\overline{x}, \overline{y})$ of problem* (1.4) *is a local optimal solution of this problem iff $(\overline{x}, \overline{y}, \overline{\lambda})$ is a local optimal solution of problem* (3.2) *for each $\overline{\lambda} \in \Lambda(\overline{x}, \overline{y})$.*

Proof If $(\overline{x}, \overline{y})$ is a local optimal solution of problem (1.4) then, since the Karush-Kuhn-Tucker conditions are sufficient and necessary optimality conditions and the objective function of (3.2) does not depend on the Lagrange multiplier the point $(\overline{x}, \overline{y}, \overline{\lambda})$ is a local optimum of (1.4) for all $\lambda \in \Lambda(\overline{x}, \overline{y})$.
Let $(\overline{x}, \overline{y}, \overline{\lambda})$ be a local optimal solution of (3.2) for all $\lambda \in \Lambda(\overline{x}, \overline{y})$ and assume that $(\overline{x}, \overline{y})$ is not a local optimal solution of (1.4). Then, there exists a sequence $\{(x^k, y^k)\}_{k=1}^{\infty}$ of feasible points for (1.4) converging to $(\overline{x}, \overline{y})$ such that $F(x^k, y^k) < F(\overline{x}, \overline{y})$ for all k. Since the KKT conditions are necessary optimality conditions there exists a sequence $\{\lambda^k\}_{k=1}^{\infty}$ with $\lambda^k \in \Lambda(x^k, y^k)$. Obviously, (x^k, y^k, λ^k) is feasible for (3.2) for all k. Since the mapping $(x, y) \mapsto \Lambda(x, y)$ is upper semicontinuous (Robinson [270]), the sequence $\{\lambda^k\}$ has an accumulation point $\widehat{\lambda} \in \Lambda(\overline{x}, \overline{y})$ and $(\overline{x}, \overline{y}, \widehat{\lambda})$ is feasible for (3.2). This violates the assumption. Hence, the theorem is proved. □

The following example shows that it is essential to consider all Lagrange multipliers:

Example 3.1 (*Dempe and Dutta* [55]) Consider the linear lower level problem

$$\min_y \{-y : x + y \leq 1,\ -x + y \leq 1\} \tag{3.3}$$

having the unique optimal solution $y(x)$ and the set $\Lambda(x, y)$ of Lagrange multipliers

$$y(x) = \begin{cases} x+1 & \text{if } x \leq 0 \\ -x+1 & \text{if } x \geq 0 \end{cases}, \quad \Lambda(x, y) = \begin{cases} \{(1,0)\} & \text{if } x > 0 \\ \{(0,1)\} & \text{if } x < 0 \\ \operatorname{conv}\{(1,0),(0,1)\} & \text{if } x = 0 \end{cases}$$

where conv A denotes the convex hull of the set A. This example is illustrated in Fig. 3.1.
The bilevel optimization problem

$$\min\{(x-1)^2 + (y-1)^2 : (x, y) \in \mathbf{gph}\, \Psi\} \tag{3.4}$$

has the unique optimal solution $(\overline{x}, \overline{y}) = (0.5, 0.5)$ and no local optimal solutions. The points $(\overline{x}, \overline{y}, \overline{\lambda}_1, \overline{\lambda}_2) = (0.5, 0.5, 1, 0)$ and $(x^0, y^0, \lambda_1^0, \lambda_2^0) = (0, 1, 0, 1)$ are global or local optimal solutions, resp., of problem (3.2). To see that the point $(x^0, y^0, \lambda_1^0, \lambda_2^0) = (0, 1, 0, 1)$ is locally optimal remark that in a not too large open neighborhood V of this point we have $\lambda_2 > 0$ implying that the second

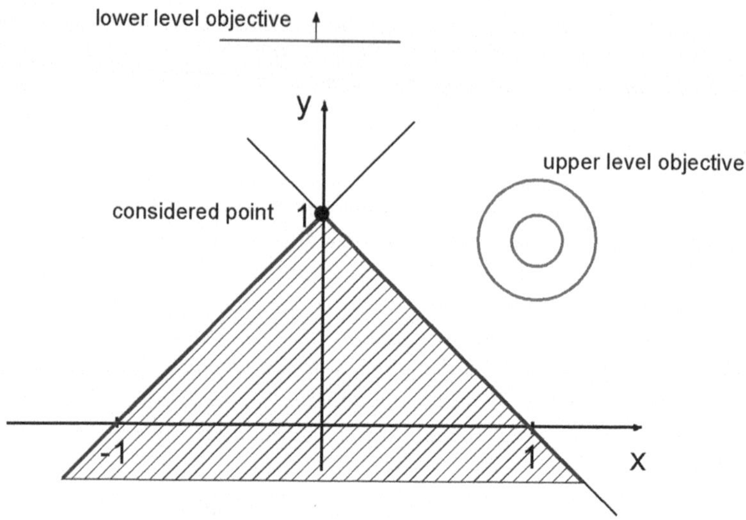

Fig. 3.1 The considered stationary solution in Example 3.1

constraint in problem (3.3) is active, i.e. $y = x + 1$ which by feasibility implies $x \leq 0$. Substituting this into the upper level objective function gives

$$(x-1)^2 + (y-1)^2 = (x-1)^2 + x^2 \geq 1$$

for each feasible point of the mathematical program with equilibrium constraints corresponding to the bilevel problem (3.4) in V. Together with $F(0, 1) = 1$ we derive that $(x^0, y^0, \lambda_1^0, \lambda_2^0)$ is indeed a local optimum of the mathematical program with equilibrium constraints corresponding to (3.4). □

If the upper level objective function is replaced with a linear function it can also be shown that a local optimal solution of the MPEC corresponding to a linear bilevel optimization problem needs not to be related to a local optimum of the original problem.
The result in Theorem 3.1 is not correct if Slater's conditions is violated at some points, see Dempe and Dutta [55].

Example 3.2 Consider the convex lower level problem

$$\min\{y_1 : y_1^2 - y_2 \leq x,\ y_1^2 + y_2 \leq 0\} \tag{3.5}$$

Then, for $x = 0$, Slater's condition is violated, $y = 0$ is the only feasible point for $x = 0$. The optimal solution for $x \geq 0$ is

$$y(x) = \begin{cases} (0,0)^\top & \text{for } x = 0 \\ (-\sqrt{x/2}, -x/2)^\top & \text{for } x > 0 \end{cases}$$

The Lagrange multiplier is $\lambda(x)$ with $\lambda_1(x) = \lambda_2(x) = \frac{1}{4\sqrt{x/2}}$ for $x > 0$. For $x = 0$, the problem is not regular and the Karush-Kuhn-Tucker conditions are not satisfied.

Let us consider the bilevel optimization problem

$$\min\{x : x \geq 0, \; y \in \Psi(x)\} \tag{3.6}$$

where $\Psi(x)$ is the solution set mapping of problem (3.5).

Then, the unique (global) optimal solution of the bilevel optimization problem (3.6) is $x = 0, \; y = 0$ and there do not exist local optimal solutions different from the global solution.

Consider the corresponding MPEC. Then, $(x, y(x), \lambda(x))$ is feasible for the MPEC for $x > 0$ and the objective function value converges to zero for $x \to 0$. However, an optimal solution of this problem does not exist, since for $x = 0$ the only optimal solution of the lower level problem is $y = 0$ and there does not exist a corresponding feasible solution of the MPEC.

Note that this example can be used to show that the MPEC need not to have any global optimal solution even if its feasible set is not empty and bounded. □

The next result shows, that under a mild assumption the assertion in Theorem 3.1 is true iff $(\overline{x}, \overline{y})$ is globally optimal for problem (1.4).

Theorem 3.2 (Dempe and Dutta [55]) *Let $(\overline{x}, \overline{y}, \overline{\lambda})$ be a global optimal solution of problem* (3.2)*, assume that the functions $y \mapsto f(x, y)$, $y \mapsto g_i(x, y)$, $i = 1, \ldots, p$ are convex and that Slater's constraint qualification is satisfied for the lower level problem* (1.1) *for each $x \in X$. Then, $(\overline{x}, \overline{y})$ is a global optimal solution of the bilevel optimization problem.*

The main part of the proof in Dempe and Dutta [55] is based on independence of the optimal function value on the vector $\lambda \in \Lambda(x, y)$.

The following example is borrowed from Dempe and Dutta [55]:

Example 3.3 Consider the lower level problem

$$\Psi_L(x) := \operatorname*{Argmin}_{y,z}\{-y - z : x + y \leq 1, \; -x + y \leq 1, \; 0 \leq z \leq 1\}$$

and the bilevel optimization problem

$$\min\{0.5x - y + 3z : (y, z) \in \Psi_L(x)\}.$$

Then, we derive $\Psi_L(x) = \{(y(x), z(x)) : z(x) = 1, \; x \in \mathbb{R}\}$ with

$$y(x) = \begin{cases} 1 - x & \text{if } x \geq 0 \\ 1 + x & \text{if } x \leq 0. \end{cases}$$

Substituting this solution into the upper level objective function we obtain

$$0.5x - y(x) + 3z(x) = \begin{cases} 0.5x - 1 + x + 3 = 2 + 1.5x \geq 2 & \text{if } x \geq 0 \\ 0.5x - 1 - x + 3 = 2 - 0.5x \geq 2 & \text{if } x \leq 0. \end{cases}$$

Hence, $(\overline{x}, \overline{y}, \overline{z}) = (0, 1, 1)$ is the global (and unique local) optimal solution of the problem. At this point, three constraints in the lower level problem are active and the linear independence constraint qualification is violated.
Note that small smooth perturbations of the data of both the lower and the upper level problems will have no impact on this property. To see this, consider e.g. the case when the right-hand side of the first constraint $x + y \leq 1$ in the follower's problem is perturbed and the new lower level problem reads as follows:

$$\Psi_L(x) := \operatorname*{Argmin}_{y,z}\{-y - z : x + y \leq 1 + \alpha,\ -x + y \leq 1,\ 0 \leq z \leq 1\}.$$

Then, the optimal solution of the lower level problem becomes $z = 1$ and

$$y(x) = \begin{cases} 1 - x + \alpha & \text{if } x \geq 0.5\alpha \\ 1 + x & \text{if } x \leq 0.5\alpha. \end{cases}$$

Substituting this solution into the upper level objective function gives

$$0.5x - y(x) + 3z(x) = 0.5x - 1 + x - \alpha + 3 = 2 + 1.5x - \alpha \geq 2 - \alpha/4$$

if $x \geq 0.5\alpha$ and

$$0.5x - y(x) + 3z(x) = 0.5x - 1 - x + 3 = 2 - 0.5x \geq 2 - \alpha/4$$

if $x \leq 0.5\alpha$. The optimal objective function value is then $2 - \alpha/4$ with optimal solution $(x, y, z) = (\alpha/2, 1 + \alpha/2, 1)$ and again three constraints of the lower level problem are active.
The other perturbations can be treated analogously. □

Generic properties of bilevel optimization problems and mathematical programs with equilibrium constraints can be found in the papers Allende and Still [2], Jongen et al. [158], Jongen and Shikhman [159], Jongen et al. [160].
Due to its importance with respect to solution algorithms for bilevel optimization problems two main observations should be repeated:

1. Solution algorithms for mathematical programs with equilibrium constraints compute stationary points as e.g. C-stationary or M-stationary points (see page 64). Under additional assumptions these points are local minima of the MPEC. Due to Theorem 3.1 such a solution needs not to be related to a local optimum of the bilevel optimization problem if the Lagrangian multiplier of the lower level problem is not unique.

2. The last example shows that the linear independence constraint qualification is not generically satisfied in the lower level problem at the (global) optimal solution of the bilevel optimization problem. Hence, the Lagrange multiplier for the lower level problem needs not to be unique in general. This is in general not related to the MPEC-LICQ defined on page 68. This is a bit surprising since (LICQ) is generically satisfied at optimal solutions of smooth optimization problems.

Optimal value transformation: Recall the definition of the optimal value function $\varphi(x)$ on page 2. Then, a third reformulation is

$$\begin{aligned} F(x,y) &\to \min \\ G(x) &\leq 0 \\ f(x,y) &\leq \varphi(x) \\ g(x,y) &\leq 0,\ y \in T, \\ x &\in X. \end{aligned} \tag{3.7}$$

This, again is a nonsmooth optimization problem since the optimal value function is in general not differentiable, even if all the constraint functions and the objective function in the lower level problem are smooth. We call this the *optimal value transformation*. Problem (3.7) is again fully equivalent to problem (1.4).

3.2 Parametric Optimization Problems

Dependence of optimal solutions and of the optimal function value of optimization problems on perturbation of the functions describing it is an important issue. We will need results of this type at different places.

Consider an optimization problem

$$\min_y \{f(x,y) : g(x,y) \leq 0\}, \tag{3.8}$$

where $f : \mathbb{R}^n \times \mathbb{R}^m \to \mathbb{R}$, $g : \mathbb{R}^n \times \mathbb{R}^m \to \mathbb{R}^p$ are sufficiently smooth functions and let

$$Y(x) := \{y : g(x,y) \leq 0\}$$

denote the *feasible set mapping*,

$$\varphi(x) := \min_y \{f(x,y) : g(x,y) \leq 0\}$$

be the *optimal value function*, and

$$\Psi(x) := \{y \in Y(x) : f(x,y) = \varphi(x)\}$$

be the *solution set mapping*.

Note that the investigation of stability and sensitivity properties for optimization problems can be done for more general problems as for instance in the monograph Guddat et al. [133]. For the investigation of bilevel optimization problems it is sufficient to investigate problem (3.8).

Definition 3.1 A point-to-set mapping $\Gamma : \mathbb{R}^n \rightrightarrows \mathbb{R}^m$ (mapping points $x \in \mathbb{R}^n$ to subsets in $\mathbb{R}^m$) is called *upper semicontinuous* at a point $x^0 \in \mathbb{R}^n$ if, for each open set $V \supseteq \Gamma(x^0)$ there is an open set $U \ni x^0$ with $\Gamma(x) \subseteq V$ for all $x \in U$.

Γ is called *lower semicontinuous* at x^0, if for each open set V with $\Gamma(x^0) \cap V \neq \emptyset$ there is an open set $U \ni x^0$ such that $\Gamma(x) \cap V \neq \emptyset$ for all $x \in U$.

A slightly weaker condition than lower semicontinuity is inner semicontinuity.

Definition 3.2 (*Mordukhovich* [241]) A point-to-set mapping $\Gamma : \mathbb{R}^n \rightrightarrows \mathbb{R}^m$ is said to be *inner semicompact* at $\overline{z}$ if, for each sequence $\{z^k\}_{k=1}^\infty$ converging to $\overline{z}$ there is a sequence $\{\alpha^k\}_{k=1}^\infty$, $\alpha^k \in \Gamma(z^k)$ that contains a convergent subsequence. It is called *inner semicontinuous* at $(\overline{z}, \overline{\alpha}) \in$ **gph** Γ provided that, for each sequence $\{z^k\}_{k=1}^\infty$ converging to $\overline{z}$ there is a sequence $\{\alpha^k\}_{k=1}^\infty$, $\alpha^k \in \Gamma(z^k)$ converging to $\overline{\alpha}$.

Clearly, if $\{(x, y) : g(x, y) \leq 0\}$ is not empty and compact, then the mappings $x \to Y(x)$ and $x \to \Psi(x)$ are inner semicompact.

In the following theorem, basic continuity results of the mapping of Lagrange multipliers

$$\Lambda(x, y) = \{\lambda \in \mathbb{R}^p_+ : 0 = \nabla_y f(x, y) + \lambda^\top \nabla_y g(x, y),\ \lambda^\top g(x, y) = 0\},$$

the set of stationary points

$$SP(x) = \{y : \Lambda(x, y) \neq \emptyset\},$$

and the optimal value function are shown assuming that the *Mangasarian-Fromovitz constraint qualification* (*MFCQ*) is satisfied for problem (3.8) at the point $(\overline{x}, \overline{y})$:

(**MFCQ**) There exists $d \in \mathbb{R}^m$ with $\nabla_y g_i(\overline{x}, \overline{y})d < 0$ for all $i : g_i(\overline{x}, \overline{y}) = 0$.

Theorem 3.3 (Bank et al. [8]; Robinson [270]) *Consider problem* (3.8) *at* $x = \overline{x}$, *let* $\{y : g(\overline{x}, y) \leq 0\} \neq \emptyset$ *be compact, the functions* f, g_i *be at least continuously differentiable for all* i, *and let* (*MFCQ*) *be satisfied at all points* $y \in Y(\overline{x})$. *Then, the point-to-set mappings* $(x, y) \mapsto \Lambda(x, y)$ *and* $x \mapsto SP(x)$ *are upper semicontinuous at* $(x, y) = (\overline{x}, \overline{y})$ *respectively* $x = \overline{x}$. *Moreover, the function* $x \mapsto \varphi(x)$ *is continuous at* $x = \overline{x}$.

Proof 1. First we show that the set $\Lambda(\cdot, \cdot)$ is locally bounded at $(\overline{x}, \overline{y})$ with $\overline{y} \in SP(\overline{x})$ which means that there exist an open neighborhood $W(\overline{x}, \overline{y})$ and a compact set K such that

$$\Lambda(x, y) \subseteq K \quad \forall (x, y) \in W(\overline{x}, \overline{y}).$$

Arguing from a contradiction and assuming that there is a sequence $\{\lambda^k\}_{k=1}^\infty$ with $\lambda^k \in \Lambda(x^k, y^k)$, (x^k, y^k) converging to $(\overline{x}, \overline{y})$ such that $\lim_{k\to\infty} \|\lambda^k\| = \infty$ it is not difficult to show that $\{\lambda^k/\|\lambda^k\|\}_{k=1}^\infty$ converges without less of generality to a point κ such that

$$\lim_{k\to\infty} \left\{ \frac{\nabla_y f(x^k, y^k)}{\|\lambda^k\|} + \frac{\lambda^k}{\|\lambda^k\|}^\top \nabla_y g(x^k, y^k) \right\} = \kappa^\top \nabla_y g(\overline{x}, \overline{y}) = 0$$

and $\kappa \geq 0$, $\|\kappa\| = 1$. Then, since by (MFCQ) there exists a direction d such that $\nabla_y g_i(\overline{x}, \overline{y})d < 0$ for all i with $g_i(\overline{x}, \overline{y}) = 0$ we derive

$$0 = \kappa^\top \nabla_y g(\overline{x}, \overline{y})d < 0$$

which is not possible.

2. Now, if $\{(x^k, y^k)\}_{k=1}^\infty$ converges to $(\overline{x}, \overline{y})$ and $\lambda^k \in \Lambda(x^k, y^k)$, due to local boundedness of $\Lambda(\cdot, \cdot)$ and continuity of all functions involved, the sequence $\{\lambda^k\}_{k=1}^\infty$ has an accumulation point $\overline{\lambda} \in \Lambda(\overline{x}, \overline{y})$. This also implies that $\overline{y} \in SP(\overline{x})$. Hence, both $\Lambda(\cdot, \cdot)$ and $SP(\cdot)$ are upper semicontinuous at $x = \overline{x}$.
3. Let $\{(x^k, y^k)\}_{k=1}^\infty$ be a sequence converging to $(\overline{x}, \overline{y})$ with $y^k \in Y(x^k)$ and $\varphi(x^k) = f(x^k, y^k)$. Then, by upper semicontinuity of $SP(\cdot)$, $\overline{y} \in SP(\overline{x}) \subseteq Y(\overline{x})$, and we derive
$$\varphi(\overline{x}) \leq f(\overline{x}, \overline{y}) = \liminf_{x\to\overline{x}} \varphi(x).$$

   On the other hand, if $\overline{y} \in Y(\overline{x})$ with $\varphi(\overline{x}) = f(\overline{x}, \overline{y})$ then, due to validity of the (MFCQ) the mapping $x \mapsto Y(x)$ is lower semicontinuous (see Bank et al. [8]) and, for each sequence $\{x^k\}_{k=1}^\infty$ there exists a sequence $\{y^k\}_{k=1}^\infty$ with $y^k \in Y(x^k)$ converging to $\overline{y}$. This implies
$$\varphi(\overline{x}) = f(\overline{x}, \overline{y}) = \lim_{k\to\infty} f(x^k, y^k) \geq \limsup_{x\to\overline{x}} \varphi(x).$$

   Hence, the function $\varphi(\cdot)$ is continuous at $x = \overline{x}$. □

Remark 3.1 Since $\Psi(x) \subseteq SP(x)$, it is possible to show that the solution set mapping is upper semicontinuous, too.

Remark 3.2 Let the function f be continuous. If the problem

$$\min_y \{f(x, y) : y \in Y(x)\}$$

is considered at some parameter value $\overline{x}$ where the feasible set mapping $x \mapsto Y(x)$ is only assumed to be upper semicontinuous, then the optimal value function

$$\varphi(x) := \min_y \{f(x, y) : y \in Y(x)\}$$

is lower semicontinuous. On the other hand, if $x \mapsto Y(x)$ is lower semicontinuous at $\overline{x}$, then, $\varphi(\cdot)$ is upper semicontinuous at $\overline{x}$. The lower semicontinuity assumption can be weakened to inner semicontinuity at the point $(\overline{x}, \overline{y}) \in \mathbf{gph}\, \Psi$.

Theorem 3.3 with the subsequent remarks has important implications concerning the existence of optimal solutions of the bilevel optimization problem. For this note that upper semicontinuity of the solution set mapping implies that the set $\{(x, y) : G(x) \leq 0,\ (x, y) \in \mathbf{gph}\, \Psi,\ x \in X\}$ is closed provided that the function G is continuous and the set X is closed.

Theorem 3.4 *Let the functions F, G be continuous and the functions f, g be continuously differentiable. Assume that the set $\{(x, y) : G(x) \leq 0,\ g(x, y) \leq 0\}$ is not empty and bounded, $X = \mathbb{R}^n$, and let (MFCQ) be satisfied at all points $(x, y) \in \mathbf{gph}\, Y$. Then, problem* (1.4) *has an optimal solution.*

Problem (1.4) is a nonconvex optimization problem. Under the assumptions of Theorem 3.4 it has a global optimal solution. But, in general, it can also have local optima.

If the point-to-set mapping Ψ is upper semicontinuous, the functions F, G are continuous and the set $\{x : G(x) \leq 0\}$ is not empty and compact then, according to Remark 3.2, the optimistic bilevel optimization problem (1.3) has an optimal solution.

For existence of an optimal solution of the pessimistic bilevel optimization problem (1.6) we need lower semicontinuity (or at least inner semicontinuity at an optimal solution) of the point-to-set mapping Ψ (see Lucchetti et al. [207]).

Definition 3.3 (*Klatte and Kummer* [189]) A point-to-set mapping $\Gamma : \mathbb{R}^n \rightrightarrows \mathbb{R}^m$ is called *pseudo-Lipschitz continuous* at $(\overline{x}, \overline{y})$ if there are open neighborhoods U and V of $\overline{x}$ and $\overline{y}$ and a finite number L_Γ such that, given $(x, y) \in (U \times V) \cap \mathbf{gph}\, \Gamma$ and $x' \in U$

there exists $y' \in \Gamma(x')$ such that $\|y - y'\| \leq L_\Gamma \|x - x'\|$.

In other references this property is called Lipschitz-like or Aubin property. It can also be written as

$$\rho(y, \Gamma(x')) := \min_z\{\|y - z\| : z \in \Gamma(x')\} \leq L_\Gamma \|x - x'\| \quad \forall\, y \in \Gamma(x) \cap V$$

for each $x, x' \in U$. Here, $\rho(\cdot, \Gamma(\cdot))$ denotes the distance function.

It has been shown in Robinson [268] that the mapping $\{y : g(y) \leq x^1,\ h(y) = x^2\}$ with continuously differentiable functions g, h is pseudo-Lipschitz continuous if and only if the Mangasarian-Fromovitz constraint qualification is satisfied. The idea of linear transformations (Rockafellar [273]) can be used to apply this result to mappings defined with nonlinearly perturbed functions. One method for this is the following:

Consider the set

$$Y(x) = \{y : g(x, y) \leq 0\}$$

and relate it to the set

$$\widehat{Y}(x) := \{(y, z) : g(z, y) \leq 0,\ z = x\}.$$

Then, if the Mangasarian-Fromovitz constraint qualification is satisfied at $\overline{y} \in Y(\overline{x})$ it is also satisfied at the point $(\overline{z}, \overline{y}) \in \widehat{Y}(\overline{x})$. Using the result of Robinson [268] pseudo-Lipschitz continuity is derived for the mapping $x \mapsto \widehat{Y}(x)$ which then implies pseudo-Lipschitz continuity for the mapping $x \mapsto Y(x)$, too.

Theorem 3.5 (Klatte and Kummer [188]; Mordukhovich and Nam [244]) *Consider problem* (1.1) *at* $x = \overline{x}$ *with* $T = \mathbb{R}^m$, *and let (MFCQ) be satisfied at every point* $y \in Y(\overline{x})$, *assume* $f, g_i \in C^1(\mathbb{R}^n \times \mathbb{R}^m, \mathbb{R})$ *for all* $i = 1, \dots, p$ *and let* $\{(x, y) : g(x, y) \leq 0\}$ *be nonempty and compact. Then, the function* $\varphi(\cdot)$ *is locally Lipschitz continuous.*

Proof Continuously differentiable functions are locally Lipschitz continuous, let $L_f < \infty$ be a Lipschitz constant for f. Take x, x' sufficiently close to $\overline{x}$, $y' \in \Psi(x')$, $y \in \Psi(x)$. Take $y'' \in Y(x')$ such that $\|y - y''\| \leq L_Y \|x - x'\|$ with finite L_Y. Then, since $f(x', y') \leq f(x', y'')$ we have

$$\begin{aligned} \varphi(x') - \varphi(x) = f(x', y') - f(x, y) \leq f(x', y'') - f(x, y) \leq \\ L_f \|(x', y'') - (x, y)\| \leq L_f(\|x - x'\| + \|y - y''\|) \leq \\ L_f(1 + L_Y)\|x - x'\|. \end{aligned}$$ □

Locally Lipschitz continuous functions have *generalized derivatives* and we will need a formula for this in what follows.

Theorem 3.6 (Gauvin and Dubeau [127]) *Consider problem* (1.1) *with* $T = \mathbb{R}^m$, *let the set* $\{(x, y) : g(x, y) \leq 0\}$ *be nonempty and compact,* $f, g_i \in C^1(\mathbb{R}^n \times \mathbb{R}^m, \mathbb{R})$ *for all* $i = 1, \dots, p$ *and assume that (MFCQ) is satisfied for* $x = \overline{x}$ *and all* $y \in Y(\overline{x})$. *Then, the function* $\varphi(\cdot)$ *is locally Lipschitz continuous at* $\overline{x}$ *and*

$$\partial^{Cl} \varphi(\overline{x}) \subseteq \operatorname{conv} \bigcup_{y \in \Psi(\overline{x})} \bigcup_{\lambda \in \Lambda(\overline{x}, y)} \nabla_x L(\overline{x}, y, \lambda). \tag{3.9}$$

Proof The function $\varphi(\cdot)$ is locally Lipschitz continuous by Theorem 3.5. By Rademacher's theorem (see e.g. Clarke [42])

$$\partial^{Cl} \varphi(\overline{x}) = \operatorname{conv} \{ \lim_{k \to \infty} \nabla_x \varphi(x^k) : \lim_{k \to \infty} x^k = \overline{x},\ \nabla_x \varphi(x^k) \text{ exists } \forall k \}. \tag{3.10}$$

Using the upper Dini directional derivative

$$D^+ \varphi(\overline{x}; r) := \limsup_{t \downarrow 0} \frac{\varphi(\overline{x} + tr) - \varphi(\overline{x})}{t}$$

it can be shown (see e.g. Dempe [52, Theorem 4.15]) that

$$\nabla_x \varphi(x^k) r = D^+ \varphi(x^k; r) \leq \nabla_x L(x^k, y^k, \lambda^k) r$$

for some $y^k \in \Psi(x^k)$, $\lambda^k \in \Lambda(x^k, y^k)$ provided that $\nabla \varphi(x^k)$ exists. By Theorem 3.3 the mappings $x \mapsto \Psi(x)$ and $(x, y) \mapsto \Lambda(x, y)$ are upper semicontinuous. For each $\zeta \in \partial^{Cl} \varphi(\overline{x})$ there exists $\overline{y} \in \Psi(\overline{x})$ and $\overline{\lambda} \in \Lambda(\overline{x}, \overline{y})$ such that

$$\zeta r \leq \nabla_x L(\overline{x}, \overline{y}, \overline{\lambda}) r \leq \max\{\eta r : \eta \in \bigcup_{y \in \Psi(\overline{x})} \bigcup_{\lambda \in \Lambda(\overline{x}, y)} \nabla_x L(\overline{x}, y, \lambda)\}. \quad (3.11)$$

Now, the result follows from the fact, that $A \subseteq \text{cl conv } B \subseteq \mathbb{R}^n$ if and only if

$$\max\{\alpha r : \alpha \in A\} \subseteq \max\{\beta r : \beta \in B\} \; \forall \, r \in \mathbb{R}^n,$$

$\Psi(\overline{x})$ and $\Lambda(\overline{x}, \overline{y})$ are (locally) compact and, hence, the right-hand side set in (3.9) is closed. □

Using the assumptions in Theorem 3.6, from Mordukhovich [242, Theorem 5.2] it follows that

$$\partial^M \varphi(\overline{x}) \subseteq \bigcup_{y \in \Psi(\overline{x})} \bigcup_{\lambda \in \Lambda(\overline{x}, y)} \nabla_x L(\overline{x}, y, \lambda).$$

3.3 Convex Quadratic Lower Level Problem

Vicente and Calamai investigate in [370] the bilevel optimization problem

$$\begin{aligned} F(x, y) &\rightarrow \min_{x,y} \\ y &\in \operatorname*{Argmin}_z \{q_x(z) : Ax + By \leq c\} \end{aligned} \quad (3.12)$$

with $F : \mathbb{R}^n \times \mathbb{R}^m$, $q_x(y) = \frac{1}{2} y^\top Q y + y^\top R x + r^\top x$, matrices A, B, Q, R and vectors r of appropriate dimensions, $c \in \mathbb{R}^p$. The matrix Q is assumed to symmetric and positive definite. The lower level problem is in this case a strictly convex quadratic optimization problem, its optimal solution exists and is unique whenever the feasible set of this problem is not empty.

Denote $\Psi_Q(x) = \operatorname{Argmin}_z \{q_x(z) : Ax + By \leq c\}$.

Let $\overline{x}, d \in \mathbb{R}^n$ be given such that $Y(\overline{x} + td) := \{y : A(\overline{x} + td) + By \leq c\} \neq \emptyset$ for $t > 0$ sufficiently small. Then, since the objective function $q_x(\cdot)$ is strictly convex and quadratic, $\Psi_Q(\overline{x} + td) \neq \emptyset$ and the optimal solution is unique. Let $\{y(\overline{x} + td)\} = \Psi_Q(\overline{x} + td)$, $y(\overline{x} + td$ converges to $\overline{y}$ for t tending to zero. Since the constraints are linear, the Karush-Kuhn-Tucker conditions are necessary and sufficient optimality conditions, i.e. there exist $\lambda(\overline{x}+td) \in \Lambda(\overline{x}+td, y(\overline{x}+td))$. Let $\lambda(\overline{x}+td)$ be a vertex of the set $\Lambda(\overline{x}+td, y(\overline{x}+td))$. $\lambda(\overline{x}+td)$ is a nonnegative solution of the system of linear equations:

$$Qy(\overline{x}+td) + R(\overline{x}+td) + r + B^\top\lambda = 0,\ \lambda^\top(A(\overline{x}+td) + By(\overline{x}+td) - c) = 0. \tag{3.13}$$

It follows that the rows B_i of the matrix B with $\lambda(\overline{x}+td)_i > 0$ are linearly independent. Denote

$$\mathscr{I}(\overline{x}+td) := \{i : \lambda(\overline{x}+td)_i > 0\}.$$

By upper semicontinuity of the set of Lagrange multipliers $\Lambda(\cdot,\cdot)$, $\lambda(\overline{x}+td)$ converges to a vertex $\lambda^0 \in \Lambda(\overline{x}, y(\overline{x}))$ for $t \downarrow 0$.

Let $\widehat{\mathscr{I}}$ be the upper limit of the sets $\mathscr{I}(\overline{x}+td)$:

$$\widehat{\mathscr{I}} := \widehat{\mathscr{I}}(\overline{x}, d, \lambda^0) = \{i : \exists\ \{t_k\}_{k=1}^\infty \downarrow 0 \text{ with } \lambda(\overline{x}+t_kd)_i > 0\ \forall\ k\}. \tag{3.14}$$

Clearly,

$$\{i : \lambda_i^0 > 0\} \subseteq \widehat{\mathscr{I}} \subseteq \{i : (A\overline{x} + B\overline{y} - c)_i = 0\}. \tag{3.15}$$

Let $B_{\widehat{\mathscr{I}}}$ be the matrix with rows $B_i,\ i \in \widehat{\mathscr{I}}$, of the matrix B and consider the system

$$\begin{aligned} Qr + B_{\widehat{\mathscr{I}}}^\top\gamma &= -Rd \\ B_{\widehat{\mathscr{I}}}r &= -A_{\widehat{\mathscr{I}}}d \end{aligned} \tag{3.16}$$

Since the matrix

$$\begin{pmatrix} Q & B_{\widehat{\mathscr{I}}}^\top \\ B_{\widehat{\mathscr{I}}} & 0 \end{pmatrix}$$

is invertible, system (3.16) has a unique solution (r, γ) and, hence, r equals the directional derivative $y'(\overline{x}; d)$ of the solution function $y(\cdot)$ at $\overline{x}$ into direction d, see Bigelow and Shapiro [22].

If some submatrix B_I of B is used and the system (3.16) has a solution, this does not need to be true since this system is not suitable to the direction. This means that a wrong system of active constraints has been taken (see Theorem 3.10). A criterion which can be used to verify if a solution of the system (3.16) equals $(y'(\overline{x}; d), \gamma)$ for some direction d is

$$A_id + B_ir \le 0 \text{ for } i \text{ with } A_i\overline{x} + B_i\overline{y} - c_i = 0$$

and

$$\gamma_i \ge 0 \text{ for } i \text{ with } \lambda_i^0 = 0.$$

A careful investigation of the system (3.13) can be used to show that the solution of the lower level problem in (3.12) is piecewise affine-linear. This implies that it is a Lipschitz continuous function, its *directional derivative*

$$d \mapsto y'(\overline{x}; d) = \lim_{t \downarrow 0} \frac{1}{t}[y(\overline{x} + td) - y(\overline{x})]$$

is also Lipschitz continuous.

Vicente and Calamai [307] call a direction $(d, y'(\overline{x}; d))$ for which there exists $\overline{t} > 0$ such that $(\overline{x} + td, y(\overline{x}) + ty'(\overline{x}; d)) \in \mathbf{gph}\, \Psi_Q \ \forall\, t \in [0, \overline{t}]$ an *induced region direction*. The set of all induced region directions is $T(\overline{x}, \overline{y} = y(\overline{x}))$.

Let $T_{\widehat{\mathscr{I}}}$ be the set of all pairs $(d, y'(\overline{x}; d))$ computed by solving system (3.16) for the index set of active constraints $\widehat{\mathscr{I}}$. This set is a convex cone. To find a formula for the cone $T_{\mathbf{gph}\, \Psi_Q}(\overline{x}, \overline{y})$ we use the union over all cones $T_{\widehat{\mathscr{I}}}$ for sets $\widehat{\mathscr{I}}$ satisfying (3.15) and all $\lambda \in \Lambda(\overline{x}, \overline{y})$.

The number of cones used in this union is finite but can be arbitrarily large.

This result can now be used to derive necessary and sufficient optimality conditions.

Theorem 3.7 (Vicente and Calamai [307]) *Consider the problem* (3.12) *and let the function F be continuously differentiable. If* $(\overline{x}, \overline{y})$ *is a local optimal solution of this problem, then*

$$\nabla F_x(\overline{x}, \overline{y})d + \nabla_y F(\overline{x}, \overline{y})r \geq 0 \ \forall\, (d, r) \in T_{\mathbf{gph}\, \Psi_Q}(\overline{x}, \overline{y}).$$

If the assertion of this theorem is satisfied, the point $(\overline{x}, \overline{y})$ is called a stationary solution.

Theorem 3.8 (Vicente and Calamai [307]) *If F is a twice continuously differentiable function,* $(\overline{x}, \overline{y})$ *is stationary and*

$$(d\ r)\nabla^2 F(\overline{x}, \overline{y}) \begin{pmatrix} d^\top \\ r^\top \end{pmatrix} > 0 \ \forall\, (d, r) \in T_{\mathbf{gph}\, \Psi_Q}(\overline{x}, \overline{y}),\ \nabla F(\overline{x}, \overline{y}) \begin{pmatrix} d^\top \\ r^\top \end{pmatrix} = 0$$

then $(\overline{x}, \overline{y})$ *is a strict local minimum of problem* (3.12).

3.4 Unique Lower Level Optimal Solution

Assume throughout this section that $T = \mathbb{R}^m$ and that all the functions defining the lower level problem are sufficiently smooth and convex with respect to the lower level variables y. Then, we can use the notion of a strongly stable optimal solution of problem (1.1) in the sense of Kojima [191] for which convexity is not necessary. But, we use global optimal solutions in the lower level problem and without convexity the global optimal solutions need not to be continuous mappings of the parameter (Jongen and Weber [161]). Then, an optimal solution y^* of problem (1.1) is called *strongly stable* at the parameter value x^* provided there is an open neighborhood V of x^* such that problem (1.1) has a unique optimal solution $y(x)$ for each $x \in V$ with $y(x^*) = y^*$ which is continuous at x^*. In this case, the bilevel optimization problem (1.2) can be replaced with

$$\min_x \{\mathscr{F}(x) := F(x, y(x)) : G(x) \leq 0,\ x \in X\} \tag{3.17}$$

locally around the point x^*. The objective function $x \mapsto F(x, y(x))$ of this problem is only implicitly determined, in general not convex and not differentiable. Using suitable assumptions it can be shown that this function is directionally differentiable and locally Lipschitz continuous.

Here a function $\alpha : \mathbb{R}^n \to \mathbb{R}$ is *directionally differentiable* at a point $\widehat{z}$ in direction d provided the limit

$$\alpha'(\widehat{z}; d) := \lim_{t \downarrow 0} \frac{1}{t}[\alpha(\widehat{z} + td) - \alpha(\widehat{z})]$$

exists and is finite. The function α is *locally Lipschitz continuous* at $\widehat{z}$ if there is an open neighborhood V of $\widehat{z}$ and a number $0 \leq L < \infty$ such that

$$|\alpha(z) - \alpha(z')| \leq L\|z - z'\| \quad \forall z, z' \in V.$$

A subclass of directionally differentiable, locally Lipschitz continuous functions are PC^1-functions, which are the subject of the next subsection.

3.4.1 Piecewise Continuously Differentiable Functions

Definition 3.4 A function $\alpha : \mathbb{R}^n \mapsto \mathbb{R}^q$ is called *PC^1-function* (or *piecewise continuously differentiable function*) at $\widehat{z}$ if there are an open neighborhood V of $\widehat{z}$ and a finite number of continuously differentiable functions $\alpha_i : V \mapsto \mathbb{R}^q$, $i = 1, \ldots, k$, such that α is continuous at $\widehat{z}$ and

$$\alpha(z) \in \{\alpha_i(z) : i = 1, \ldots, k\} \; \forall z \in V.$$

Let the functions f, g_i be sufficiently smooth. Recall that the Mangasarian-Fromovitz constraint qualification (MFCQ) is satisfied for problem (1.1) at the point $(\overline{x}, \overline{y})$ if

(**MFCQ**) there exists $d \in \mathbb{R}^m$ with $\nabla_y g_i(\overline{x}, \overline{y})d < 0$ for all $i : g_i(\overline{x}, \overline{y}) = 0$.

Clearly, the Mangasarian-Fromovitz constraint qualification is satisfied at any point in the feasible set of a convex, differentiable optimization problem iff the Slater's condition is satisfied. Let

$$L(x, y, \lambda) = f(x, y) + \lambda^\top g(x, y)$$

denote the Lagrange function of problem (1.1) with $T = \mathbb{R}^m$. The *strong sufficient optimality condition of second order* (*SSOSC*) for problem (1.1) at the stationary point $(\overline{x}, \overline{y})$ with Lagrange-multiplier

$$\lambda \in \Lambda(\overline{x}, \overline{y}) := \{\lambda \geq 0 : \lambda^\top g(\overline{x}, \overline{y}) = 0,\ \nabla_y L(\overline{x}, \overline{y}, \lambda) = 0\}$$

reads as follows:

(SSOSC) For each direction $d \neq 0$ with $\nabla g_i(\overline{x}, \overline{y})d = 0$ for each $i : \lambda_i > 0$ we have

$$d^\top \nabla^2_{yy} L(\overline{x}, \overline{y}, \lambda)d > 0.$$

The last assumption we need here is the constant rank constraint qualification (CRCQ):

(CRCQ) The *constant rank constraint qualification* (*CRCQ*) is satisfied at the point $(\overline{x}, \overline{y})$ for the problem (1.1) if there exists an open neighborhood U of $(\overline{x}, \overline{y})$ such that, for each subset $I \subseteq \{i : g_i(\overline{x}, \overline{y}) = 0\}$ the family of gradient vectors $\{\nabla_y g_i(x, y) : i \in I\}$ has constant rank on U.

Theorem 3.9 (Ralph and Dempe [265]) *Consider problem* (1.1) *for parameter value* $\overline{x}$ *at a stationary point* $\overline{y}$ *and assume that the assumptions* (*MFCQ*), (*SSOSC*), *and* (*CRCQ*) *are satisfied there. Then, there exist open neighborhoods* $V \subset \mathbb{R}^n$ *of* $\overline{x}$ *and* $U \subset \mathbb{R}^m$ *of* $\overline{y}$ *as well as a unique function* $y : V \to \mathbb{R}^m$ *such that* $y(x)$ *is the unique (local) minimum of problem* (1.1) *in* U *for each* $x \in V$. *Moreover, this function is a* PC^1*-function, it is directionally differentiable at the point* $\overline{x}$ *and locally Lipschitz continuous.*

For convex lower level problems, $y(x)$ is, of course, a global optimum. Then, the transformation to (3.17) is possible and the objective function is at least directionally differentiable if $F \in C^1(\mathbb{R}^n \times \mathbb{R}^m, \mathbb{R})$. To use this, either a formula for computing the directional derivative or the generalized derivative in Clarke's sense is needed.

For a PC^1-function $\alpha : \mathbb{R}^n \mapsto \mathbb{R}^q$ let

$$\text{Supp}(\alpha, \alpha_i) := \{x : \alpha(x) = \alpha_i(x)\}$$

denote the set of all points, where the PC^1-function coincides with one of its members.

The result in Theorem 3.9 has been illustrated in Dempe [52] using the following example:

Example 3.4 (*Dempe* [52]) Consider the problem

$$\begin{aligned} -y &\to \min_y \\ y &\leq 1, \\ y^2 &\leq 3 - x_1^2 - x_2^2, \\ (y - 1.5)^2 &\geq 0.75 - (x_1 - 0.5)^2 - (x_2 - 0.5)^2, \end{aligned}$$

with two parameters x_1 and x_2. Then, y is a continuous selection of three continuously differentiable functions $y^1 = y^1(x)$, $y^2 = y^2(x)$, $y^3 = y^3(x)$ in an open neighborhood of the point $x^0 = (1, 1)^\top$:

$$y(x) = \begin{cases} y^1 = 1, & x \in \mathrm{Supp}(y, y^1), \\ y^2 = \sqrt{3 - x_1^2 - x_2^2}, & x \in \mathrm{Supp}(y, y^2), \\ y^3 = 1.5 - \sqrt{0.75 - (x_1 - 0.5)^2 - (x_2 - 0.5)^2}, & x \in \mathrm{Supp}(y, y^3), \end{cases}$$

where

$$\begin{aligned} \mathrm{Supp}(y, y^1) &= \{x : x_1^2 + x_2^2 \le 2,\ (x_1 - 0.5)^2 + (x_2 - 0.5)^2 \ge 0.5\}, \\ \mathrm{Supp}(y, y^2) &= \{x : 2 \le x_1^2 + x_2^2 \le 3\}, \\ \mathrm{Supp}(y, y^3) &= \{x : (x_1 - 0.5)^2 + (x_2 - 0.5)^2 \le 0.5\}. \end{aligned}$$

The sets $Y_{\{i\}} = \mathrm{Supp}(y, y^i)$ and the function y are illustrated in Figs. 3.2 and 3.3, respectively. □

Using Rademacher's theorem, the generalized derivative of a locally Lipschitz continuous function $f : \mathbb{R}^n \mapsto \mathbb{R}$ can be defined as

$$\partial^{Cl} f(\overline{x}) := \mathrm{conv}\,\{\lim_{k\to\infty} \nabla f(x^k) : \lim_{k\to\infty} x^k = \overline{x},\ \nabla f(x^k) \text{ exists } \forall\, k\}.$$

The same definition can be used for the generalized Jacobian of a locally Lipschitz continuous function $f : \mathbb{R}^n \mapsto \mathbb{R}^m$. It has been shown in Kummer [194] and Scholtes [283] that, under the assumptions of Theorem 3.9, we have

$$\partial^{Cl} y(\overline{x}) = \mathrm{conv}\,\{\nabla y^i(\overline{x}) : \overline{x} \in \mathrm{cl\ int\ Supp}(y, y^i)\}.$$

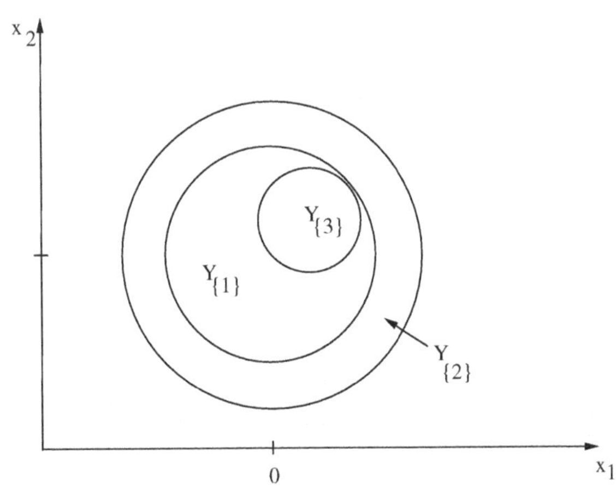

Fig. 3.2 The sets $\mathrm{Supp}(y, y^i)$ in Example 3.4

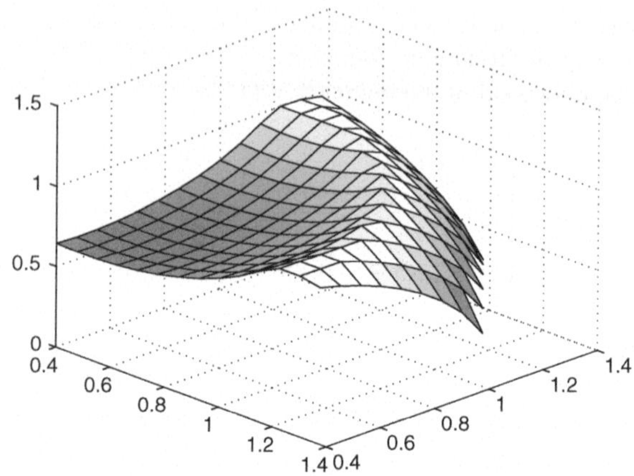

Fig. 3.3 The optimal solution in Example 3.4

Here int A denotes the set if inner points of some set A. Moreover, PC^1-functions are semismooth (Chaney [41]).

PC^1-functions are directionally differentiable and it is easy to see that

$$\alpha'(\overline{x}; d) \in \{\nabla\alpha^i(\overline{x})d : \overline{x} \in \text{cl int Supp}(\alpha, \alpha^i)\}.$$

For the optimal solution function of a parametric optimization problem (1.1) the directional derivative can be computed by solving a quadratic optimization problem.

Theorem 3.10 (Ralph and Dempe [265]) *Consider problem* (1.1) *at a point* $x = \overline{x}$ *and let* $\overline{y}$ *be a local optimal solution of this problem where the assumptions* (*MFCQ*), (*SSOSC*), *and* (*CRCQ*) *are satisfied. Then the directional derivative of the function* $y(\cdot)$ *at* $\overline{x}$ *in direction* r *coincides with the unique optimal solution of the convex quadratic optimization problem* $QP(\lambda^0, r)$

$$0.5 d^\top \nabla^2_{yy} L(\overline{x}, \overline{y}, \lambda^0) d + d^\top \nabla^2_{xy} L(\overline{x}, \overline{y}, \lambda^0) r \rightarrow \min_d$$

$$\nabla_x g_i(\overline{x}, \overline{y}) r + \nabla_y g_i(\overline{x}, \overline{y}) d \begin{cases} = 0, \; \textit{if}\, \lambda^0_i > 0, \\ \leq 0, \; \textit{if}\, g_i(\overline{x}, \overline{y}) = \lambda^0_i = 0 \end{cases} \tag{3.18}$$

for an arbitrary Lagrange multiplier $\lambda^0 \in \Lambda(\overline{x}, \overline{y})$ *solving*

$$\nabla_x L(\overline{x}, \overline{y}, \lambda) r \rightarrow \max_{\lambda \in \Lambda(\overline{x}, \overline{y})}. \tag{3.19}$$

The quadratic optimization problem (3.18) can be replaced equivalently with its Karush-Kuhn-Tucker conditions. Using linear optimization duality it can be shown that problem (3.18) has a feasible solution if and only if λ solves problem (3.19).

3.4.2 Necessary and Sufficient Optimality Conditions

Applying the results in Sect. 3.4.1 to problem (3.17) necessary optimality conditions for the bilevel optimization problem can be obtained.

Theorem 3.11 (Dempe [50]) *Let $(\overline{x}, \overline{y})$ be a local optimal solution of the bilevel optimization problem* (1.1), (1.4) *and assume that the lower level problem* (1.1) *is a convex parametric optimization problem satisfying the conditions* (*MFCQ*), (*SSOC*), *and* (*CRCQ*) *at $(\overline{x}, \overline{y})$. Then the following optimization problem has a nonnegative optimal objective function value:*

$$
\begin{aligned}
\alpha &\to \min_{\alpha,r} \\
\nabla_x F(\overline{x}, \overline{y})r + \nabla_y F(\overline{x}, \overline{y})y'(\overline{x}; r) &\leq \alpha \\
\nabla G_i(\overline{x})r &\leq \alpha, \forall\, i : G_i(\overline{x}) = 0 \\
\|r\| &\leq 1.
\end{aligned}
\tag{3.20}
$$

Moreover, if there exists a direction r such that $\nabla G_i(\overline{x})r < 0$ for all $i : G_i(\overline{x}) = 0$, problem (3.20) *can be replaced by*

$$
\begin{aligned}
\nabla_x F(\overline{x}, \overline{y})r + \nabla_y F(\overline{x}, \overline{y})y'(\overline{x}; r) &\to \min_{r} \\
\nabla G_i(\overline{x})r &\leq 0, \ \forall\, i : G_i(\overline{y}) = 0 \\
\|r\| &\leq 1.
\end{aligned}
\tag{3.21}
$$

Proof Under the assumptions of this theorem, due to Theorem 3.9, the bilevel optimization problem (1.4) can be replaced with (3.17). Then, using the definition of the directional derivative, a necessary optimality condition is that the feasible set of problem (3.20) with $\alpha < 0$ is empty.

Applying the (MFCQ),

$$\{r : \nabla G_i(\overline{x})r \leq 0, \ \forall\, i : G_i(\overline{y}) = 0\} = \mathrm{cl}\, S$$

with

$$S = \{r : \nabla G_i(\overline{x})r < 0, \ \forall\, i : G_i(\overline{y}) = 0\}$$

and, for each $r \in S$ we have $\mathscr{F}'(\overline{x}; r) \geq 0$ by the first part of the theorem. By Theorem 3.9 and Bector et al. [14, Theorem 5.1.1] the mapping $r \mapsto \mathscr{F}'(\overline{x}; r)$ is continuous and, hence, the optimal function value of problem (3.21) must be nonnegative. □

If the condition $\|r\| \leq 1$ in problem (3.21) is replaced with $\|r\| = 1$ and the resulting problem has a strictly positive optimal objective function value $v_1 > 0$, then the feasible solution $(\overline{x}, \overline{y})$ is a strict local optimal solution of the bilevel optimization problem (1.1), (1.4), see Dempe [50].

3.4.3 Solution Algorithm

The optimization problem (3.17) is under the conditions of Theorem 3.9 a Lipschitz continuous optimization problem with an implicitly determined objective function. To solve this problem bundle-trust region algorithms (Outrata et al. [259], see also Dempe [51]) can be used provided that $G(x) \equiv 0$. For the computation of elements of the generalized gradient see Dempe and Vogel [79] or Dempe and Pallaschke [73], see Theorem 3.12.

The following algorithm has been described in Dempe and Schmidt [76].

Descent algorithm for the bilevel problem:
Input: Bilevel optimization problem (3.17).
Output: A Clarke stationary solution.

Algorithm: Step 1: Select x^0 satisfying $G(x^0) \leq 0$, set $k := 0$, choose $\varepsilon \in (0, 1)$.

Step 2: Compute a direction r^k, $\|r^k\| \leq 1$, satisfying

$$\mathscr{F}'(x^k; r^k) \leq s^k, \nabla G_i(x^k) r^k \leq -G_i(x^k) + s^k,\ i = 1, \ldots, q,$$

and $s^k < 0$.

Step 3: Choose a step-size t_k such that

$$\mathscr{F}(x^k + t_k r^k) \leq \mathscr{F}(x^k) + \varepsilon t_k s^k,\ G(x^k + t_k r^k) \leq 0.$$

Step 4: Set $x^{k+1} := x^k + t_k r^k$, compute the optimal solution $y^{k+1} = y(x^{k+1}) \in \Psi(x^{k+1})$, set $k := k + 1$.

Step 5: If a stopping criterion is satisfied stop, else goto step 2.

Using Theorem 3.9 and continuous differentiability of the function F we derive

$$\mathscr{F}'(x; r) = \nabla_x F(x, y(x)) r + \nabla_y F(x, y(x)) y'(x; r),$$

where the directional derivative of the vector-valued function $y(\cdot)$ at the point x in direction r can be computed solving the quadratic optimization problem (3.18) or, alternatively, a system of quadratic equations. Note that, due to the complementarity constraints, this is a combinatorial problem. This, mainly, is the reason for the following Theorem 3.13, where Clarke stationarity of the computed solution is claimed.

We need the assumption

(**FRR**) For each vertex $\lambda^0 \in \Lambda(x, y)$ the matrix

$$\mathscr{M} := \begin{pmatrix} \nabla^2_{yy} L(x, y, \lambda^0) & \nabla_y^\top g_{J(\lambda^0)}(x, y) & \nabla^2_{xy} L(x, y, \lambda^0) \\ \nabla_y g_{I_0}(x, y) & 0 & \nabla_x g_{I_0}(x, y) \end{pmatrix}$$

has full row rank $m + |I_0|$.

Here $I_0 = \{i : g_i(x, y) = 0\}$ and $J(\lambda) = \{i : \lambda_i > 0\}$.

The following theorem completes the characterization of the generalized gradient in the sense of Clarke of the optimal solution function $y(x)$.

Theorem 3.12 (Dempe and Pallaschke [73]) *Consider the lower level problem* (1.1), *assume that it is a convex parametric optimization problem and that (FRR), (MFCQ), (CRCQ), and (SSOC) are satisfied at a point* $(\overline{x}, \overline{y}) \in \mathbf{gph}\,\Psi$. *Then, for any vertex* $\overline{\lambda} \in \Lambda(\overline{x}, \overline{y})$ *and each index set* I *satisfying* $J(\overline{\lambda}) \subseteq I \subseteq I(\overline{x}, \overline{y})$ *such that the gradients* $\{\nabla_y g_i(\overline{x}, \overline{y}) : i \in I\}$ *are linearly independent, the matrix* $\nabla y^I(\overline{x})$ *with*

$$\begin{pmatrix} \nabla^2_{yy} L(\overline{x}, \overline{y}, \overline{\lambda}) & \nabla^\top_y g_I(\overline{x}, \overline{y}) \\ \nabla_y g_I(\overline{x}, \overline{y}) & 0 \end{pmatrix} \begin{pmatrix} \nabla y^I(\overline{x}) \\ w \end{pmatrix} = \begin{pmatrix} -\nabla^2_{xy} L(\overline{x}, \overline{y}, \overline{\lambda}) \\ -\nabla_x g_I(\overline{x}, \overline{y}) \end{pmatrix} \tag{3.22}$$

belongs to $\partial^{Cl} y(\overline{x})$.

The matrix w in formula (3.22) is related to a generalized gradient of a certain Lagrange multiplier of problem (1.1).

An algorithm for computing elements of the generalized gradient of the PC^1-function $y(\cdot)$ can be found in Dempe and Vogel [79]. In this article the interested reader can also find an example showing that the assumption (FRR) is essential.

Theorem 3.13 (Dempe and Schmidt [76]) *Consider problem* (3.17), *where* $y(x)$ *is an optimal solution of the lower level problem* (1.1) *which is a convex parametric optimization problem. Let* $T = \mathbb{R}^m$, *assume that all functions* F, f, G, g *are sufficiently smooth and the set* $\{(x, y) : G(x) \leq 0,\ g(x, y) \leq 0\}$ *is nonempty and bounded. Let the assumptions (FRR), (MFCQ), (CRCQ), and (SSOC) for all* $(x, y),\ y \in \Psi(x),\ G(x) \leq 0$ *be satisfied for problem* (1.1) *and (MFCQ) is satisfied with respect to* $G(x) \leq 0$ *for all* x. *Then, each accumulation point* $(\overline{x}, \overline{y})$ *of the sequence of iterates* $\{(x^k, y^k)\}_{k=1}^\infty$ *computed with the above algorithms is Clarke stationary.*

To motivate Theorem 3.13 we first see that the above descent algorithm is a realization of the method of feasible directions (Bazaraa and Shetty [12]) in the modification by Topkis and Veinott [301]. Then, using standard argumentation the algorithm converges to a point $(\overline{x}, \overline{y})$, where the system

$$\begin{aligned} \mathscr{F}'(\overline{x}; r) &< 0 \\ \nabla G_i(\overline{x}) &< 0,\ \forall i : G_i(\overline{x}) = 0 \end{aligned}$$

does not have a solution. Since $y(\cdot)$ is a PC^1-function, this means that there is some selection function $y^j(\cdot)$ such that

$$\mathscr{F}'(x^k; r^k) = \nabla_x F(x^k, y^k) r^k + \nabla_y F(x^k, y^k) \nabla y^j(x^k) r^k$$

and $y^k = y^j(x^k)$ for an infinite subsequence of the sequence $\{(x^k, y^k, r^k, s^k)\}_{k=1}^\infty$ of points generated by the descent algorithm and some of the selection functions of $y(\cdot)$ at $x = \overline{x}$. Hence, the system

$$\begin{aligned} \nabla_x F(\overline{x}, \overline{y}) r + \nabla_y F(\overline{x}, \overline{y}) \nabla y^j(\overline{x}) r &< 0 \\ \nabla G_i(\overline{x}) < 0,\ \forall i : G_i(\overline{x}) &= 0 \end{aligned}$$

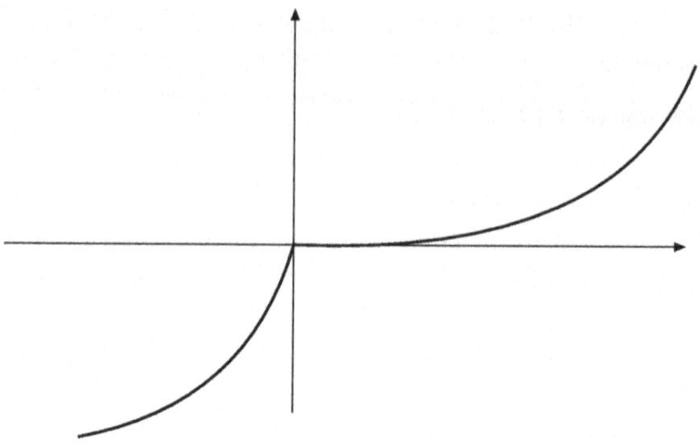

Fig. 3.4 Convergence to a Clarke stationary point at the origin

has no solution. This implies that the point $(\overline{x}, \overline{y})$ is a Karush-Kuhn-Tucker point for the problem

$$\min_x \{F(x, y^j(x)) : G(x) \leq 0\}.$$

Then,

$$0 \in \partial_x^{Cl} F(\overline{x}, y^j(\overline{x})) + N_M^{Cl}(\overline{x}),$$

where $M = \{x : G(x) \leq 0\}$ and $N_M^{Cl}(\overline{x})$ is the Clarke normal cone to M in $\overline{x}$. Since, by (FRR), $\nabla y^j(\overline{x}) \in \partial^{Cl} y(\overline{x})$ (Dempe and Vogel [79]; Dempe and Pallaschke [73]) we derive

$$0 \in \partial^{Cl} \mathscr{F}(\overline{x}) + N_M^{Cl}(\overline{x})$$

implying that $\overline{x}$ is a Clarke stationary point.

Unfortunately, Clarke stationary points need not to be local optimal solutions. This can be seen in the Fig. 3.4 borrowed from Dempe [52].

To circumvent the undesired convergence to a Clarke stationary point and guarantee convergence to a Bouligand stationary point, i.e. a point where there does not exist a feasible direction of descent, a modification of the direction finding problem in Step 2 of the above algorithm is needed. The interested reader is referred to the paper Dempe and Schmidt [76] for the respective results.

3.5 The Classical KKT Transformation

3.5.1 *Stationary Solutions*

The classical KKT transformation is often used in literature. Assume for simplicity throughout this subsection that $T = \mathbb{R}^m$ and $X = \mathbb{R}^n$. Moreover, the lower level problem needs to be convex throughout this subsection, and a regularity condition as Slater's condition (or, equivalently, the (MFCQ)) needs to be satisfied at all feasible points. If the lower level problem is not convex for fixed parameter value, the set of feasible solutions is enlarged by adding local optimal as well as stationary solutions of the lower level problem to it. This can imply that a local (or global) optimal solution of the bilevel problem (1.4) needs not to be stationary for the resulting problem (3.2), see Mirrlees [232]. Moreover, if Slater's condition is violated at some feasible points of problem (1.1) the set of feasible points for problem (3.2) needs not to be closed, see Dempe and Dutta [55]. The classical KKT transformation reduces to the problem

$$\begin{aligned} F(x,y) &\to \min \\ G(x) &\leq 0 \\ 0 &\in \partial_y f(x,y) + \lambda^\top \partial_y g(x,y) \\ \lambda &\geq 0,\ g(x,y) \leq 0,\ \lambda^\top g(x,y) = 0. \end{aligned} \tag{3.23}$$

This is a mathematical optimization problem with complementarity constraints (MPEC or MPCC for short), see Luo et al. [208] and Outrata et al. [259]. The following result is due to Scheel and Scholtes [280]:

Theorem 3.14 *The Mangasarian-Fromovitz constraint qualification is violated at every feasible point of the problem* (3.23).

This makes the solution of the problem and also the formulation of (necessary and sufficient) optimality conditions difficult. On the one hand this resulted in the formulation of stationary solutions of different types. On the other hand, for solving the problem both the application of adapted algorithms in nonlinear optimization as well as regularization algorithms have been suggested.

Necessary optimality conditions can be defined for problem (3.2) as usual for mathematical programs with complementarity constraints.

Assume in this subsection that all the functions F, f, g_i, G_j are sufficiently smooth. Use the following sets:

1. $I_G(x) = \{j : G_j(x) = 0\}$,
2. $I_{-0}(x,y,\lambda) = \{i : g_i(x,y) < 0,\ \lambda_i = 0\}$,
3. $I_{00}(x,y,\lambda) = \{i : g_i(x,y) = 0,\ \lambda_i = 0\}$,
4. $I_{0+}(x,y,\lambda) = \{i : g_i(x,y) = 0,\ \lambda_i > 0\}$.

If the set $I_{00}(\overline{x},\overline{y},\overline{\lambda}) = \emptyset$, problem (3.2) reduces to an optimization problem without complementarity constraints and can be treated as a classical optimization problem locally around the point $(\overline{x},\overline{y},\overline{\lambda})$. This is a consequence of continuity. Hence, the set $I_{00}(\overline{x},\overline{y},\overline{\lambda})$ is the interesting one. It is called the set of bi-active constraints.

To simplify the notation let $\mathscr{L}(x,y,\lambda) = \nabla_y f(x,y) + \lambda^\top \nabla_y g(x,y)$. Moreover, $\nabla_x \mathscr{L}(x,y,\lambda)\gamma = \nabla_x(\mathscr{L}(x,y,\lambda)\gamma)$ and $\nabla_y \mathscr{L}(x,y,\lambda)\gamma = \nabla_y(\mathscr{L}(x,y,\lambda)\gamma)$.

Main part of the optimality conditions are the following equations resulting from Karush-Kuhn-Tucker conditions for problem (3.23):

$$\nabla_x F(x,y) + \alpha^\top \nabla G(x) + \beta^\top \nabla_x g(x,y) + \nabla_x \mathscr{L}(x,y,\lambda)\gamma = 0, \tag{3.24}$$

$$\nabla_y F(x,y) + \beta^\top \nabla_y g(x,y) + \nabla_y \mathcal{L}(x,y,\lambda)\gamma = 0, \tag{3.25}$$

$$\alpha \geq 0,\ \alpha^\top G(x) = 0, \tag{3.26}$$

$$\nabla_y g_{I_{0+}(x,y,\lambda)}(x,y)\gamma = 0,\ \beta_{I_{-0}(x,y,\lambda)} = 0, \tag{3.27}$$

where $t_I = 0$ for a system of inequalities $t_i \geq 0$ and an index set I means $\{t_i = 0\,\forall\, i \in I\}$.

Definition 3.5 A feasible solution $(\overline{x}, \overline{y}, \overline{\lambda})$ for the problem (3.23) for which there exists a vector $(\alpha, \beta, \gamma) \in \mathbb{R}^q \times \mathbb{R}^p \times \mathbb{R}^m$ such that

1. Conditions (3.24)–(3.27) are satisfied is called *weakly stationary*.
2. Conditions (3.24)–(3.27) together with

$$\beta_i \geq 0 \text{ or } \nabla_y g_i(\overline{x},\overline{y})^\top \gamma \geq 0\ \forall\, i \in I_{00}(\overline{x},\overline{y},\overline{\lambda})$$

   are satisfied, is an *A-stationary solution*.
3. Equations (3.24)–(3.27) together with

$$\beta_i \nabla_y g_i(\overline{x},\overline{y})^\top \gamma \geq 0\ \forall\, i \in I_{00}(\overline{x},\overline{y},\overline{\lambda})$$

   hold, is *C-stationary*.
4. Equations (3.24)–(3.27) hold together with

$$(\beta_i > 0 \wedge \nabla_y g_i(\overline{x},\overline{y})^\top \gamma > 0) \vee \beta_i \nabla_y g_i(\overline{x},\overline{y})^\top \gamma = 0\ \forall\, i \in I_{00}(\overline{x},\overline{y},\overline{\lambda}),$$

   is a *M-stationary solution*.
5. Conditions (3.24)–(3.27) and also

$$\beta_i \geq 0 \text{ and } \nabla_y g_i(\overline{x},\overline{y})^\top \gamma \geq 0\ \forall\, i \in I_{00}(\overline{x},\overline{y},\overline{\lambda})$$

hold, is a *S-stationary solution*.

Related results for mathematical programs with complementarity (equilibrium) constraints can be found in Flegel and Kanzow [112], the PhD thesis by Flegel [111], the papers Pang and Fukushima [262] as well as Scheel and Scholtes [280].

As examples for necessary optimality conditions we will consider two results, one for a M-stationary and one for a C-stationary solution. More results can be found especially in the PhD thesis of Zemkoho [328] and in Ye [323], respective results for MPEC's have been developed in the PhD thesis of Flegel [111].

In the following theorem we need the (basic) normal cone in the sense of variational analysis, see Mordukhovich [241, 242]. Let $A \subseteq \mathbb{R}^p$ be a closed set and $\overline{z} \in A$. Then,

$$\hat{N}_A(\overline{z}) := \{d : d^\top (z - \overline{z}) \leq o(\|z - \overline{z}\|)\ \forall\, z \in A\}$$

is the Fréchet normal cone to the set A at $\overline{z}$. Here, $o(t)$ is a function satisfying $\lim_{t\to 0} o(t)/t = 0$. The Mordukhovich normal cone to A at $\overline{z}$ is the

Kuratowski-Painlevé upper limit of the Fréchet normal cone, i.e.

$$N_A^M(\overline{z}) = \{r : \exists\, \{z^k\}_{k=1}^\infty \subseteq A,\ \exists\{r^k\}_{k=1}^\infty \text{ with } \lim_{k\to\infty} z^k = \overline{z},$$
$$\lim_{k\to\infty} r^k = r,\ r^k \in \hat{N}_A(z^k)\ \forall\, k\}.$$

The *Mordukhovich subdifferential* of a lower semicontinuous function $f : \mathbb{R}^n \to \overline{\mathbb{R}}$ at some point $\overline{x} \in \operatorname{dom} f$ is

$$\partial^M f(\overline{x}) := \{z^* \in \mathbb{R}^n : (z^*, -1) \in N_{epi f}(\overline{x}, f(\overline{x}))\},$$

where epi $f := \{(x, \alpha) : f(x) \le \alpha\}$ is the epigraph of the function f.

We also need the following theorem borrowed from Mordukhovich [240, Proposition 2.10]:

Theorem 3.15 *Let $f : \mathbb{R}^p \to \mathbb{R}^q$ be Lipschitz continuous around $\overline{x}$ and $g : \mathbb{R}^q \to \overline{\mathbb{R}}$ be Lipschitz continuous around $\overline{y} = f(\overline{x}) \in \operatorname{dom} g$. Then,*

$$\partial^M (g \circ f)(\overline{x}) \subseteq \bigcup \{\partial^M \langle w, f(\overline{x})\rangle : w \in \partial^M g(\overline{y})\}.$$

The following theorem can be found in Mordukhovich [238, Corollary 4.6]:

Theorem 3.16 *Let the functions $f, g : \mathbb{R}^n \to \mathbb{R}$ be locally Lipschitz continuous around $\overline{z} \in \mathbb{R}^n$. Then,*

$$\partial^M (f + g)(\overline{z}) \subseteq \partial^M f(\overline{z}) + \partial^M g(\overline{z}).$$

Equality holds if one of the functions is continuously differentiable.

The last ingredient for the necessary optimality condition comes from Rockafellar and Wets [274, Theorem 8.15]:

Theorem 3.17 *Let $f : \mathbb{R}^n \to \mathbb{R}$ be a locally Lipschitz continuous function and $A \subseteq \mathbb{R}^n$ a closed set. If $\overline{z} \in A$ is a local minimizer of the function f over A, then*

$$0 \in \partial^M f(\overline{z}) + N_A^M(\overline{z}).$$

Theorem 3.18 (Dempe and Zemkoho [82]) *Let $(\overline{x}, \overline{y}, \overline{\lambda})$ be a local optimal solution of problem* (3.23) *and assume that the constraint qualification*

$$\left.\begin{array}{l} \alpha^\top \nabla G(x) + \beta^\top \nabla_x g(x, y) + \nabla_x \mathscr{L}(x, y, \lambda)\gamma = 0, \\ \beta^\top \nabla_y g(x, y) + \nabla_y \mathscr{L}(x, y, \lambda)\gamma = 0, \\ \alpha \ge 0,\ \alpha^\top G(x) = 0, \\ \nabla_y g_{I_{0+}(x,y,\lambda)}(x, y)\gamma = 0,\ \beta_{I_{-0}(x,y,\lambda)} = 0, \\ (\beta_i > 0 \wedge \nabla_y g_i(\overline{x}, \overline{y})\gamma > 0) \vee \beta_i \nabla_y g_i(\overline{x}, \overline{y})\gamma = 0 \\ \hfill \forall\, i \in I_{00}(\overline{x}, \overline{y}, \overline{\lambda}) \end{array}\right\} \Rightarrow \left\{\begin{array}{l} \alpha = 0 \\ \beta = 0 \\ \gamma = 0 \end{array}\right. \tag{3.28}$$

is satisfied there. Then, there exists (α, β, γ) with $\|(\alpha, \beta, \gamma)\| \leq r$ for some $r < \infty$ such that the point is a M-stationary solution.

Proof Let

$$\Theta = \{(a, b) \in \mathbb{R}^{2p} : a \geq 0,\ b \geq 0,\ a^\top b = 0\}.$$

Then, it can be shown, see Flegel and Kanzow [113], that

$$N_\Theta^M(\overline{u}, \overline{v}) = \left\{ (u^*, v^*) : \begin{bmatrix} u_i^* = 0, & \forall\, i : \overline{u}_i > 0 = \overline{v}_i \\ v_i^* = 0, & \forall\, i : \overline{u}_i = 0 < \overline{v}_i \\ (u_i^* < 0 \wedge v_i^* < 0) \vee u_i^* v_i^* = 0, & \forall\, i : \overline{u}_i = \overline{v}_i = 0 \end{bmatrix} \right\}.$$

Let $(\overline{x}, \overline{y}, \overline{\lambda})$ be a local optimal solution of problem (3.23).

Set

$$\Gamma(x, y, \lambda, v) = (G(x), g(x, y) + v, \mathcal{L}(x, y, \lambda))^\top,$$

$$\Xi = \mathbb{R}^q_- \times \{0_{p+m}\},\ \Omega = \mathbb{R}^n \times \mathbb{R}^m \times \Theta.$$

Then, it is easy to see that there is a vector $\overline{v}$ such that $(\overline{x}, \overline{y}, \overline{\lambda}, \overline{v})$ is a local optimal solution of the problem

$$\min_{x,y,\lambda,v} \{F(x, y) : (x, y, u, v) \in \Omega \cap \Gamma^{-1}(\Xi)\}. \tag{3.29}$$

In this case,

$$\begin{aligned} N_\Omega^M(\overline{x}, \overline{y}, \overline{\lambda}, \overline{v}) &= \{0_{m+n}\} \times N_\Theta(\overline{u}, \overline{v}), & (3.30)\\ N_\Xi^M(\Gamma(\overline{x}, \overline{y}, \overline{\lambda}, \overline{v})) &= \{(\alpha, \beta, \gamma) : \alpha \geq 0,\ \alpha^\top G(\overline{x}) = 0\}, & (3.31)\\ \nabla\Gamma(x, y, \lambda, v)(\alpha, \beta, \gamma) &= \begin{pmatrix} A(\alpha, \beta, \gamma) \\ \beta \end{pmatrix}, & (3.32) \end{aligned}$$

where

$$A(\alpha, \eta, \gamma) = \begin{pmatrix} \alpha^\top \nabla G(\overline{x}) + \beta^\top \nabla_x g(\overline{x}, \overline{y}) + \nabla_x \mathcal{L}(\overline{x}, \overline{y}, \overline{\lambda})\gamma \\ \beta^\top \nabla_y g(\overline{x}, \overline{y}) + \nabla_y \mathcal{L}(\overline{x}, \overline{y}, \overline{\lambda})\gamma \\ \nabla_y g(\overline{x}, \overline{y})\gamma \end{pmatrix}.$$

Using Theorem 3.17 we derive

$$0 \in \partial^M F(\overline{x}, \overline{y}) \times (0,\ 0)^\top + N_W^M(\overline{x}, \overline{y}, \overline{\lambda}, \overline{v}),$$

where

$$\begin{aligned} W &= \{(x, y, \lambda, v) : (x, y, \lambda, v) \in \Omega \cap \Gamma^{-1}(\Xi)\} \\ &= \{(x, y, \lambda, v) \in \Omega : \Gamma(x, y, \lambda, v) \in \Xi\}. \end{aligned}$$

Assumption (3.28) implies the basic constraint qualification (Mordukhovich [239]; Rockafellar and Wets [274])

$$\left.\begin{array}{r} 0 \in \partial^M \langle u^*, \Gamma(x, y, \lambda, v)\rangle + N^M_\Omega(x, y, \lambda, v) \\ u^* \in N^M_\Xi(\Gamma(x, y, \lambda, v)) \end{array}\right\} \Rightarrow u^* = 0$$

is satisfied. Hence,

$$\begin{aligned} & N^M_W(x, y, \lambda, v) \\ & \quad = \{z^* : \exists\, u^* \in N^M_\Xi(\Gamma(x, y, \lambda, v) \text{ with} \\ & \qquad z^* = \partial^M \langle u^*, \Gamma(x, y, \lambda, v)\rangle + N^M_\Omega(x, y, \lambda, v)\}. \end{aligned}$$

Thus, there exists $\mu > 0$ such that, for all $r \geq \mu$, we can find

$$u^* \in N^M_\Xi(\Gamma(\overline{x}, \overline{y}, \overline{\lambda}, \overline{v})), \ \|u^*\| \leq r,$$

with

$$0 \in \partial^M F(\overline{x}, \overline{y}) \times (0, 0) + \partial^M \langle u^*, \Gamma\rangle(\overline{x}, \overline{y}, \overline{\lambda}, \overline{v}) + N^M_\Omega(\overline{x}, \overline{y}, \overline{\lambda}, \overline{v})$$

(Dempe and Zemkoho [80]). This implies the existence of a vector

$$(\alpha, \beta, \gamma), \ \|(\alpha, \beta, \gamma)\| \leq r$$

such that (3.24)–(3.26) together with

$$(-\nabla_y g(\overline{x}, \overline{y})\gamma, -\beta) \in N_\Theta(\overline{\lambda}, -g(\overline{x}, \overline{y})).$$

Here, we used that the upper level objective F does not depend on λ, v. The result then follows from the formula for $N_\Theta(\overline{\lambda}, -g(\overline{x}, \overline{y}))$. □

To derive optimality conditions using the approaches for mathematical programs with equilibrium constraints (MPECs) we can adapt the corresponding regularity conditions. For this, consider the tightened problem

$$\begin{aligned} F(x, y) &\to \min \\ G(x) &\leq 0 \\ 0 &= \nabla_y f(x, y) + \lambda^\top \nabla_y g(x, y) \\ \lambda_i &= 0, \ \forall\, i \in I_{-0}(\overline{x}, \overline{y}, \overline{\lambda}) \cup I_{00}(\overline{x}, \overline{y}, \overline{\lambda}) \\ g_i(x, y) &= 0, \ \forall\, i \in I_{0+}(\overline{x}, \overline{y}, \overline{\lambda}) \cup I_{00}(\overline{x}, \overline{y}, \overline{\lambda}) \\ \lambda_i &\geq 0, \ i \in I_{0+}(\overline{x}, \overline{y}, \overline{\lambda}) \\ g_i(x, y) &\leq 0, \ i \in I_{-0}(\overline{x}, \overline{y}, \overline{\lambda}). \end{aligned} \tag{3.33}$$

around the point $(\overline{x}, \overline{y}, \overline{\lambda})$.

Definition 3.6 The problem (3.23) is said to satisfy at the point $(\overline{x}, \overline{y}, \overline{\lambda})$ the

1. (**MPEC-LICQ**) if the (LICQ) is satisfied for the problem (3.33).
2. (**MPEC-MFCQ**) if the (MFCQ) is satisfied for the problem (3.33).

It has been shown in Scholtes and Stöhr [284] that the (MPEC-LICQ) is a generic regularity condition for MPEC's which means, that, if it is not satisfied, then small (but unknown) perturbations of the data of the problem can be used to transform it into a problem where this condition is satisfied. The following theorem goes back to Scheel and Scholtes [280] as well as Zemkoho [328].

Theorem 3.19 *Let* $(\overline{x}, \overline{y}, \overline{\lambda})$ *be a local optimal solution of problem* (3.23) *and assume that the* (*MPEC-MFCQ*) *is satisfied there. Then,* $(\overline{x}, \overline{y}, \overline{\lambda})$ *is a C-stationary point of* (3.23).

Proof Problem (3.23) can be written as

$$\begin{aligned} F(x,y) &\to \min \\ G(x) &\le 0 \\ \mathscr{L}(x,y,\lambda) &= 0 \\ \min\{\lambda_i, -g_i(x,y)\} &= 0, \ \forall i = 1,\dots,p. \end{aligned}$$

This is a Lipschitz optimization problem. Hence, there exist $\lambda_0 \ge 0,\ \alpha, \beta, \gamma$, not all vanishing, such that

$$0 \in \lambda_0 \nabla_x F(\overline{x},\overline{y}) + \alpha^\top \nabla G(\overline{x}) - \sum_{i=1}^p \beta_i \partial_x^{Cl} s_i(\overline{x},\overline{y},\overline{\lambda}) + \nabla_x \mathscr{L}(\overline{x},\overline{y},\overline{\lambda})\gamma, \quad (3.34)$$

$$0 \in \lambda_0 \nabla_y F(\overline{x},\overline{y}) - \sum_{i=1}^p \beta_i \partial_y^{Cl} s_i(\overline{x},\overline{y},\overline{\lambda}) + \nabla_y \mathscr{L}(\overline{x},\overline{y},\overline{\lambda})\gamma, \quad (3.35)$$

$$\alpha \ge 0, \alpha^\top G(\overline{x}) = 0, \quad (3.36)$$

$$0 \in -\sum_{i=1}^p \beta_i \partial_\lambda^{Cl} s_i(\overline{x},\overline{y},\overline{\lambda}) + \nabla_y g(\overline{x},\overline{y})\gamma, \quad (3.37)$$

where $s_i(x,y,\lambda) = \min\{\lambda_i, -g_i(x,y)\}$ and $\partial_x^{Cl} f(x,y)$ is the Clarke generalized derivative of the function f at (x,y) with respect to x.

The (MPEC-MFCQ) can be used to show that $\lambda_0 \ne 0$. Let $\lambda_0 = 1$ without loss of generality.

Moreover,

$$\partial^{Cl} s_i(\overline{x},\overline{y},\overline{\lambda}) = \begin{cases} -(\nabla g_i(\overline{x},\overline{y}), 0), & \text{if } i \in I_{0+}(\overline{x},\overline{y},\overline{\lambda}), \\ (0, e^{i\top}), & \text{if } i \in I_{-0}(\overline{x},\overline{y},\overline{\lambda}), \\ \operatorname{conv}\{-(\nabla g_i(\overline{x},\overline{y}), 0), (0, e^{i\top})\}, & \text{if } i \in I_{00}(\overline{x},\overline{y},\overline{\lambda}). \end{cases}$$

For $i \in I_{0+}(\overline{x}, \overline{y}, \overline{\lambda})$, Eq. (3.37) implies $\nabla_y g_i(\overline{x}, \overline{y})\gamma = 0$.

For $i \in I_{-0}(\overline{x}, \overline{y}, \overline{\lambda})$, $\partial_x^{Cl} s_i(\overline{x}, \overline{y}, \overline{\lambda}) = 0$ and $\partial_y^{Cl} s_i(\overline{x}, \overline{y}, \overline{\lambda}) = 0$. This implies that $\beta_i = 0$ can be taken in Eqs. (3.34) and (3.35).

If $i \in I_{00}(\overline{x}, \overline{y}, \overline{\lambda})$ then

$$\partial^{Cl} s_i(\overline{x}, \overline{y}, \overline{\lambda}) = \{\xi_i = -\mu \nabla g_i(\overline{x}, \overline{y}) \times \{0\} + (1-\mu)(\{0\} \times \{0\} \times e^{i\top}) : 0 \leq \mu \leq 1\}$$

and

$$\beta_i(1-\mu) = \nabla_y g_i(\overline{x}, \overline{y})\gamma$$

by Eq. (3.37). Hence,

$$\beta_i^2(1-\mu) = \beta_i \nabla_y g_i(\overline{x}, \overline{y})\gamma \geq 0 \text{ for } i \in I_{00}(\overline{x}, \overline{y}, \overline{\lambda}). \qquad \square$$

It has been shown in Flegel [111] that (MPEC-MFCQ) even implies M-stationarity.

Theorem 3.20 (Scheel and Scholtes [280]; Dempe and Franke [59]) *If a point* $(\overline{x}, \overline{y}, \overline{\lambda})$ *is a local optimal solution of problem* (3.23) *where the* (MPEC-LICQ) *is satisfied then it is a S-stationary point.*

3.5.2 Solution Algorithms

The difficulties in solving the bilevel optimization problem on the basis of the classical KKT reformulation result, on the first hand, from violation of mostly used constraint qualifications (cf. Theorem 3.14). On the other hand, this is a result of Theorem 3.1 showing that global optimal solutions need to be considered for equivalence of problems (1.4) and (3.2). Problem (3.2) is a nonconvex optimization problem. For such problems most algorithms compute stationary or local optimal solutions. A number of different algorithms have been suggested for solving MPECs, some of them are presented in the following. At least two can be used for solving the bilevel optimization problem locally, too.

Algorithms to compute a global optimum of the MPEC and of bilevel optimization problems can be based on enumeration plus approximation principles (Meyer and Floudas [229]) or reverse convex optimization (Horst and Tuy [147]).

3.5.2.1 Reduction to a Mixed-Integer Nonlinear Optimization Problem

Using Boolean variables z_i, $i = 1, \ldots, p$, and a sufficiently large positive number Q, the problem

$$\begin{aligned} F(x, y) &\to \min_{x,y} \\ G(x) &\leq 0 \\ 0 &\in \partial_y f(x, y) + \lambda^\top \partial_y g(x, y) \end{aligned}$$

$$\begin{aligned} \lambda &\geq 0,\ g(x,y) \leq 0,\ \lambda^\top g(x,y) = 0 \\ x &\in X. \end{aligned}$$

can be replaced by

$$\begin{aligned} F(x,y) &\to \min_{x,y} \\ G(x) &\leq 0 \\ 0 &\in \partial_y f(x,y) + \lambda^\top \partial_y g(x,y) \\ 0 \leq \lambda_i &\leq Qz_i,\ i = 1,\dots,p, \\ 0 \geq g_i(x,y) &\geq -Q(1-z_i),\ i = 1,\dots,p, \\ x &\in X,\ z_i \in \{0,1\},\ i = 1,\dots,p. \end{aligned} \tag{3.38}$$

The number Q is bounded if the Mangasarian-Fromovitz constraint qualification is satisfied for the lower level problem (see the proof of Theorem 3.3) and the set $\{(x,y) : g(x,y) \leq 0,\ G(x) \leq 0,\ x \in X\}$ is compact. Problem (3.38) is a nonconvex mixed-integer optimization problem and can be solved using e.g. the approaches in Fletcher and Leyffer [114] and Leyffer [200]. For linear bilevel optimization problems this idea has been suggested in Hansen et al. [136]. Due to the nature of approaches to mixed-integer optimization (as for instance branch-and-bound algorithms) these methods aim to find global optimal solutions of the problem (3.38). Hence, applying them the bilevel optimization problem can also be solved globally.

3.5.2.2 Metaheuristics

Over the years a large number of heuristic approaches such as genetic algorithm (Hejazi et al. [142]), simulated annealing (Sahin and Ciric [278]), particle swarm optimization (Jiang et al. [157]) and other approaches have been suggested for linear bilevel optimization problems. Due to the nature of bilevel optimization problems, in many cases, the metaheuristic algorithms (local perturbation, mutation, etc.) are applied to the upper level variables only and the lower level variables are computed by solving the lower level problem itself. A collection of recent metaheuristics can be found in the edited volume Talbi [297].

3.5.2.3 Regularization of the Classical KKT Transformation

Due to failure of regularity conditions for the problem (3.2) (see Theorem 3.14) finding a direct method solving this problem is in general not easy. However, the problem can be regularized which has been done e.g. in Scholtes [282], Demiguel et al. [48] and in Mersha [226]. For more regularization approaches as well as for a numerical comparison of different approaches, the interested reader is referred to Hoheisel et al. [146].

An Exact Algorithm

Let $T = \mathbb{R}^m$, $X = \mathbb{R}^n$ and assume that the functions f, g_i are differentiable. Scholtes [282] replaced problem (3.2) with

$$\begin{aligned} F(x, y) &\rightarrow \min \\ G(x) &\leq 0 \\ 0 &= \nabla_y f(x, y) + \lambda^\top \nabla_y g(x, y) \\ \lambda &\geq 0,\ g(x, y) \leq 0, \\ -\lambda^\top g(x, y) &\leq \varepsilon \end{aligned} \tag{3.39}$$

and solved this problem for $\varepsilon \downarrow 0$. Assume that the set $\mathscr{M} := \{(x, y) : g(x, y) \leq 0, G(x) \leq 0\}$ is not empty and bounded. Let the lower level problem be a convex parametric optimization problem and suppose that Slater's condition is satisfied for each x with $(x, y) \in \mathscr{M}$ for some y. Then, the feasible set of problem (1.4) is not empty since there exists $(\widehat{x}, \widehat{y}) \in \mathscr{M}$ and $\{\widehat{x}\} \times \Psi(\widehat{x}) \subset \mathscr{M}$. Due to Theorem 3.3, the feasible set of problem (1.4) is compact and, thus, this problem has an optimal solution, see Theorem 3.4. Then, due to Theorem 3.1, problem (3.2) has also a (global) optimal solution $(\overline{x}, \overline{y}, \overline{\lambda})$. Let $\{\varepsilon^k\}_{k=1}^\infty$ be a sequence of positive numbers converging to zero and let $\{(x^k, y^k, \lambda^k)\}_{k=1}^\infty$ be a sequence of (global) optimal solutions of problem (3.39). Then, comparing the feasible sets of problems (3.39) and (3.2) it is easy to see that

$$F(x^k, y^k) \leq F(\overline{x}, \overline{y})\ \forall\, k.$$

Moreover, the sequence $\{(x^k, y^k, \lambda^k)\}_{k=1}^\infty$ has an accumulation point and each accumulation point $(\overline{x}, \overline{y})$ is a global optimal solution of problem (1.4), cf. Theorem 3.2.

An Interior Point Method

Using a diffeomorphism as e.g. in Guddat et al. [133], the problem (3.2) can be reduced to its essential form

$$\min\{f(x) : c(x) = 0,\ x_1 \geq 0,\ x_2 \geq 0,\ x_1^\top x_2 = 0,\ x_0 \geq 0\} \tag{3.40}$$

where $x = (x_0, x_1, x_2) \in \mathbb{R}^{p+n+n}$, $f : \mathbb{R}^{p+2n} \rightarrow \mathbb{R}$, $c : \mathbb{R}^{p+2n} \rightarrow \mathbb{R}^m$. Without using a diffeomorphism this transformation is also possible with the aid of slack variables to replace inequalities by equations and introducing new variables (being equal to the left-hand side of the equations). This problem has been considered in Demiguel et al. [48]. We will only give a short description of main ideas of this approach and ask the reader interested in details to consult the original work. The authors of the article Demiguel et al. [48] apply a regularization approach to problem (3.40) which is more general than the one in problem (3.39). This transforms problem (3.40) into

$$\min\{f(x) : c(x) = 0,\ x_1 \geq -\delta_1,\ x_2 \geq -\delta_2,\ x_1^\top x_2 \leq \delta_0,\ x_0 \geq 0\} \tag{3.41}$$

with $\delta_0, \delta_1, \delta_2 > 0$. Let

$$\mathscr{L}(x, \lambda) := f(x) + \lambda^\top c(x)$$

denote a certain Lagrangian function for problem (3.40), $\overline{x}$ be a feasible point for problem (3.40) and let

$$\begin{aligned}\mathscr{X}_1(x) &= \{j : x_{1j} = 0 < x_{2j}\},\\ \mathscr{X}_2(x) &= \{j : x_{1j} > 0 = x_{2j}\},\\ \mathscr{B}(x) &= \{j : x_{1j} = 0 = x_{2j}\}.\end{aligned}$$

A relaxed problem related to (3.41) is

$$\begin{aligned} f(x) &\to \min\\ c(x) &= 0\\ x_0 &\geq 0\\ x_{1j} = 0,\ x_{2j} &\geq 0,\ j \in \mathscr{X}_1(\overline{x})\\ x_{1j} \geq 0,\ x_{2j} &= 0,\ j \in \mathscr{X}_2(\overline{x})\\ x_{1j} \geq 0,\ x_{2j} &\geq 0,\ j \in \mathscr{B}(\overline{x}). \end{aligned} \tag{3.42}$$

The *linear independence constraint qualification* (LICQ) is satisfied for problem (1.1) at the point $(\overline{x}, \overline{y})$ if:

(LICQ) the gradients $\nabla_y g_i(\overline{x}, \overline{y})$ are linearly independent for all i with $g_i(\overline{x}, \overline{y}) = 0$.

Similarly, this assumption can be posed for problem (3.42). This condition is again called (MPEC-LICQ), and is equivalent to the (MPEC-LICQ) on page 68. It has been shown in Scheel and Scholtes [280] that, if (LICQ) is satisfied at $\overline{x}$ for (3.42), its Karush-Kuhn-Tucker conditions are also necessary optimality conditions for (3.40). This leads to the following stationarity concept for MPECs: A point $(\overline{x}, \overline{\lambda}, \overline{z})$ is *strongly stationary* for problem (3.40) if it satisfies the Karush-Kuhn-Tucker conditions for a problem which is strongly related to problem (3.42):

$$\begin{aligned} \nabla_x \mathscr{L}(\overline{x}, \overline{\lambda}) &= \overline{z},\\ c(\overline{x}) &= 0,\\ \min\{\overline{x}_0, \overline{z}_0\} &= 0,\\ \min\{\overline{x}_1, \overline{x}_2\} &= 0,\\ \overline{x}_{1j}\overline{z}_{1j} &= 0\ \forall\ j,\\ \overline{x}_{2j}\overline{z}_{2j} &= 0\ \forall\ j,\\ \overline{z}_{1j}, \overline{z}_{2j} &= 0\ \forall\ j \text{ with } \overline{x}_{1j} = \overline{x}_{2j} = 0. \end{aligned} \tag{3.43}$$

It satisfies the *weak strict complementarity slackness condition* (MPEC-WSCS) if it is strongly stationary and $\max\{\overline{x}_{0j}, \overline{z}_{0j}\} > 0$ for all j. Moreover, the *strong second-order sufficient optimality condition* (MPEC-SSOSC) is satisfied for problem (3.40) at $(\overline{x}, \overline{\lambda}, \overline{z})$ if

$$d^\top \nabla^2_{xx} \mathscr{L}(\overline{x}, \overline{\lambda}) d > 0$$

for all $d \neq 0$ belonging to a certain tangent cone to the feasible set of problem (3.42).

If for problem (3.42) (LICQ), the weak strict complementarity slackness condition and a sufficient optimality condition of second order [or (MPEC-LICQ), (MPEC-WSCS) and (MPEC-SSOSC) for problem (3.40)] are satisfied, then an interior point algorithm (see e.g. Nesterov and Nemirovskii [252]) can be applied to (3.42) to compute a strongly stationary solution of problem (3.40). This has been done in Demiguel et al. [48]. Moreover, if the solution algorithm starts in a sufficiently small neighborhood of a strongly stationary solution $\overline{w} := (\overline{x}, \overline{\lambda}, \overline{z})$ of problem (3.40), this algorithm converges Q-superlinearly to $(\overline{x}, \overline{\lambda}, \overline{z})$, i.e. there exists a positive constant σ such that

$$\|w_{k+1} - w^*\| \leq \sigma \|w_k - w^*\|^{1+\tau}$$

for some small positive τ and all k.

Bouligand Stationary Solution

In the PhD thesis of Mersha [226] another regularization is developed guaranteeing convergence to Bouligand stationary solutions under some appropriate assumptions, see also Mersha and Dempe [228]. This approach uses the following regularization:

$$\begin{aligned} F(x, y) &\to \min \\ G(x) &\leq 0 \\ \|\nabla_y L(x, y, \lambda)\|_\infty &\leq \varepsilon \\ \lambda &\geq 0,\ g(x, y) \leq 0, \\ -\lambda_i g_i(x, y) &\leq \varepsilon,\ i = 1, \ldots, p, \\ x &\in X, \end{aligned} \tag{3.44}$$

where $L(x, y, \lambda) = f(x, y) + \lambda^\top g(x, y)$ is the Lagrangian of the lower level problem (1.1). Let $\{(x^k, y^k, \lambda^k)\}_{k=1}^\infty$ be a sequence of globally optimal solutions of this problem for $\varepsilon = \varepsilon^k > 0$ and $\{\varepsilon^k\}_{k=1}^\infty$ converging to zero. Assume that $T = \mathbb{R}^m$, the set $M := \{(x, y) : x \in X,\ G(x) \leq 0,\ g(x, y) \leq 0\}$ is not empty and compact, that Slater's condition is satisfied for all $(x, y) \in M$ and the lower level problem (1.1) is a convex parametric optimization problem. Then, using the same ideas as in the preceding subsections we find that each accumulation point $(\overline{x}, \overline{y}, \lambda)$ of the sequence $\{(x^k, y^k, \lambda^k)\}_{k=1}^\infty$ is a global optimal solution of problem (3.2).

But, since problem (3.44) is again a nonconvex optimization problem, assuming that (x^k, y^k, λ^k) is a global optimal solution of this problem is too restrictive.

In the next theorem we will see, that using a sequence $\{(x^k, y^k)\}_{k=1}^{\infty}$ of local optimal solutions of problem (3.44) for $\varepsilon^k \downarrow 0$ we can compute local optimal solutions of problem (1.4), provided that the optimal solution of the lower level problem (1.1) is strongly stable in the sense of Kojima [191]. The main ideas of this approach are again only shortly explained and the interested reader is referred to the original paper [228] of Mersha and Dempe.

Theorem 3.21 (Mersha and Dempe [228]) *Let* $(\overline{x}, \overline{y}, \overline{\lambda})$ *be feasible for problem* (3.44) *for* $\varepsilon = 0$, *assume that the lower level problem is convex with respect to* y *for every fixed* x, $X = \mathbb{R}^n$, *and that the assumptions* (*CRCQ*) *and* (*SSOSC*) *are satisfied for the lower level problem* (1.1) *for all feasible points of* (1.4). *Assume that*

$$\left\{ (d, r) \;\middle|\; \begin{array}{c} \nabla_x G_j(\overline{x})d < 0, \; \forall\, j : G_j(\overline{x}) = 0 \\ \nabla g_i(\overline{x}, \overline{y}) \begin{pmatrix} d \\ r \end{pmatrix} < 0, \; \forall\, i : g_i(\overline{x}, \overline{y}) = 0 \end{array} \right\} \neq \emptyset.$$

If $(\overline{x}, \overline{y})$ *is not a Bouligand stationary point of problem* (1.4) *then there exist a direction* d *and a positive number* $\delta > 0$ *such that*

$$\nabla_x G_i(x)d < 0 \text{ for all } i \text{ with } G_i(\overline{x}) = 0$$

and

$$\nabla_x F(x, y)d + \nabla_y F(x, y)\nabla y^j(x)d < 0,$$

for all $(x, y) \in B_\delta(\overline{x}, \overline{y})$ *and for some of the selection functions* y^j *composing the* PC^1 *function* $\{y(x)\} = \Psi(x)$ *locally around* $\overline{x}$.

Here, $B_\delta(\overline{x}, \overline{y})$ is an open ball of radius $\delta > 0$ around $(\overline{x}, \overline{y})$.

Proof Due to the assumptions and Theorem 3.9 there are an open neighborhood $U_\gamma(\overline{x})$ with $\gamma > 0$ and a PC^1-function $y : U_\gamma(\overline{x}) \mapsto \mathbb{R}^m$ such that $\Psi(x) = \{y(x)\}\ \forall\, x \in U_\gamma(\overline{x})$. If $(\overline{x}, \overline{y})$ is not a Bouligand stationary point of problem (1.4), then there exists a direction d such that

$$\nabla_x G_i(\overline{x})d \leq 0 \text{ for all } i \text{ with } G_i(\overline{x}) = 0$$

and

$$\nabla_x F(\overline{x}, \overline{y})d + \nabla_y F(\overline{x}, \overline{y})y'(\overline{x}; d) < 0,$$

cf. Theorem 3.11. Due to $y'(\overline{x}; d) \in \{\nabla y^j(\overline{x})d : j \in \mathrm{Supp}(y, y^j),\ j = 1, \ldots, k\}$, there is some selection function $y^j(x)$ for the PC^1-function $y(\cdot)$ such that

$$\nabla_x G_i(\overline{x})d \leq 0 \text{ for all } i \text{ with } G_i(\overline{x}) = 0$$

and

$$\nabla_x F(\overline{x}, \overline{y})d + \nabla_y F(\overline{x}, \overline{y})\nabla y^j(\overline{x})d < 0.$$

Since the (MFCQ) is satisfied for the upper level constraints $G(x) \leq 0$ at $x = \overline{x}$ the direction d can be perturbed to $\widehat{d}$ such that

$$\nabla_x G_i(\overline{x})\widehat{d} < 0 \text{ for all } i \text{ with } G_i(\overline{x}) = 0$$

and

$$\nabla_x F(\overline{x}, \overline{y})\widehat{d} + \nabla_y F(\overline{x}, \overline{y})\nabla y^j(\overline{x})\widehat{d} < 0.$$

The proof now follows by smoothness of the functions G_i, y^j, F. □

Now, let $\varepsilon^k \downarrow 0$ be a sequence of positive numbers converging to zero and $\{(x^k, y^k, \lambda^k)\}_{k=1}^{\infty}$ be a sequence of feasible solutions of problem (3.44) for $\varepsilon = \varepsilon^k$, $k = 1, 2, \ldots$, converging to $(\overline{x}, \overline{y}, \overline{\lambda})$. Assume that $(\overline{x}, \overline{y})$ is not a Bouligand stationary solution of problem (1.4). Then, there exists a direction $\overline{d}$ such that

$$\nabla_x G_i(\overline{x})\overline{d} \leq 0 \text{ for all } i \text{ with } G_i(\overline{x}) = 0$$

and

$$\nabla_x F(\overline{x}, \overline{y})\overline{d} + \nabla_y F(\overline{x}, \overline{y}) y'(\overline{x}; \overline{d}) < 0.$$

If the assumptions of Theorem 3.21 are satisfied, due to Theorem 3.10, system

$$\begin{aligned}
\nabla_x F(\overline{x}, \overline{y})d + \nabla_y F(\overline{x}, \overline{y})r &< 0 \\
\nabla_x G_i(\overline{x})d &\leq 0 \quad \text{for all } i \text{ with } G_i(\overline{x}) = 0 \\
\left\| \nabla(\nabla_y L(\overline{x}, \overline{y}, \overline{\lambda})) \begin{pmatrix} d \\ r \\ \gamma \end{pmatrix} \right\|_\infty &= 0 \\
\nabla_x g_i(\overline{x}, \overline{y})d + \nabla_y g_i(\overline{x}, \overline{y})r &\leq 0 \quad \text{for all } i \text{ with } g_i(\overline{x}, \overline{y}) = 0 \\
\gamma_i &\geq 0 \quad \text{for all } i \text{ with } \overline{\lambda}_i = 0 \\
-\nabla(\overline{\lambda}_i g_i(\overline{x}, \overline{y})) \begin{pmatrix} d \\ r \\ \gamma \end{pmatrix} &= 0
\end{aligned} \tag{3.45}$$

has a solution (d, r, γ). For instance, $d = \overline{d}$, $r = y'(\overline{x}; \overline{d})$ and γ standing for a "directional derivative of the multiplier" is one solution. Using again the assumptions of Theorem 3.21 we obtain that the following system has a solution at each point (x, y, λ) in an open neighborhood around $(\overline{x}, \overline{y}, \overline{\lambda})$ for arbitrary small $\varepsilon > 0$:

$$\begin{aligned}
\nabla_x F(x, y)d + \nabla_y F(x, y)r &< 0 \\
G_i(x) + \nabla_x G_i(x)d &< 0 \quad \text{for all } i \\
\left\| \nabla(\nabla_y L(x, y, \lambda)) \begin{pmatrix} d \\ r \\ \gamma \end{pmatrix} \right\|_\infty &\leq \varepsilon/2
\end{aligned} \tag{3.46}$$

$$g_i(x, y) + \nabla_x g_i(x, y)d + \nabla_y g_i(x, y)r < 0 \quad \text{for all } i$$
$$\lambda_i + \gamma_i > 0 \quad \text{for all } i$$
$$-\nabla(\lambda_i g_i(x, y)) \begin{pmatrix} d \\ r \\ \gamma \end{pmatrix} \leq \varepsilon/2$$

This implies the existence of a direction of descent within the tangent cone to the feasible set of problem (3.44) at (x, y, λ) near $(\overline{x}, \overline{y}, \lambda)$ and, therefore, the point (x, y, λ) is not a stationary solution of problem (3.44) for $\varepsilon > 0$.

Theorem 3.22 (Mersha and Dempe [228]) *Under the assumptions of Theorem* 3.21, *if* $\{(x^k, y^k, \lambda^k)\}_{k=1}^\infty$ *is a sequence of stationary solutions of problem* (3.44) *converging to* $\{\overline{x}, \overline{y}, \overline{\lambda}\}$ *for* $\{\varepsilon^k\}_{k=1}^\infty$ *tending to zero, then* $(\overline{x}, \overline{y})$ *is a Bouligand stationary solution of problem* (1.4).

3.5.2.4 Approaches Using NCP Functions

Another approach to solve the problem (3.2) is to replace the complementarity constraints
$$\lambda^\top g(x, y) = 0,\ g(x, y) \leq 0,\ \lambda \geq 0$$
by the help of nonlinear complementarity (NCP) functions. Using auxiliary functions $a_i = -g_i(x, y),\ i = 1, \ldots, p$ and $b_i = \lambda_i,\ i = 1, \ldots, p$ the complementarity conditions can be reduced to

$$a_i \geq 0,\ b_i \geq 0,\ a_i b_i = 0,\ i = 1, \ldots, p. \tag{3.47}$$

These conditions can then be replaced using so called NCP functions, see Galántai [122] where over 30 different NCP functions are formulated. In Kadrani [163], Kadrani et al. [164] and Hoheisel et al. [146] various NCP functions are used to solve MPECs.

Fukushima and Pang [121] use the Fischer-Burmeister function

$$\kappa_i(a, b) = a_i + b_i - \sqrt{a_i^2 + b_i^2}$$

and replace Eq. (3.47) with the equivalent conditions $\kappa_i(a, b) = 0$ respectively the inequalities $\kappa_i(a, b) \leq 0$ for all i.

Theorem 3.23 *We have* $a \geq 0,\ b \geq 0,\ ab = 0$ *for* $a, b \in \mathbb{R}$ *if and only if* $a + b - \sqrt{a^2 + b^2} = 0$.

Proof $a + b - \sqrt{a^2 + b^2} = 0$ implies $(a + b)^2 = a^2 + b^2$ or $ab = 0$. Hence, $a = 0$ or $b = 0$ which by $a + b \geq 0$ leads to $a \geq 0,\ b \geq 0$. On the other hand, let without loss of generality $b = 0$ and $a \geq 0$. Then, this implies $a = \sqrt{a^2}$. □

The function $\kappa_i(a, b)$ is not differentiable at $a_i = b_i = 0$. The regularized function $\kappa_i^\varepsilon(a, b) = a_i + b_i - \sqrt{a_i^2 + b_i^2 + \varepsilon}$ is differentiable for $\varepsilon > 0$.

It can be shown similar to the proof of Theorem 3.23 that

$$a + b - \sqrt{a^2 + b^2 + \varepsilon} = 0 \Leftrightarrow a > 0,\ b > 0,\ ab = \varepsilon/2.$$

Assume that the functions F, G_i, f, g_i are sufficiently smooth, $T = \mathbb{R}^m$, $X = \mathbb{R}^n$. Using the function $\kappa_i^\varepsilon(\cdot, \cdot)$, Fukushima and Pang [121] replace problem (3.2) by

$$\begin{aligned} F(x, y) &\to \min \\ G(x) &\le 0, \\ \nabla_y f(x, y) + \lambda^\top \nabla_y g(x, y) &= 0, \\ \Phi^\varepsilon(x, y, \lambda) &= 0, \end{aligned} \tag{3.48}$$

where

$$\Phi^\varepsilon(x, y, \lambda) = \begin{pmatrix} \kappa_1^\varepsilon(\lambda, -g(x, y)) \\ \vdots \\ \kappa_p^\varepsilon(\lambda, -g(x, y)) \end{pmatrix}.$$

Problem (3.48) is solved for $\varepsilon \downarrow 0$. For a practical realization of this, a sequence $\{\varepsilon_k\}_{k=1}^\infty$ is (carefully) selected, problem (3.48) is solved for $\varepsilon = \varepsilon_k$ and all k. Thus, a sequence of optimal solutions $\{(x^k, y^k, \lambda^k)\}_{k=1}^\infty$ is computed. If this sequence has an accumulation point $(\overline{x}, \overline{y}, \overline{\lambda})$, properties of this accumulation point need to be investigated. The formulation $(x^\varepsilon, y^\varepsilon, \lambda^\varepsilon) \to (\overline{x}, \overline{y}, \overline{\lambda})$ is just an abbreviation for such a process.

Theorem 3.24 (Fukushima and Pang [121]) *Let $(x^\varepsilon, y^\varepsilon, \lambda^\varepsilon)$ be feasible for problem* (3.48) *for $\varepsilon > 0$. If $(x^\varepsilon, y^\varepsilon, \lambda^\varepsilon) \to (\overline{x}, \overline{y}, \overline{\lambda})$ for $\varepsilon \downarrow 0$, then $(\overline{x}, \overline{y}, \overline{\lambda})$ is feasible for problem* (3.2).

The proof follows from the properties of the function $\Phi^\varepsilon(x, y, \lambda)$.

Let

$$\mathscr{F}^\varepsilon(x, y, \lambda, \alpha, \beta, \gamma) = F(x, y) + \alpha^\top G(x) + \beta^\top \mathscr{L}(x, y, \lambda) - \gamma^\top \Phi^\varepsilon(x, y, \lambda)$$

denote some Lagrangian function for problem (3.48) with

$$\mathscr{L}(x, y, \lambda) = \nabla_y f(x, y) + \lambda^\top \nabla_y g(x, y).$$

If some regularity conditions are satisfied at a local optimal solution $(x^\varepsilon, y^\varepsilon, \lambda^\varepsilon)$ of problem (3.48), then Lagrange multipliers $(\alpha, \beta, \gamma) = (\alpha^\varepsilon, \beta^\varepsilon, \gamma^\varepsilon)$ exist such that the following Karush-Kuhn-Tucker conditions are satisfied at this point:

$$\begin{aligned} \nabla \mathscr{F}^{\varepsilon}(x, y, \lambda, \alpha, \beta, \gamma) &= 0 \\ \Phi^{\varepsilon}(x, y, \lambda) &= 0 \\ G(x) \le 0, \ \alpha \ge 0 \quad \alpha^{\top} G(x) &= 0 \\ \mathscr{L}(x, y, \lambda) &= 0. \end{aligned} \tag{3.49}$$

The sufficient optimality condition of second order for problem (3.48) is:

$$d^{\top} \nabla^2_{x,y,\lambda} \mathscr{F}(x, y, \lambda, \alpha, \beta, \gamma) d > 0 \ \forall \, d \in \mathscr{T}^{\varepsilon}(x, y, \lambda), \tag{3.50}$$

where

$$\begin{aligned} \mathscr{T}^{\varepsilon}(x, y, \lambda) = \{d : \nabla \Phi^{\varepsilon}(x, y, \lambda) d = 0, \ \nabla \mathscr{L}(x, y, \lambda) d = 0, \\ (\nabla G_{I_\varepsilon}(x), 0, 0) d = 0\} \end{aligned}$$

with $I_\varepsilon = \{j : G_j(x^\varepsilon) = 0\}$.

We need one further assumption. Suppose that $(x^\varepsilon, y^\varepsilon, \lambda^\varepsilon) \to (\overline{x}, \overline{y}, \overline{\lambda})$ for $\varepsilon \downarrow 0$. Then it follows from the formula of the generalized Jacobian $\partial^{Cl} \Phi^{\varepsilon}(x, y, \lambda)$ of locally Lipschitz continuous functions and Rademacher's theorem (see Clarke [42]) that each accumulation point $\overline{r}_i$ of $\nabla \Phi_i^{\varepsilon}(x^\varepsilon, y^\varepsilon, \lambda^\varepsilon)$ belongs to the set $\partial^{Cl} \Phi_i^0(\overline{x}, \overline{y}, \overline{\lambda})$ and is, hence, represented by

$$\overline{r}_i = \xi_i (0 \, 0 \, e^{i\top}) - \eta_i (\nabla g_i(\overline{x}, \overline{y}) \, 0),$$

where e^i is the ith unit vector and $(1 - \xi_i)^2 + (1 - \eta_i)^2 \le 1$. We say that $(x^\varepsilon, y^\varepsilon, \lambda^\varepsilon)$ is *asymptotically weakly nondegenerate* if neither ξ_i nor η_i vanishes for any accumulation point $\overline{r}_i$. Clearly, $(x^\varepsilon, y^\varepsilon, \lambda^\varepsilon)$ is asymptotically weakly nondegenerate if $I_{00}(\overline{x}, \overline{y}, \overline{\lambda}) = \emptyset$ (see page 63 for the definition of the set $I_{00}(\overline{x}, \overline{y}, \overline{\lambda})$).

Theorem 3.25 (Fukushima and Pang [121]) *Let* $(x^\varepsilon, y^\varepsilon, \lambda^\varepsilon)$ *together with multipliers* $(\alpha^\varepsilon, \beta^\varepsilon, \gamma^\varepsilon)$ *satisfy the Karush-Kuhn-Tucker conditions* (3.49) *and the sufficient optimality conditions of second order* (3.50) *for problem* (3.48). *Furthermore, let* $\{(x^\varepsilon, y^\varepsilon, \lambda^\varepsilon, \alpha^\varepsilon, \beta^\varepsilon, \gamma^\varepsilon)\}_{\varepsilon \downarrow 0}$ *converge to* $(\overline{x}, \overline{y}, \overline{\lambda}, \overline{\alpha}, \overline{\beta}, \overline{\gamma})$ *as* $\varepsilon \downarrow 0$. *If (MPEC-LICQ) holds at* $(\overline{x}, \overline{y}, \overline{\lambda})$ *for problem* (3.2) *and* $(x^\varepsilon, y^\varepsilon, \lambda^\varepsilon)$ *is asymptotically weakly nondegenerate, then* $(\overline{x}, \overline{y}, \overline{\lambda})$ *is a B-stationary point of problem* (3.2).

Proof First, using the Karush-Kuhn-Tucker conditions for problem (3.48) and the formula for the generalized derivative in the sense of Clarke we obtain

$$\begin{aligned} &(\nabla F(\overline{x}, \overline{y}) \, 0) + \alpha^{\top} (\nabla G(\overline{x}) \, 0 \, 0) + \beta^{\top} \nabla \mathscr{L}(\overline{x}, \overline{y}, \overline{\lambda}) \\ &\qquad - \sum_{i \in I_{00}(\overline{x}, \overline{y}, \overline{\lambda})} \xi_i \gamma_i (0 \, 0 \, e^{i\top}) + \sum_{i \in I_{00}(\overline{x}, \overline{y}, \overline{\lambda})} \eta_i \gamma_i (\nabla g_i(\overline{x}, \overline{y}) \, 0) \\ &\qquad - \sum_{i \in I_{-0}(\overline{x}, \overline{y}, \overline{\lambda})} \gamma_i (0 \, 0 \, e^{i\top}) + \sum_{i \in I_{0+}(\overline{x}, \overline{y}, \overline{\lambda})} \gamma_i (\nabla g_i(\overline{x}, \overline{y}) \, 0) = 0. \end{aligned}$$

Moreover, the point $(\overline{x}, \overline{y}, \overline{\lambda})$ is feasible and the complementarity slackness conditions with respect to the constraints $G(x) \leq 0$ hold. Here, all gradients are taken with respect to x, y, λ. To show that these conditions imply the Karush-Kuhn-Tucker conditions of the relaxed problem

$$\begin{aligned} F(x, y) &\to \min \\ G(x) &\leq 0 \\ 0 &= \nabla_y f(x, y) + \lambda^\top \nabla_y g(x, y) \\ \lambda_i &= 0, \ \forall\, i \in I_{-0}(\overline{x}, \overline{y}, \overline{\lambda}) \\ g_i(x, y) &= 0, \ \forall\, i \in I_{0+}(\overline{x}, \overline{y}, \overline{\lambda}) \\ \lambda_i &\geq 0, \ i \in I_{0+}(\overline{x}, \overline{y}, \overline{\lambda}) \cup I_{00}(\overline{x}, \overline{y}, \overline{\lambda}) \\ g_i(x, y) &\leq 0, \ i \in I_{-0}(\overline{x}, \overline{y}, \overline{\lambda}) \cup I_{00}(\overline{x}, \overline{y}, \overline{\lambda}) \end{aligned} \tag{3.51}$$

we need to verify that $\nu_i = \eta_i \gamma_i \geq 0$ and $\zeta_i = \xi_i \gamma_i \geq 0$ for all $i \in I_{00}(\overline{x}, \overline{y}, \overline{\lambda})$. This can be shown by contradiction using (LICQ) together with $\xi_i > 0$, $\eta_i > 0$ due to asymptotical weak nondegeneracy of $(x^\varepsilon, y^\varepsilon, \lambda^\varepsilon)$. For details see Fukushima and Pang [121].

B-stationarity of the point $(\overline{x}, \overline{y}, \overline{\lambda})$ follows by Luo et al. [208, Proposition 4.3.7], since this point is a Karush-Kuhn-Tucker point for the relaxed problem (3.51) and the (MPEC-LICQ) is satisfied. □

3.5.2.5 Application of Algorithms for Smooth Nonlinear Optimization

Convergence of algorithms for smooth optimization problems is usually shown using regularity assumptions. Unfortunately, as mentioned in Theorem 3.14 standard constraint qualifications in nonlinear optimization are violated in all feasible points of the problem (3.2). Nevertheless, computational experiments and also theoretical investigations show encouraging convergence properties of some algorithms for smooth optimization, see Fletcher et al. [115, 116]. Anitescu [3] applies a sequential quadratic optimization algorithm to problem (3.2) and mentions that the direction finding problems often have an empty feasible set if the Mangasarian-Fromovitz constraint qualification is not satisfied at the solution. Hence, this algorithm is applied to a relaxation of problem (3.2). To explain this idea, consider again a problem with simple complementarity constraints [cf. problem (3.40)]:

$$\min\{f(x) : c(x) = 0,\ w(x) \leq 0,\ x_1 \geq 0,\ x_2 \geq 0,\ x_1^\top x_2 = 0\}, \tag{3.52}$$

where $f : \mathbb{R}^{2n} \to \mathbb{R}$, $c : \mathbb{R}^{2n} \to \mathbb{R}^q$ and $w : \mathbb{R}^{2n} \to \mathbb{R}^p$ are sufficiently smooth functions throughout this subsection and $x = (x_1\ x_2)^\top$. It has been shown in Scheel and Scholtes [280], that (MPEC-LICQ) at a local optimal solution $\overline{x}$ of (3.40) implies (LICQ) at $\overline{x}$ in the relaxed problem (3.42) and, hence, this problem has then unique Lagrange multipliers at $\overline{x}$.

To derive conditions guaranteeing these assumptions, Anitescu [3] applies a penalty function approach to problem (3.52). Let

$$p_\infty(x) := \sum_{i=1}^{p} \max\{0, w_i(x)\} + \sum_{i=1}^{q} |c_i(x)| + |x_1^\top x_2|$$

be this penalty function. Then, the penalty function approach for this problem consists of solving

$$\min\{f(x) + \zeta_\infty p_\infty(x) : x \geq 0\} \tag{3.53}$$

for $0 \leq \zeta_\infty < \infty$. A realization of this approach clearly means that a sequence $\{\zeta_k\}_{k=1}^\infty$ is (carefully) selected and problem (3.53) is solved for each $\zeta_\infty = \zeta_k$, $k = 1, \ldots, \infty$. Thus, a sequence $\{x^k\}_{k=1}^\infty$ is computed. If this sequence has an accumulation point $\overline{x}$, properties of this accumulation point need to be investigated. It will be shown in the following that (3.53) is an exact penalty function, which means that ζ_k does not need to go to infinity, the sequence $\{x^k\}_{k=1}^\infty$ is finite.

Consider the following

Condition (A1) Let $\widehat{x}$ be an optimal solution of problem (3.52). The set $\widehat{\Lambda}(\widehat{x}) = \{(\lambda, \mu, \gamma) : \nabla\widehat{L}(\widehat{x}, \lambda, \mu, \gamma) = 0,\ \lambda \geq 0,\ \lambda^\top w(\widehat{x}) = 0\}$ of Lagrange multipliers for problem (3.52) is not empty, where $\widehat{L}(x, \lambda, \mu, \gamma) = f(x) + \lambda^\top w(x) + \mu^\top c(x) + \gamma x_1^\top x_2$.

This assumption is satisfied e.g. under (MPEC-LICQ), see Scheel and Scholtes [280].

Theorem 3.26 (Anitescu [3]) *If condition (A1) is satisfied, problem* (3.53) *is an exact penalty function approach to problem* (3.52), *i.e. there exists* $\zeta_\infty^0 < \infty$ *such that, for each* $\zeta_\infty \geq \zeta_\infty^0$, *each optimal solution of problem* (3.52) *is also an optimal solution of problem* (3.53).

Proof Let $\widetilde{x}$ be an optimal solution of problem (3.53), take $(\lambda, \mu, \gamma) \in \widehat{\Lambda}(\widehat{x})$ and

$$\zeta_\infty^0 \geq \max\{\lambda_1, \ldots, \lambda_p, |\mu_1|, \ldots, |\mu_q|, |\gamma|\}$$

as well as $\overline{\mu_i} = |\mu_i|$. Then,

$$\begin{aligned} f(\widetilde{x}) + \zeta_\infty p_\infty(\widetilde{x}) &\leq f(\widehat{x}) = \widehat{L}(\widehat{x}, \lambda, \mu, \gamma) \leq \widehat{L}(x, \lambda, \mu, \gamma) \\ &= f(x) + \lambda^\top w(x) + \mu^\top c(x) + \gamma x_1^\top x_2 \\ &\leq f(x) + \lambda^\top w(x) + \overline{\mu}^\top |c(x)| + |\gamma||x_1^\top x_2| \\ &\leq f(x) + \zeta_\infty^0 p_\infty(x) \leq f(x) + \zeta_\infty p_\infty(x) \end{aligned}$$

for all optimal solutions $\widehat{x}$ of problem (3.52), $x \in \mathbb{R}^{2n}$. □

Problem (3.53) can equivalently be written as

$$
\begin{aligned}
f(x) + \zeta_\infty \theta &\to \min \\
w_i(x) &\le \theta, \; i = 1, \dots, p \\
|c_i(x)| &\le \theta, \; i = 1, \dots, q \\
|x_1^\top x_2| &\le \theta \\
x, \theta &\ge 0.
\end{aligned}
\tag{3.54}
$$

Consider the *quadratic growth condition*

(**QGC**) There exists σ such that

$$\max\{f(x) - f(\overline{x}), p_\infty(x)\} \ge \sigma \|x - \overline{x}\|^2, \; \forall\, x \text{ sufficiently close to } \overline{x}$$

is satisfied at each (local) optimal solution $\overline{x}$ of problem (3.52).

For smooth problems the quadratic growth condition is equivalent to a sufficient optimality condition of second order, see e.g. Bonnans and Shapiro [23].

Theorem 3.27 (Anitescu [3]) *Let assumption (A1) together with the quadratic growth condition be satisfied for problem* (3.52) *at an optimal solution* $\overline{x}$. *Then,* $(\overline{x}, 0)$ *is an isolated local minimum of problem* (3.54) *and (MFCQ) as well as the quadratic growth condition are satisfied there.*

Problem (3.54) can be solved using sequential quadratic optimization.

The sequential quadratic optimization algorithm for an optimization problem with equality and inequality constraints

$$\min\{\widetilde{f}(x) : \widetilde{g}(x) \le 0, \; \widetilde{h}(x) = 0\} \tag{3.55}$$

with sufficiently smooth functions $\widetilde{f} : \mathbb{R}^n \to \mathbb{R}$, $\widetilde{g} : \mathbb{R}^n \to \mathbb{R}^p$, $\widetilde{h} : \mathbb{R}^n \to \mathbb{R}^q$ solves quadratic direction finding problems

$$
\begin{aligned}
\nabla \widetilde{f}(x^k)(x - x^k) + \frac{1}{2}(x - x^k)^\top H_k (x - x^k) &\to \min \\
\widetilde{g}_i(x^k) + \nabla \widetilde{g}_i(x^k)(x - x^k) &\le 0, \; i = 1, \dots, p \\
\widetilde{h}_i(x^k) + \nabla \widetilde{h}_i(x^k)(x - x^k) &\le 0, \; i = 1, \dots, q
\end{aligned}
\tag{3.56}
$$

in all iterations. Here, H_k is in general a quadratic, positive definite matrix approximating the Hessian matrix of the Lagrangian of problem (3.55). This algorithm can be formulated as follows, if $H_k = \nabla^2_{xx} \widetilde{L}(x^k, \lambda^k, \mu^k)$ is used (see e.g. Geiger and Kanzow [128]):

Algorithm: [SQP algorithm solving problem (3.55)]

Step 0: `Select` $(x^0, \lambda^0, \mu^0) \in \mathbb{R}^n \times \mathbb{R}^p \times \mathbb{R}^q$, `set` $k := 0$.
Step 1: `If` (x^k, λ^k, μ^k) `satisfies the Karush-Kuhn-Tucker conditions of problem (3.55), stop.`
Step 2: `Solve problem (3.56). Let` $(x^{k+1}, \lambda^{k+1}, \mu^{k+1})$ `be a KKT point of this problem.`
`If this problem has more than one KKT point take one whose distance`

$$\|(x^{k+1}, \lambda^{k+1}, \mu^{k+1}) - (x^k, \lambda^k, \mu^k)\|$$

`is minimal.`
Step 4: `Set` $k := k + 1$, `goto Step 1.`

If this algorithm is applied to problem (3.55) we obtain

Theorem 3.28 (Geiger and Kanzow [128]) *Let* $(\overline{x}, \overline{\lambda}, \overline{\mu})$ *be a KKT point of problem* (3.55) *and let the following assumptions be satisfied:*

1. *We have* $\widetilde{g}_i(\overline{x}) + \overline{\lambda}_i \neq 0,\ i = 1, \ldots, p$ *(strict complementarity slackness),*
2. *(LICQ) is satisfied,*
3. *The sufficient optimality condition of second order is valid at* $\overline{x}$*.*

Then there is $\varepsilon > 0$ *such that the following conditions are satisfied, provided* $\|(x^0, \lambda^0, \mu^0) - (\overline{x}, \overline{\lambda}, \overline{\mu})\| \le \varepsilon$*:*

1. *The sequence* $\{(x^k, \lambda^k, \mu^k)\}_{k=1}^{\infty}$ *converges superlinearly to* $(\overline{x}, \overline{\lambda}, \overline{\mu})$*, i.e. there exists a sequence* $\{\varepsilon_k\}_{k=1}^{\infty}$ *converging to zero from above such that*

$$\|(x^{k+1}, \lambda^{k+1}, \mu^{k+1}) - (\overline{x}, \overline{\lambda}, \overline{\mu})\| \le \varepsilon_k \|(x^k, \lambda^k, \mu^k) - (\overline{x}, \overline{\lambda}, \overline{\mu})\|$$

   for all k.
2. *If the Hessian matrices of all functions* $\widetilde{f}, \widetilde{g}_i, \widetilde{h}_j$ *are locally Lipschitz continuous, the convergence rate is quadratic, which means that there exists* $0 < C < \infty$ *such that*

$$\|(x^{k+1}, \lambda^{k+1}, \mu^{k+1}) - (\overline{x}, \overline{\lambda}, \overline{\mu})\| \le C \|(x^k, \lambda^k, \mu^k) - (\overline{x}, \overline{\lambda}, \overline{\mu})\|^2$$

   for all k.

The difficulty in applying this result to solving mathematical programs with equilibrium constraints is that the assumptions cannot be satisfied, because the linear independence constraint qualification is violated in every feasible point. To circumvent this difficulty, Anitescu [3] suggests solving the penalized problem (3.53) or an equivalent problem (3.54). The direction finding problem for problem (3.53) is (3.57):

$$
\begin{aligned}
0.5\,d^\top \widehat{W} d + \nabla f(x)d + \zeta_\infty(\theta + d_\theta) &\to \min_{d,d_\theta}, \\
|c_i(x) + \nabla c_i(x)d| &\le \theta + d_\theta,\ i = 1, \dots, q, \\
w_i(x) + \nabla w_i(x) &\le \theta + d_\theta,\ i = 1, \dots, p \\
x_{2t} + d_{2t} &\ge 0,\ t = 1, \dots, n, \\
x_{1t} + d_{1t} &\ge 0,\ t = 1, \dots, n, \\
x_1^\top d_2 + x_2^\top d_1 + x_1^\top x_2 &\le \theta + d_\theta, \\
\theta + d_\theta &\ge 0.
\end{aligned}
\tag{3.57}
$$

The direction finding problem for problem (3.52) reads as

$$
\begin{aligned}
\nabla f(x)d + \frac{1}{2} d^\top \widehat{W} d &\to \min \\
w_i(x) + \nabla w_i(x)d &\le 0,\ i = 1, \dots, p, \\
c_i(x) + \nabla c_i(x)d &= 0,\ i = 1, \dots, q \\
x_{1t} + d_{1t} &\le 0,\ t = 1, \dots, n \\
x_{2t} + d_{2t} &\le 0,\ t = 1, \dots, n \\
x_1^\top x_2 + x_1^\top d_2 + x_2^\top d_1 &\le 0.
\end{aligned}
\tag{3.58}
$$

This direction finding problem can be used whenever it has a feasible solution.

Then, the idea in Anitescu [3] is to use the following algorithm:

Algorithm: Start with x^0, $\zeta_\infty < \infty$, $k = 1$.

1. Compute a KKT point of problem (3.58).
2. If the norm of the multipliers is not larger than the penalty parameter and problem (3.58) has a KKT point, set $x^{k+1} = x^k + d^k$, $k := k + 1$ and goto Step 1.
3. Otherwise compute a KKT point of problem (3.57), set $x^{k+1} = x^k + d^k$, $\theta^{k+1} = \theta^k + d_\theta^k$, $k := k + 1$.
4. If the norm of (d^k, d_θ^k) is not too large, enlarge ζ_∞. Goto Step 3.

For constructing the matrices $\widehat{W}$ either the Hessian of the Lagrangian of the respective optimization problem or an estimate (which needs to be a positive definite matrix) of it (using an estimate of the Lagrange multiplier of this problem) can be used. Step 3 is called the elastic mode in the following theorem.

The sufficient optimality condition of second order for an MPEC (MPEC-SOSC) demands that the Hessian matrix of the Lagrangian function of the relaxed MPEC problem (where the complementarity constraint is dropped) with respect to x is positive definite (which means $d^\top \nabla^2_{xx} \mathscr{L}(x, \lambda, \mu)d > 0$) for all directions d in the critical cone of this problem.

Theorem 3.29 (Anitescu [3]) *Let (MPEC-LICQ) and (MPEC-SOSC) be satisfied as well as strict complementarity slackness (i.e.* $\mu_{1t} > 0$ *and* $\mu_{2t} > 0$ *for all* $t \in \{1, \dots, n\}$ *with* $\overline{x}_{1t} = \overline{x}_{2t} = 0$ *where* μ_{it} *are the multipliers to* x_{it} *and* $w_i(\overline{x}) + \overline{\lambda}_i \neq 0$ *for all* $i = 1, \dots, p$*) be satisfied near a solution* $\overline{x}$ *of problem* (3.40)*. Assume that the starting point* x^0 *is sufficiently close to* $\overline{x}$*, the Lagrange multipliers used for computing* $\widehat{W}$ *are sufficiently close to* $(\overline{\lambda}, \overline{\mu}, \overline{\gamma})$*, and* $\theta > 0$ *is sufficiently small. Then,* $\{x^k\}_{k=1}^{\infty}$ *converges to* $\overline{x}$ *superlinearly if the elastic mode is never used and* $\{(x^k, \theta^k)\}_{k=1}^{\infty}$ *converges superlinearly to* $(\overline{x}, 0)$ *in the other case, when the elastic mode is used in all iterations beginning at iteration* k_0.

The application of the elastic mode SQP algorithm is closely related to the use of problem (3.44) to solve the bilevel optimization problem. The condition

$$\nabla_y L(x, y, \lambda) = \nabla_y f(x, y) + \lambda^\top \nabla_y g(x, y) = 0$$

is a nonlinear equation in the classical KKT transformation of the bilevel optimization problem (3.2). In problem (3.40) it is part of the constraints $c(x) = 0$. The use of descent algorithms as the feasible directions method in the Paragraph "Bouligand stationary solution" starting on page 73 or the SQP algorithm can lead to direction finding problems which do not have a solution or to convergence of the computed sequence $\{x^k, y^k, \lambda^k\}_{k=1}^{\infty}$ to a point $(\overline{x}, \overline{y}, \overline{\lambda})$ which is a stationary solution for the MPEC but is not related to a stationary point of the bilevel optimization problem. The constraint $\nabla_y L(x, y, \lambda) = 0$ was relaxed to $\|\nabla_y L(x, y, \lambda)\| \leq \varepsilon$ in the Paragraph "Bouligand stationary solution" to circumvent this situation. The same is done in the elastic mode SQP in this section. Hence, the use of the elastic mode SQP approach is also possible for solving the bilevel optimization problem. Applied at the same point both direction finding problems will compute a direction of descent at the same time. Under the used assumptions both approaches avoid convergence to a stationary solution of the MPEC which is not related to a stationary solution of the bilevel optimization problem.

In Coulibaly and Orban [46] an elastic interior point algorithm is suggested converging to a strongly stationary solution of problem (3.40) under certain assumptions. As a special feature the algorithm detects if the (MPEC-MFCQ) is violated at the limit point of the sequence of points generated.

3.6 The Optimal Value Transformation

In this section the optimal value function

$$\varphi(x) := \min_y \{f(x, y) : g(x, y) \leq 0, \; y \in T\}$$

is used to transform the bilevel optimization problem (1.4) into problem (3.7):

$$\begin{aligned} F(x,y) &\to \min \\ G(x) &\leq 0 \\ f(x,y) &\leq \varphi(x) \\ g(x,y) &\leq 0, \ y \in T, \\ x &\in X. \end{aligned}$$

Recall that both formulations (1.4) and (3.7) are fully equivalent and that problem (3.7) is a nonsmooth, nonconvex optimization problem.

Theorem 3.30 (Ye and Zhu [325]; Pilecka [264]) *If the function $\varphi(\cdot)$ is locally Lipschitz continuous and the functions F, G, f, g are at least differentiable, $T \subseteq \mathbb{R}^n$ is convex, then the (nonsmooth) Mangasarian-Fromovitz constraint qualification is violated at every feasible point of the problem* (3.7).

Proof Problem (3.7) is a Lipschitz optimization problem and we can apply necessary optimality conditions from Lipschitz optimization, see Clarke [42]. Let (x^0, y^0) be any feasible point of problem (3.7). Then, this point is a global optimal solution of the problem

$$\min_{x,y}\{f(x,y) - \varphi(x) : g(x,y) \leq 0, \ y \in T\},$$

since, by definition of the optimal value function $\varphi(\cdot)$ of the lower level optimization problem, the optimal function value of this problem is zero. This value is attained by feasibility of the point (x^0, y^0) to (1.2). Thus, there exist $\lambda_0 \geq 0, \ \lambda \in \mathbb{R}^p_+$ such that

$$0 \in \lambda_0 \nabla f(x^0, y^0) + \lambda^\top \nabla g(x^0, y^0) + \{0\} \times N_T(y^0).$$

Hence, there exists an abnormal Lagrange multiplier for problem (3.7) which is equivalent to violation of the (nonsmooth) Mangasarian-Fromovitz constraint qualification, see Ye and Zhu [325]. □

3.6.1 Necessary Optimality Conditions

Problem (3.7) can be transformed into an optimization problem where the nonsmooth function appears in the objective if it is partially calm.

Definition 3.7 (*Ye and Zhu* [325]) Let (x^0, y^0) be a (local) optimal solution of problem (3.7). Then, this problem is called *partially calm* at (x^0, y^0) if there exists $\kappa > 0$ and an open neighborhood $W(x^0, y^0, 0) \subset \mathbb{R}^n \times \mathbb{R}^m \times \mathbb{R}$ such that the following condition is satisfied: For each feasible solution (x', y') of the problem

$$\begin{aligned} F(x, y) &\to \min \\ G(x) &\le 0 \\ f(x, y) - \varphi(x) + u &= 0 \\ g(x, y) &\le 0, \ y \in T, \\ x &\in X. \end{aligned} \tag{3.59}$$

for $u = u'$ with $(x', y', u') \in W(x^0, y^0, 0)$ the inequality

$$F(x', y') - F(x^0, y^0) + \kappa |u'| \geq 0$$

is satisfied.

Using partial calmness it is easy to see that the (local) optimal solution (x^0, y^0) is also a (local) optimal solution of the problem

$$\begin{aligned} F(x, y) + \gamma(f(x, y) - \varphi(x)) &\to \min_{x,y} \\ G(x) &\le 0 \\ g(x, y) &\le 0, \ y \in T, \\ x &\in X. \end{aligned} \tag{3.60}$$

for some finite $\gamma > 0$.

This problem can now be used to derive necessary conditions for a local optimal solution of the optimistic bilevel optimization problem (1.4).

Theorem 3.31 (Ye and Zhu [325]) *Assume that all the functions F, G, f, g are sufficiently smooth, the sets $T = \mathbb{R}^m$, $X = \mathbb{R}^n$ and that the Mangasarian-Fromovitz constraint qualification is satisfied for the sets $\{x : G(x) \le 0\}$ and $\{y : g(x^0, y) \le 0\}$ at all points in these sets, where x^0 with $G(x^0) \le 0$ is arbitrary. Let problem* (3.7) *be partially calm at a local optimal solution $(\overline{x}, \overline{y})$ and let the set $\{(x, y) : g(x, y) \le 0,\ G(x) \le 0\}$ be compact. Then, there exist $\gamma > 0$, $\alpha \in \mathbb{R}^p_+$, $\beta \in \mathbb{R}^q_+$ such that*

$$\begin{aligned} &0 \in \nabla F(\overline{x}, \overline{y}) + \gamma(\nabla f(\overline{x}, \overline{y}) - \partial^{Cl} \varphi(\overline{x}) \times \{0\}) + \alpha^\top \nabla g(\overline{x}, \overline{y}) + \beta^\top \nabla G(\overline{x}) \times \{0\} \\ &0 = \alpha^\top g(\overline{x}, \overline{y}) \\ &0 = \beta^\top G(\overline{x}). \end{aligned} \tag{3.61}$$

Proof Due to partial calmness, $(\overline{x}, \overline{y})$ is a local optimal solution of problem (3.60) for some $\gamma > 0$. By Theorem 3.5, this problem is a Lipschitz optimization problem. Due to the Mangasarian-Fromovitz constraint qualifications, there does not exist an abnormal multiplier. Then the result follows by applying the necessary optimality conditions for Lipschitz optimization problems as given in Clarke [42]. □

Using Theorem 3.6 it is possible to formulate optimality conditions using the functions F, G, f, g only. Necessary for local optimality of the point $(\overline{x}, \overline{y})$ is the existence of $\gamma > 0$, $\alpha \in \mathbb{R}^p_+$, $\beta \in \mathbb{R}^q_+$, $\zeta \in \mathbb{R}^{n+1}_+$, $\sum_{k=1}^{n+1} \zeta_k = 1$ and $y^k \in \Psi(\overline{x})$, $\lambda^k \in \Lambda(\overline{x}, y^k)$, $k = 1, \ldots, n+1$ such that the following conditions are satisfied:

$$\nabla_x F(\overline{x}, \overline{y}) + \gamma(\nabla_x f(\overline{x}, \overline{y}) - \sum_{k=1}^{n+1} \zeta_k \nabla_x L(\overline{x}, y^k, \lambda^k)) + \alpha^\top \nabla_x g(\overline{x}, \overline{y}) + \beta^\top \nabla_x G(\overline{x}) = 0$$
$$\nabla_y F(\overline{x}, \overline{y}) + \gamma \nabla_y f(\overline{x}, \overline{y}) + \alpha^\top \nabla_y g(\overline{x}, \overline{y}) = 0$$
$$\alpha^\top g(\overline{x}, \overline{y}) = 0$$
$$\beta^\top G(\overline{x}) = 0.$$

Here we used Caratheodory's theorem to treat the convex hull on the right-hand side of the inclusion (3.9).

If $\Psi(\overline{x})$ reduces to a singleton $\{\overline{y}\} = \Psi(\overline{x})$ then we need only one multiplier $\zeta = 1$ since the set $\Lambda(\overline{x}, \overline{y})$ is convex. Then, the first equation of the last system of equations reduces to

$$\nabla_x F(\overline{x}, \overline{y}) + (\alpha - \gamma\lambda)^\top \nabla_x g(\overline{x}, \overline{y}) + \beta^\top \nabla_x G(\overline{x}) = 0. \tag{3.62}$$

This result has been shown in Dempe et al. [56] under the weaker assumption of inner semicontinuity of the solution set mapping $\Psi(\cdot)$ at the point $(\overline{x}, \overline{y})$.

In the case when the functions f, g_i are jointly convex in both x and y the optimal value function $\varphi(\cdot)$ is also convex. Then the union over all optimal solutions in the lower level problem (1.1) is not necessary in the formula (3.9) for the generalized gradient of $\varphi(\cdot)$, see e.g. Shimizu et. al [288]. Since an arbitrary optimal solution of the lower level problem can then be used in the formula (3.9) an analogous result to (3.62) can be obtained.

3.6.2 Solution Algorithms

3.6.2.1 Jointly Convex Lower Level Problem

First consider the case when the functions f, g_i in the lower level problem (1.1) are jointly convex and $T = \mathbb{R}^m$. The optimal value function $\varphi(x) = \min_y\{f(x, y) : g(x, y) \leq 0\}$ is also convex in this case. Let X be a polytop, i.e. a bounded polyhedron and let $\mathscr{X} = \{x^i : i = 1, \ldots, s\}$ be the set of its vertices:

$$X = \operatorname{conv} \mathscr{X}.$$

For each $x \in X$ there exist $\mu_i \geq 0$, $\sum_{i=1}^s \mu_i = 1$ such that $x = \sum_{i=1}^s \mu_i x^i$ and we can apply a variable transformation substituting the variables x in problem (1.4) by μ. Moreover, by convexity,

$$\varphi(x) \leq \sum_{i=1}^{s} \mu_i \varphi(x^i).$$

Note that for a fixed x the vector μ with $x = \sum_{i=1}^{s} \mu_i x^i$ is in general not unique and that this formula is correct for all $\mu \geq 0,\ \sum_{i=1}^{s} \mu_i = 1$. The best bound for $\varphi(\cdot)$ is the function

$$\xi(x) = \min_{\mu} \left\{ \sum_{i=1}^{s} \mu_i \varphi(x^i) : \mu \geq 0,\ \sum_{i=1}^{s} \mu_i = 1,\ x = \sum_{i=1}^{s} \mu_i x^i \right\}. \tag{3.63}$$

This implies that the feasible set of problem (3.7) is a subset of the feasible set of the following optimization problem:

$$\begin{aligned} F(x, y) &\to \min_{x,y} \\ G(x) &\leq 0 \\ f(x, y) &\leq \xi(x) \\ g(x, y) &\leq 0 \\ x &\in X. \end{aligned} \tag{3.64}$$

Thus, the optimal objective function value of problem (3.64) is not larger than the optimal objective function value of problem (3.7). If (x^0, y^0) is an optimal solution of problem (3.64) it is also optimal for problem (3.7) provided it is feasible for this problem. In the opposite case, adding x^0 to the set $\mathscr{X}$ we obtain a better approximation of the function $\varphi(\cdot)$ and can, hence, proceed with solving problem (3.64) using an updated function $\xi(\cdot)$.

This leads to the following

Algorithm for bilevel optimization problems with jointly convex lower level problems:

Algorithm: Start `Compute the set` $\mathscr{X}$ `of all vertices of the set` X, `compute the function` $\xi(\cdot)$, $t := 1$.

Step 1 `Solve problem (3.64). Let` (x^t, y^t) `be a global optimal solution.`

Step 2 `If` $y^t \in \Psi(x^t)$, `stop,` (x^t, y^t) `is the solution of the problem (3.7). Otherwise, set` $\mathscr{X} := \mathscr{X} \cup \{x^t\}$, `compute` $\varphi(x^t)$, `update` $\xi(\cdot)$, `set` $t := t + 1$ `and goto Step 1.`

Theorem 3.32 *Assume that the above algorithm computes an infinite sequence* $\{(x^t, y^t)\}_{t=1}^{\infty}$, *that the set* $\{(x, y) : G(x) \leq 0,\ g(x, y) \leq 0,\ x \in X\}$ *is not empty and compact and that* $\forall\, \widetilde{x} \in X$ *there exists a point* $(\widetilde{x}, \widetilde{y})$ *such that* $g(\widetilde{x}, \widetilde{y}) < 0$. *Then,*

1. *the sequence* $\{(x^t, y^t)\}_{t=1}^{\infty}$ *has accumulation points* $(\overline{x}, \overline{y})$,
2. *each accumulation point of this sequence is a globally optimal solution of problem* (3.7).

Proof The assumed existence of the point $(\widetilde{x}, \widetilde{y})$ means that the Slater's condition is satisfied for all $x \in X$ with $G(x) \leq 0$ for the convex lower level problem. This implies that the optimal value function $\varphi(\cdot)$ is Lipschitz continuous due to the compactness assumption. Hence, the set $M := \{(x, y) : G(x) \leq 0,\ g(x, y) \leq 0,\ f(x, y) \leq \varphi(x),\ x \in X\}$ is also compact and the sequence $\{(x^t, y^t)\}_{t=1}^{\infty} \subset M$ has accumulation points.

Denote the set $\mathscr{X}$ in iteration t of the algorithm by $\mathscr{X}^t$ and let s^t be the number of its elements.

Let $(\overline{x}, \overline{y})$ be an accumulation point of this sequence and let

$$\xi^t(x) = \min_{\mu} \left\{ \sum_{i=1}^{s^t} \mu_i \varphi(x^i) : \mu_i \geq 0,\ i = 1, \ldots, s^t,\ \sum_{i=1}^{s^t} \mu_i = 1,\ \sum_{i=1}^{s^t} \mu_i x^i = x \right\} \tag{3.65}$$

denote the approximation of the function $\varphi(x)$ in iteration t of the algorithm. Then, the sequence $\{\xi^t(x)\}$ converges uniformly to $\overline{\xi}(x)$ and $\varphi(x^t) = \overline{\xi}(x^t) = f(x^t, \overline{y}^t)$ for all t, where $\overline{y}^t$ is computed in Step 2 of the algorithm as an optimal solution of the lower level problem. Using the same idea as in the proof of Theorem 2.8 we can show that the point $(\overline{x}, \overline{y})$ is feasible for problem (3.7). Thus, this point is also a global optimal solution of this problem. This implies that the proof of the theorem is complete. □

Problem (3.65) is a linear optimization problem and for each x there is a set of basic variables $\mu_i,\ i \in \mathscr{B}(x) \subseteq \{1, \ldots, s^t\}$ such that $x = \sum_{i \in \mathscr{B}(x)} \mu_i x^i$ with $\xi(x) = \sum_{i \in \mathscr{B}(x)} \mu_i \varphi(x^i)$. Consider the sets $\mathscr{V}(x) = \{x^i : i \in \mathscr{B}(x)\} \subseteq \mathscr{X}^t$. By parametric linear optimization, the convex hull conv $\mathscr{V}(x)$ is the so-called *region of stability*, i.e. the set of all points x for which the set of basic variables in an optimal solution of problem (3.65) remains constant. $\xi^t(\cdot)$ is affine linear over conv $\mathscr{V}(x)$. If we consider all $x \in X$, a finite number of subsets of the set $\mathscr{X}^t$ arises, the union if which equals $\mathscr{X}^t$. Denote these sets as $\mathscr{V}_{kt},\ k = 1, \ldots, w_t$.

Using the sets $\mathscr{V}_{kt}$ and linear parametric optimization, problem (3.64) can be decomposed into a number of problems

$$\begin{aligned} F(x, y) &\to \min_{\mu, y} \\ G(x) &\leq 0 \\ f(x, y) &\leq \xi(x) \\ g(x, y) &\leq 0 \\ x &= \sum_{i=1}^{|\mathscr{V}_{kt}|} \mu_i x^{ki} \in \operatorname{conv} \mathscr{V}_{kt} := \operatorname{conv} \{x^{kti},\ i = 1, \ldots, |\mathscr{V}_{kt}|\} \\ \xi(x) &= \sum_{i=1}^{|\mathscr{V}_{kt}|} \mu_i \varphi(x^{kti}) \end{aligned} \tag{3.66}$$

with $\mathcal{V}_{kt} = \{x^{kti},\ i = 1, \ldots, |\mathcal{V}_{kt}|\}$. Here, clearly, x and $\xi(x)$ are computed using the same values for μ_i. Then, an optimal solution of problem (3.64) is one of the optimal solutions of the problems (3.66), especially the best one. If it is an optimal solution of problem (3.66) with $k = \overline{k}$, then only the set $\mathcal{V}_{\overline{k}t}$ needs to be updated and decomposed into some subsets.

Summing up, this leads to an enumerative algorithm which can be improved to an algorithm of branch-and-bound type if upper bounds of the optimal objective function value of problem (3.7) in form of function values for feasible solutions (e.g. obtained by computing $\overline{y}^t \in \Psi(x^t)$ and $F(x^t, \overline{y}^t)$) are used.

The following example illustrates this algorithm.

Example 3.5 Consider the lower level problem

$$\Psi(x) = \operatorname*{Argmin}_y\{-y : x + y \leq 2,\ -x + y \leq 2,\ y \geq 0\}$$

and the bilevel optimization problem

$$\begin{aligned} (x-1)^2 + y^2 &\to \min \\ -2 \leq x &\leq 2 \\ y &\in \Psi(x) \end{aligned}$$

The feasible set and local and global optimal solutions of this problem are depicted in Fig. 3.5 In the first step of the algorithm we need to compute a global optimal solution of the problem

$$\begin{aligned} (x-1)^2 + y^2 &\to \min \\ -2 \leq x &\leq 2 \\ x + y &\leq 2, \\ -x + y &\leq 2, \\ y &\geq 0 \end{aligned}$$

Fig. 3.5 Feasible set and optimal solutions of the example

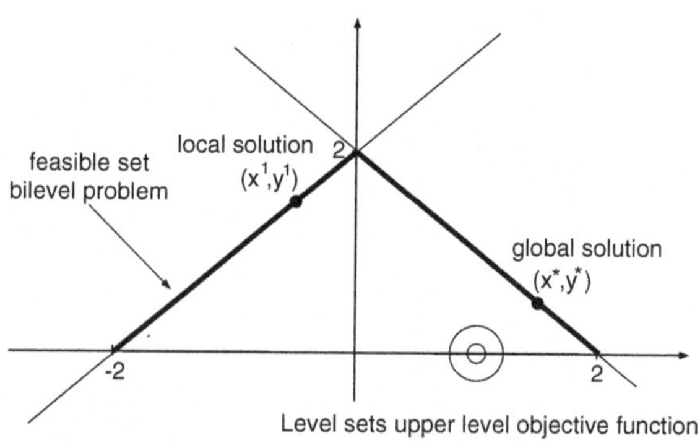

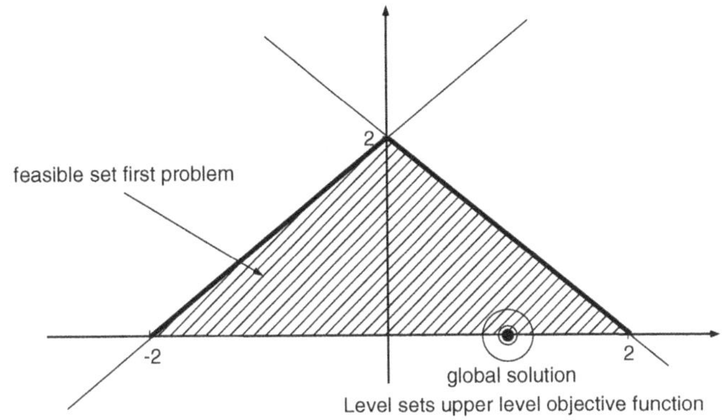

Fig. 3.6 Feasible set and optimal solutions of the first subproblem

The optimal solution of this problem is $x = 1,\ y = 0$, see Fig. 3.6 Then, the first approximation of the optimal value function $\varphi(x) := \min_y\{-y : x + y \leq 2,\ -x + y \leq 2,\ y \geq 0\}$ is

$$\xi(x) = \min\{-\frac{2}{3}x - \frac{1}{3}, x - 2\}$$

and we have to solve the problem

$$\begin{aligned} (x-1)^2 + y^2 &\to \min \\ -2 \leq x &\leq 2 \\ x + y &\leq 2, \\ -x + y &\leq 2, \\ y &\geq 0 \\ -y &\leq \min\{-\frac{2}{3}x - \frac{1}{3}, x - 2\} \end{aligned}$$

globally. This is depicted in Fig. 3.7. The global optimal solution of this problem is also globally optimal for the bilevel optimization problem. □

3.6.2.2 A Discontinuous Approximation of the Optimal Value Function

Mitsos et al. consider in [234] an algorithm which is able to approximate a global optimal solution of the bilevel optimization problem even if the lower level problem is not a convex parametric optimization problem. This is possible using the optimal value function of the lower level problem and a piecewise, yet discontinuous, approximation of the function $\varphi(\cdot)$.

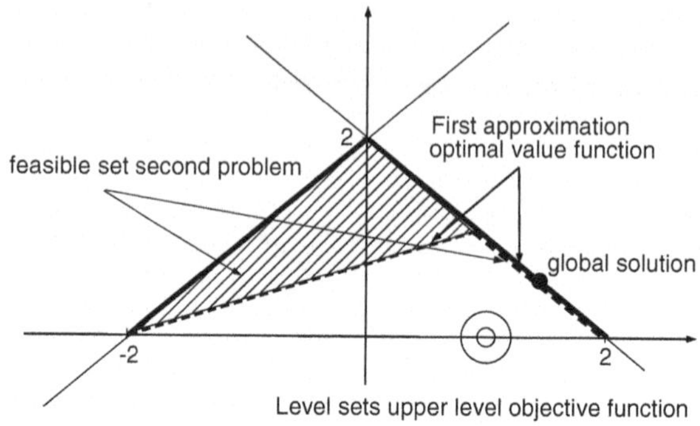

Fig. 3.7 Feasible set and optimal solutions of the second subproblem

For that, consider the optimistic bilevel optimization problem (1.1), (1.4) with $T = \mathbb{R}^m$ in its transformation using the optimal value function of the lower level problem (3.7). Let $(\overline{x}, \overline{y}) \in X \times \mathbb{R}^m$ be a feasible point for problem (3.7) and $\widehat{y}^1 \in \mathbb{R}^m$ be such that $g_i(\overline{x}, \widehat{y}^1) < 0$ for $i = 1, \ldots, p$ and $f(\overline{x}, \widehat{y}^1) \leq \varphi(\overline{x}) + \varepsilon$ for some (fixed) small $\varepsilon > 0$.

Such a point exists e.g. if the Mangasarian-Fromovitz constraint qualification is satisfied at $(\overline{x}, \overline{y})$ for the lower level problem, all functions f, g_i are sufficiently smooth and

$$W := \{(x, y) : g(x, y) \leq 0,\ G(x) \leq 0,\ x \in X\}$$

is not empty and compact. Then, there is a neighborhood V_1 of $\overline{x}$ with

$$\overline{x} \in \operatorname{int} V_1 \cap \{x \in X : G(x) \leq 0\}$$

such that

$$\widehat{y}^1 \in Y(x) := \{y : g(x, y) \leq 0\},\ \forall x \in V_1. \tag{3.67}$$

To find the point $\widehat{y}^1$, the problem

$$\min_{z,u}\{u : g_i(\overline{x}, z) \leq u, i = 1, \ldots, p,\ f(\overline{x}, z) \leq \varphi(\overline{x}) + \varepsilon + u\} \tag{3.68}$$

can be solved. Then, since $\varphi(\cdot)$ is continuous under the above assumptions (see Theorem 3.3) around $\overline{x}$, the set V_1 exists. Mitsos et al. [234] compute an inner approximation of the largest possible set V_1. Denote this set again by V_1. Hence, the optimal function value of the problem

$$\begin{aligned} F(x, y) &\to \min_{x,y} \\ G(x) &\leq 0 \\ x \in V_1 &\Rightarrow f(x, y) \leq f(x, \widehat{y}^1) \\ g(x, y) &\leq 0 \\ x &\in X \end{aligned} \tag{3.69}$$

is not larger than that of problem (3.7). Note that the second constraint in this problem is an implication: if $x \in V_1$, then $f(x, y) \leq f(x, \widehat{y}^1)$. It is not a constraint which needs to be satisfied for all $x \in V_1$. If an optimal solution of problem (3.69) is feasible for problem (3.7), then it is globally optimal for this problem. Otherwise let $(x, \widehat{y}^2)$ be an optimal solution of problem (3.69) and use $\widehat{y}^2$ to construct the next set V_2 as above. In that case the implication

$$x \in V_2 \Rightarrow f(x, y) \leq f(x, \widehat{y}^2)$$

is added to problem (3.69) and the process is repeated.

This implies that a sequence of problems

$$\begin{aligned} F(x, y) &\to \min_{x,y} \\ G(x) &\leq 0 \\ x \in V_k &\Rightarrow f(x, y) \leq f(x, \widehat{y}^k),\ k = 1, \ldots, q \\ g(x, y) &\leq 0 \\ x &\in X \end{aligned} \tag{3.70}$$

needs to be solved globally, where the points $\widehat{y}^k$ are computed as solutions of the problems (3.68) and the sets V^k satisfy the respective assumption in (3.67).

Assume that the algorithm computes infinite sequences of points $\{(\overline{x}^k, \overline{y}^k)\}_{k=1}^{\infty}$ as global optimal solutions of the problem (3.70) and $\{(\overline{x}^k, \widetilde{y}^k)\}_{k=1}^{\infty}$ with $\widetilde{y}^k \in \Psi(\overline{x}^k)$ for all k. Then, since $(\overline{x}^k, \overline{y}^k) \in W$, $(\overline{x}^k, \widetilde{y}^k) \in W$ and the set W is compact, both sequences have accumulation points $(\overline{x}, \overline{y})$ and $(\overline{x}, \widetilde{y})$. Moreover, the point $(\overline{x}, \widetilde{y})$ is feasible for problem (3.7) by continuity of the function $\varphi(\cdot)$.

The feasible set of problem (3.70) needs not to be closed. Due to boundedness of the feasible set of problem (3.70) the infimal objective function value ξ_q is finite and it is possible to compute points $(\overline{x}^q, \overline{y}^q)$ which are feasible for this problem having an objective function value close to ξ_q:

$$F(\overline{x}^q, \overline{y}^q) \leq \xi_q + \varepsilon_{BL}.$$

Moreover, since the feasible set of problem (3.70) becomes smaller and smaller during the iterations, the sequence $\{\xi_q\}_{q=1}^{\infty}$ is not decreasing.

It is also possible to compute the sequence of best objective function values $\{\theta_q\}_{q=1}^{\infty}$ obtained for feasible solutions for problem (3.7):

$$\theta_q = \min\{F(\overline{x}^k, \widetilde{y}^k) : k = 1, \ldots, q\}.$$

This sequence is not increasing. Moreover,

$$\xi_q \leq \theta_q \ \forall\, q.$$

If both sequences tend to the common limit κ then, κ equals the global optimal function value of the problem (3.7) and each accumulation point $(\widetilde{x}, \widetilde{y})$ of the sequence $\{(\overline{x}^q, \widetilde{y}^q)\}_{q=1}^{\infty}$ with $F(\overline{x}^q, \widetilde{y}^q) = \theta_q$ for all q is a global optimum. Feasibility of this point is a consequence of continuity of the optimal value function $\varphi(\cdot)$. This algorithm is investigated in Mitsos et al. [234], where it is shown that the algorithm computes an almost global optimal solution of the bilevel optimization problem within a finite number of iterations.

Theorem 3.33 (Mitsos et al. [234]) *Consider the bilevel optimization problem* (1.4) *with sufficiently smooth functions F, f, G, g and a closed set X. Let the set W be nonempty and compact, assume that the Mangasarian-Fromovitz constraint qualification is satisfied for the lower level problem for all $x \in X$ with $G(x) \leq 0$ and $y \in Y(x)$. Let $\varepsilon_{BL} > 0, \varepsilon > 0$ be sufficiently small. Then, the described algorithm converges to an ε-feasible, ε_{BL}-optimal solution.*

Proof We first show that the sets V_k have nonempty interior and their diameter $\max\{\|\overline{x} - x\| : \overline{x}, x \in V_q\}$ does not tend to zero. Indeed, by continuity of g_i, f, φ and $\varepsilon > 0$, compactness of W and continuity of the optimal value function of problem (3.68) with respect to $\overline{x}$ the optimal function value of this problem is strictly less than zero. Hence, again by boundenness of W there is $\overline{u} < 0$ such that this optimal function value is not larger than $\overline{u}$ for all $\overline{x}$. This implies that the set

$$\{x : g_i(x, \widehat{y}^1) \leq 0,\ f(x, \widehat{y}^1) \leq \varphi(x) + \varepsilon,\ G(x) \leq 0\}$$

contains $\overline{x} + \{z : \|z\| \leq \delta_1\}$ and δ_1 does not depend on $\overline{x}$.

Consider an infinite sequence $\{(\overline{x}^k, \overline{y}^k)\}_{k=1}^{\infty}$ of global optimal solutions of problem (3.70). Using a compactness argument it can then be shown that the infinite subsequence of $\{\overline{x}^q\}_{q=1}^{\infty}$ has the property that all its elements belong to one of the sets $V_{\overline{q}}$. Let, without loss of generality, $\{\overline{x}^q\}_{q=1}^{\infty} \subset V_{\overline{q}}$. This implies that

$$f(\overline{x}^{q+1}, \overline{y}^{q+1}) \leq f(\overline{x}^{q+1}, \widehat{y}^q) \leq \varphi(\overline{x}^q) + \varepsilon$$

by problems (3.68), (3.69). Thus,

$$f(\overline{x}, \overline{y}) \leq \varphi(\overline{x}) + \varepsilon.$$

Let $LB = \lim_{q\to\infty} \xi_q$. Then,

$$LB \geq F(\overline{x}, \overline{y}) - \varepsilon_{BL}.$$

This implies that $(\overline{x}, \overline{y})$ is a ε-feasible and ε_{BL} optimal solution of the bilevel optimization problem. □

In the paper Mitsos et al. [234] the algorithm is investigated under weaker assumptions and in combination with a branch-and-bound framework to accelerate the quality of the obtained solution. Moreover, a method of implementing the implication constraint in problem (3.70) in algorithms is described.

3.7 Primal KKT Transformation

Assume for simplicity throughout this section $T = \mathbb{R}^m$, $X = \mathbb{R}^n$ and replace the optimistic bilevel optimization problem (1.1), (1.4) with (3.1):

$$\min\{F(x, y) : G(x) \leq 0,\ 0 \in \partial_y f(x, y) + N_{Y(x)}(y)\}.$$

Following the remarks in Sect. 3.1 assume that the function $y \mapsto f(x, y)$ is convex and the feasible set mapping Y is convex-valued. For the existence of an optimal solution of problem (3.1) we need at least that its feasible set is closed, which by upper semicontinuity of the mapping $(x, y) \mapsto \partial_y f(x, y)$ (see Rockafellar [272]), continuity of the functions $G_i(x)$ and closedness of the set X is guaranteed if the mapping $(x, y) \mapsto Q(x, y) := N_{Y(x)}(y)$ is closed.

Definition 3.8 A point-to-set mapping $\Gamma : \mathbb{R}^n \rightrightarrows \mathbb{R}^m$ is called closed at a point $x^0 \in \mathbb{R}^n$ if for each sequence $\{x^k, y^k\}_{k=1}^{\infty} \subseteq \textbf{gph}\, \Gamma$ converging to (x^0, y^0), $y^0 \in \Gamma(x^0)$ follows.

Theorem 3.34 (Zemkoho [328]) *If the mapping Y is convex-valued and lower semicontinuous at (x^0, y^0), then Q is closed at this point.*

Proof Let $\{(x^k, y^k, z^k)\}_{k=1}^{\infty} \subseteq \textbf{gph}\, Q$ converge to (x^0, y^0, z^0). Since $Y(x^k)$ is convex we have $z^k \in N_{Y(x^k)}(y^k)$ iff

$$z^{k\top}(u^k - y^k) \leq 0\ \forall\, u^k \in Y(x^k).$$

Since Y is lower semicontinuous at (x^0, y^0), for each $u^0 \in Y(x^0)$ there exists a sequence $\{u^k\}_{k=1}^{\infty}$ converging to u^0 with $u^k \in Y(x^k)$ for all k. This implies $z^{0\top}(u^0 - y^0) \leq 0$. Hence, the result follows. □

Note that Y is lower semicontinuous at x^0 if the Mangasarian-Fromovitz constraint qualification (MFCQ) is satisfied at each (x^0, y) with $y \in Y(x^0)$ [see Bank et al. [8, Theorem 3.1.5)]. Hence, under the assumptions in Theorem 3.34 an optimal solution of problem (3.1) exists provided its feasible set is bounded.

Following Zemkoho [328] it is possible to derive necessary optimality conditions for problem (3.1) using the coderivative for the mapping Q. One constraint qualification which is often used in nonsmooth optimization for problems of the type

$$\min_{x,y}\{F(x, y) : x \in X,\ \rho(x, y) = 0\}$$

is the basic constraint qualification (Mordukhovich [237])

$$\partial^M \rho(\overline{x}, \overline{y}) \cap (-N_{X \times \mathbb{R}^m}(\overline{x}, \overline{y})) = \emptyset. \tag{3.71}$$

To apply it to problem (3.1) set

$$\kappa(x, y) = (x, y, -\nabla_y f(x, y)),\ \rho(x, y) = d_{\mathbf{gph}\, Q}(\kappa(x, y)),$$

where $d_A(z) := \inf\{\|z - u\| : u \in A\}$ is the distance of the point z from the set A and the equation $\rho(x, y) = 0$ is equivalent to $\kappa(x, y) \in \mathbf{gph}\, Q$ or $-\nabla_y f(x, y) \in N_{Y(x)}(y)$. Following Mordukhovich [240]

$$\partial^M(\rho(x, y)) \subseteq \cup\{\partial^M(u^\top \kappa(x, y)) : u \in \mathbb{B} \cap N_{\mathbf{gph}\, Q}(\kappa(x, y))\}, \tag{3.72}$$

provided that **gph** Q is locally closed. The next theorem shows that the basic constraint qualification (3.71) fails for problem (3.1) at every feasible point under a weak assumption.

Theorem 3.35 (Zemkoho [328]) *Let* $(\overline{x}, \overline{y})$ *be feasible for problem* (3.1), *where the lower level problem is a convex optimization problem. Assume that Slater's condition is satisfied at* $\overline{x}$ *and that equality holds in inclusion* (3.72). *Then, the Eq.* (3.71) *is not satisfied.*

Proof Slater's constraint qualification guarantees local closedness of **gph** Q thus validity of the inclusion (3.72). Since equality is assumed in this inclusion we derive that $0 \in \partial^M \rho(\overline{x}, \overline{y})$ since $0 \in N_{\mathbf{gph}\, Q}(\kappa(\overline{x}, \overline{y}))$. Using the same arguments we obtain $0 \in N_{X \times \mathbb{R}^m}(\overline{x}, \overline{y})$. Hence the result. □

As a constraint qualification we need the notion of a calm point-to-set mapping:

Definition 3.9 A point-to-set mapping $\Gamma : \mathbb{R}^n \rightrightarrows \mathbb{R}^m$ is called *calm* at a point $(x^0, y^0) \in \mathbf{gph}\, \Gamma$ if there are neighborhoods U of x^0 and V of y^0 and a finite constant L such that

$$\Gamma(x) \cap V \subseteq \Gamma(x^0) + L\|x - x^0\|\mathbb{B}\ \forall\, x \in U,$$

where $\mathbb{B}$ denotes the unit ball in $\mathbb{R}^m$.

If $V = \mathbb{R}^m$ calmness reduces to the upper Lipschitz property used by Robinson in [269].

Assume throughout this section that the functions $f(x, y), g_i(x, y)$ are continuously differentiable. Reformulate problem (3.1) as

$$\min_{x,y}\{F(x, y) : G(x) \leq 0, -\nabla_y f(x, y) \in Q(x, y)\},$$

or, setting $\xi(x, y) := (G(x), x, y, -\nabla_y f(x, y))$, $\Omega = \mathbb{R}^q_- \times \mathbf{gph}\, Q$, as

$$\min_{x,y}\{F(x, y) : \xi(x, y) \in \Omega\} = \min_{x,y}\{F(x, y) : (x, y) \in \xi^{-1}(\Omega)\}. \tag{3.73}$$

Theorem 3.36 (Zemkoho [328]) *Let* $(\overline{x}, \overline{y})$ *be a local optimal solution of problem* (3.73), *assume that the point-to-set mapping* $\Xi(u) := \{(x, y) : \xi(x, y) + u \in \Omega\}$ *is calm at* $(0, \overline{x}, \overline{y})$ *and the function* F *is locally Lipschitz continuous. Then, there* *exists* $\mu > 0$ *such that for all* $r \geq \mu$ *we can find* $\gamma = (\gamma_1, \gamma_2, \gamma_3, \gamma_4) \in N_\Omega(\xi(\overline{x}, \overline{y}))$ *with* $\|\gamma\| \leq r$ *such that*

$$0 \in \nabla F(\overline{x}, \overline{y}) + \nabla(\gamma^\top \xi)(\overline{x}, \overline{y}) + D^*(\overline{x}, \overline{y}, -\nabla f(\overline{x}, \overline{y}))(\gamma_4).$$

The proof by Zemkoho in [328] is by verifying that $(0, \overline{x}, \overline{y})$ is a local optimal solution of the problem

$$\min_{r,x,y}\{F(x, y) + r\|u\| : (x, y) \in V, (u, x, y) \in \mathbf{gph}\, \Xi\} \tag{3.74}$$

for $r \geq l_V l_F$, where V is a set and l_V a constant given by calmness of the mapping Ξ, l_F is the Lipschitz constant of F. Then, using Mordukhovich [242, Chap. 5),

$$0 \in r\mathbb{B} \times \partial F(\overline{x}, \overline{y}) + N_{\mathbf{gph}\,\Xi}(0, \overline{x}, \overline{y}).$$

Here we used that $(\overline{x}, \overline{y}) \in \operatorname{int} V$. Hence, there exist γ with $\|\gamma\| \leq r$ and a point $(\alpha, \beta) \in \partial F(\overline{x}, \overline{y})$ such that $(-\gamma, -\alpha, -\beta) \in N_{\mathbf{gph}\,\Xi}(0, \overline{x}, \overline{y})$. Using the notion of the Mordukhovich coderivative

$$D^*\Theta(\overline{a}, \overline{b})(v) := \{u : (u, -v) \in N_{\mathbf{gph}\,\Theta}(\overline{a}, \overline{b})\}$$

of a mapping $\Theta : \mathbb{R}^p \rightrightarrows \mathbb{R}^q$ at $(\overline{a}, \overline{b}) \in \mathbf{gph}\,\Theta$ and

$$\begin{aligned} D^*\Xi(0, \overline{x}, \overline{y})(v) \subseteq \{&\gamma \in N_\Omega(\xi(\overline{x}, \overline{y})) : -v \in \partial(\gamma^\top \xi)(\overline{x}, \overline{y})\} \\ &+ D^*(\overline{x}, \overline{y}, -\nabla f(\overline{x}, \overline{y}))(\gamma_4), \end{aligned}$$

see Zemkoho [328, Eq. 3.150), resulting from theorems in Mordukhovich [238], the theorem follows.

Theorem 3.37 (Theorem 3.4.5 in Zemkoho [328]) *Let* $(\overline{x}, \overline{y}) \in \mathbf{gph}\,\Psi$. *If the lower level problem* (1.1) *is a convex optimization problem for which the Mangasarian-Fromovitz constraint qualification* (*MFCQ*) *is satisfied at every point* $(x, y) \in \mathbf{gph}\, Y$, *then,*

$$D^*(\overline{x}, \overline{y}, -\nabla_y f(\overline{x}, \overline{y}))(\gamma) \subseteq \bigcup_{\lambda \in \Lambda(\overline{x}, \overline{y})} \left\{ \begin{pmatrix} \sum_{i=1}^p \lambda_i \nabla^2_{xy} g_i(\overline{x}, \overline{y})\gamma + \sum_{i=1}^p \beta_i \nabla_x g_i(\overline{x}, \overline{y}) \\ \sum_{i=1}^p \lambda_i \nabla^2_{yy} g_i(\overline{x}, \overline{y})\gamma + \sum_{i=1}^p \beta_i \nabla_y g_i(\overline{x}, \overline{y}) \end{pmatrix} \middle| \begin{array}{r} \nabla_y g_{I_{0+}(\overline{x}, \overline{y}, \lambda)}(\overline{x}, \overline{y})\gamma = 0, \beta_{I_{-0}(\overline{x}, \overline{y}, \lambda)} = 0 \\ (\beta_i > 0 \wedge \nabla_y g_i(\overline{x}, \overline{y})\gamma > 0) \vee \beta_i \nabla_y g_i(\overline{x}, \overline{y})\gamma = 0)\ \forall\, i \in I_{00}(\overline{x}, \overline{y}, \lambda) \end{array} \right\}$$

Proof Using Mordukhovich and Outrata [245] we obtain

$$D^*(\overline{x}, \overline{y}, -\nabla_y f(\overline{x}, \overline{y}))\gamma$$
$$\subseteq \bigcup_{\lambda \in \Lambda(\overline{x}, \overline{y})} \left\{ \sum_{i=1}^p \lambda_i \nabla\nabla_y g_i(\overline{x}, \overline{y})\gamma + \nabla_y g(\overline{x}, \overline{y}) D^* N_{\mathbb{R}^p_-}(g(\overline{x}, \overline{y}), \lambda)(\nabla_y g(\overline{x}, \overline{y})\gamma) \right\}.$$

Now the result follows analogously to the application of Flegel and Kanzow's result in [113] in Theorem 3.18. □

Combining Theorems 3.36 and 3.37 we obtain that a local optimal solution of the problem (3.1) is under the assumptions of Theorem 3.36 an M-stationary solution, see page 64.

Henrion and Surowiec consider in their paper [143] the special bilevel optimization problem where the upper level constraint functions g_i do not depend on the lower level variable: $g_i(x, y) = g_i(y)$ for all i. Assume that the lower level problem (1.1) is a convex parametric optimization problem and that the Slater's condition is satisfied for it. Let $G_i(x) \equiv 0$ for all i and $X = \mathbb{R}^n$. Then, the bilevel optimization problem (1.1), (1.4) can equivalently be replaced with

$$\min_{x,y} \{F(x, y) : 0 \in \nabla_y f(x, y) + N_Y(y)\}. \tag{3.75}$$

Note that this problem is fully equivalent to the bilevel optimization problem (1.1), (1.4).

Applying the ideas from Sect. 3.6, the following necessary optimality condition is derived:

Theorem 3.38 (Dempe et al. [56]) *Consider the bilevel optimization problem* (1.1), (1.4), *where the functions* $y \mapsto f(x, y)$ *and* $y \mapsto g_i(y)$, $i = 1 \ldots, p$, *are convex and there exists a point* $\widehat{y}$ *satisfying* $g_i(\widehat{y}) < 0$, $i = 1, \ldots, p$. *Let* $(\overline{x}, \overline{y})$ *be a local optimal solution, and assume that the problem* (3.7) *is partially calm at* $(\overline{x}, \overline{y})$. *Let the solution set mapping* $\Psi(\cdot)$ *of the lower level problem be inner semicontinuous or inner semicompact at* $(\overline{x}, \overline{y})$. *Then, there are real numbers* $\alpha \geq 0$, $\lambda_i \geq 0$, $i = 1, \ldots, p$, $\mu_i \geq 0$, $i = 1, \ldots, p$, *satisfying the following equations:*

$$\nabla_x F(\overline{x}, \overline{y}) = 0,$$
$$\nabla_y F(\overline{x}, \overline{y}) + \alpha \nabla_y f(\overline{x}, \overline{y}) + \sum_{i=1}^p \mu_i \nabla g_i(\overline{y}) = 0,$$
$$\nabla_y f(\overline{x}, \overline{y}) + \sum_{i=1}^p \lambda_i \nabla g_i(\overline{y}) = 0,$$
$$\mu^\top g(\overline{y}) = \lambda^\top g(\overline{y}) = 0.$$

Partial calmness is satisfied if the value function constraint qualification is satisfied:

Definition 3.10 The *value function constraint qualification* is satisfied for problem

$$\min_{x,y}\{F(x,y) : y \in Y,\ f(x,y) - \varphi(x) \leq 0\} \tag{3.76}$$

at a point $(\overline{x}, \overline{y})$ if the mapping

$$u \mapsto K(u) := \{(x,y) : y \in Y,\ f(x,y) - \varphi(x) \leq u\}$$

is calm at $(0, \overline{x}, \overline{y})$.

Theorem 3.39 (Henrion and Surowiec [143]) *If the value function constraint qualification is satisfied for problem* (3.76) *at a local optimal solution* $(\overline{x}, \overline{y})$ *then, this problem is partially calm.*

Replacing the partial calmness assumption with a calmness condition for the perturbed mapping

$$\nu \mapsto M(\nu) := \{(x,y) : \nu \in \nabla_y f(x,y) + N_Y(y)\}$$

a necessary optimality condition can be derived which uses second order derivatives of the objective function and the constraint functions of the lower level problem.

Theorem 3.40 (Suroviec [296]; Henrion and Surowiec [143]) *Let* $(\overline{x}, \overline{y})$ *be a local optimal solution of problem* (3.75), *where the lower level problem is assumed to be a convex parametric optimization problem. Assume that there exists* $\widehat{y}$ *with* $g(\widehat{y}) < 0$. *Let the constant rank constraint qualification be satisfied for the lower level at* $\overline{y}$. *Assume further that the mapping* $\nu \mapsto M(\nu) := \{(x,y) : \nu \in \nabla_y f(x,y) + N_Y(y)\}$ *is calm at* $(0, \overline{x}, \overline{y})$. *Then, there are multiplier vectors* $\lambda \geq 0, v, w$ *(of appropriate dimension) such that the following equations are satisfied:*

$$0 = \nabla_x F(\overline{x}, \overline{y}) + v^\top \nabla^2_{xy} f(\overline{x}, \overline{y}),$$

$$0 = \nabla_y F(\overline{x}, \overline{y}) + v^\top \nabla^2_{yy} f(\overline{x}, \overline{y}) + \lambda^\top \nabla^2_{yy} g(\overline{y}) + w^\top \nabla g(\overline{y}),$$

$$0 = \nabla g_i(\overline{y}) v\ \forall\, i : g_i(\overline{y}) = 0,\ \lambda_i > 0$$

$$0 = \lambda_i\ \forall\, i : g_i(\overline{y}) < 0$$

$$0 = w_i\ \forall\, i : g_i(\overline{y}) = 0,\ \lambda_i = 0,\ \nabla g_i(\overline{y}) v < 0$$

$$0 \leq w_i\ \forall\, i : g_i(\overline{y}) = 0,\ \lambda_i = 0,\ \nabla g_i(\overline{y}) v > 0$$

$$0 = \nabla_y f(\overline{x}, \overline{y}) + \lambda^\top \nabla g(\overline{y}).$$

An example is formulated by Henrion and Surowiec in [143] where partial calmness is violated but the optimality conditions of Theorem 3.40 can be used to characterize a local optimal solution.

It is also shown in [143] that the mapping $\nu \mapsto M(\nu) := \{(x, y) : \nu \in \nabla_y f(x, y) + N_Y(y)\}$ is calm at $(0, \overline{x}, \overline{y})$ in any of the following three cases:

1. The lower level constraint set Y is a polyhedron and the lower level objective function is
$$f(x, y) = y^\top(Ay + Bx)$$
for matrices A, B of appropriate size.
2. The linear independence constraint qualification LICQ and the strong sufficient optimality condition of second order (SSOSC) are satisfied at the local optimal solution.
3. The matrix $\nabla^2_{xy} f(\overline{x}, \overline{y})$ is surjective (which is the so-called 'ample parametrization').

3.8 The Optimistic Bilevel Programming Problem

3.8.1 One Direct Approach

Consider the optimistic bilevel optimization problem

$$\min\{\varphi_o(x) : G(x) \leq 0\} \tag{3.77}$$

with

$$\varphi_o(x) := \min\{F(x, y) : y \in \Psi(x)\}, \tag{3.78}$$

where $\Psi(x)$ is the set of optimal solutions of the lower level problem (1.1). As mentioned in Sect. 1.1 this problem is closely related to problem (1.4), but not equivalent.

Theorem 3.41 (Dempe et al. [71]) *If $\overline{x}$ is a local optimal solution of problem* (3.77) *and $\overline{y} \in \Psi(\overline{x})$, then $(\overline{x}, \overline{y})$ is a local optimal solution of problem* (1.4). *If $(\overline{x}, \overline{y})$ is a local optimal solution of* (1.4) *and the mapping $x \mapsto \Psi_o(x)$ with*

$$\Psi_o(x) = \{y \in \Psi(x) : F(x, y) \leq \varphi_o(x)\}$$

is inner semicontinuous at $(\overline{x}, \overline{y})$, then $\overline{x}$ is a local optimal solution of problem (3.77).

Proof Assume that the first assertion is not true, i.e. assume that there exists a sequence $\{x^k, y^k\}_{k=1}^\infty$ tending to $(\overline{x}, \overline{y})$ with $y^k \in \Psi(x^k),\ G(x^k) \leq 0,\ F(x^k, y^k) < \varphi_o(x^k)$ for all k. By the definition of the optimistic optimal value function $\varphi_o(x)$ we have $\varphi_o(x^k) \leq F(x^k, y^k) < \varphi_o(x^k)$ which is not possible.

Now assume that the second assertion is not correct. Hence, a sequence $\{x^k\}_{k=1}^\infty$ converging to $\overline{x}$ exists with $G(x^k) \leq 0$ and $\varphi_o(x^k) < \varphi_o(\overline{x})$ for all k. Since $\Psi_o(\cdot)$

is assumed to be inner semicontinuous at $(\overline{x}, \overline{y})$ there is a sequence $\{y^k\}_{k=1}^{\infty}$ with $y^k \in \Psi_(x^k)$ and $F(x^k, y^k) \leq \varphi_o(x^k) < \varphi_o(\overline{x}) \leq F(\overline{x}, \overline{y})$ for all k. The last inequality comes from the definition of the function $\varphi_o(x)$. This contradicts the assumption that $(\overline{x}, \overline{y})$ is a local optimal solution of problem (1.4). □

Note that upper semicontinuity of the point-to-set mapping $\Psi(x)$ and continuity of the function $F(x, y)$ imply that the function $\varphi_o(x)$ is lower semicontinuous, see the proof of Theorem 3.3. Hence, by the famous theorem of Weierstrass, problem (3.77) has an optimal solution if the assumptions of Theorem 3.3 are satisfied at every feasible point, the functions $F(x, y)$, $G_j(x)$ are continuous for all j and the set $\{x : G(x) \leq 0\}$ is not empty and compact.

Theorem 3.42 (Dempe et al. [71]) *Consider the function φ_o given in* (3.77) *and assume that the function F is sufficiently smooth and* **gph** Ψ *is a closed set. Then, the following assertions hold:*

1. *If the mapping $x \mapsto \Psi_o(x)$ is inner semicontinuous at $(\overline{x}, \overline{y}) \in$* **gph** *$\Psi_o$, then φ_o is lower semicontinuous at $\overline{x}$ and we have*

$$\partial^M(\overline{x}) \subseteq \nabla_x F(\overline{x}, \overline{y}) + D^*\Psi(\overline{x})(\nabla_y F(\overline{x}, \overline{y})).$$

If $x \mapsto \Psi(x)$ is Lipschitz-like around $(\overline{x}, \overline{y})$ then, the function $x \mapsto \varphi_o(x)$ is locally Lipschitz continuous.

2. *If the mapping $x \mapsto \Psi_o(x)$ is inner semicompact at $(\overline{x}, \overline{y}) \in$* **gph** *$\Psi_o$, then φ_o is lower semicontinuous at $\overline{x}$ and we have*

$$\partial^M(\overline{x}) \subseteq \bigcup_{y \in \Psi_o(\overline{x})} \left\{\nabla_x F(\overline{x}, \overline{y}) + D^*\Psi(\overline{x})(\nabla_y F(\overline{x}, \overline{y}))\right\}.$$

If $x \mapsto \Psi(x)$ is Lipschitz-like around $(\overline{x}, \overline{y})$ then, the function $x \mapsto \varphi_o(x)$ is locally Lipschitz continuous.

The proof of this theorem follows from Mordukhovich [241, Corollary 1.109] and [242, Theorem 5.2]. This theorem enables the formulation of necessary optimality conditions.

For that let $X = \{x : G(x) \leq 0\}$. Then, a necessary optimality condition for $\overline{x} \in X$ being a local optimal solution of problem (3.77) is

$$0 \in \partial^M \varphi_o(\overline{x}) + N_X(\overline{x})$$

provided that the assumptions of the Theorem 3.42 are satisfied. One difficulty is the use of the coderivative $D^*\Psi$ of the mapping Ψ. Using reformulations of the function $\varphi_o(x)$ this coderivative is also replaced.

Let

$$\mathscr{L}(x, y, \lambda) = \nabla_y f(x, y) + \lambda^\top \nabla_y g(x, y).$$

Theorem 3.43 (Dempe et al. [71]) *Assume that the functions $y \mapsto f(x, y)$, $y \mapsto g_i(x, y)$ are convex for all i and that (MFCQ) is satisfied for problem* (1.1) *for all $(x, y) \in \mathbf{gph}\, \Psi$ with $G(x) \leq 0$. Then,*

$$\varphi_o(x) = \min_{y,\lambda}\{F(x, y) : \mathscr{L}(x, y, \lambda) = 0,\ \lambda \geq 0,\ g(x, y) \leq 0,\ \lambda^\top g(x, y) = 0\}.$$

Proof On the one hand we have

$$\varphi_o(x) \geq \min_{y,\lambda}\{F(x, y) : \mathscr{L}(x, y, \lambda) = 0,\ \lambda \geq 0,\ g(x, y) \leq 0,\ \lambda^\top g(x, y) = 0\}$$

by the necessary optimality conditions for the lower level problem using (MFCQ).

On the other hand, for a global optimal solution $\overline{x}$ of the problem (3.78), we have

$$\begin{aligned}
\varphi(\overline{x}) &= F(\overline{x}, \overline{y}) \\
&\leq F(\overline{x}, y)\ \forall\, y \in \Psi(\overline{x}) \\
&= F(\overline{x}, y)\ \forall\, y \text{ with } 0 \in \nabla_y f(\overline{x}, y) + N_{Y(\overline{x})}(y) \\
&\quad \text{by convexity of } y \mapsto f(x, y)\ y \mapsto g_i(x, y) \text{ and regularity} \\
&\leq F(\overline{x}, y)\ \forall\, (y, \lambda) \text{ with } \mathscr{L}(\overline{x}, y, \lambda) = 0,\ \lambda \geq 0,\ g(x, y) \leq 0,\ \lambda^\top g(x, y) = 0
\end{aligned}$$

by the normal cone representation for solutions of systems of convex inequalities under regularity. Combining both inequalities we obtain the result. □

Hence, problem (1.1), (1.2) can be replaced by problem (3.77) of minimizing the function $\varphi_o(x)$ in Eq. (3.78) on the set $\{x : G(x) \leq 0\}$. To investigate the properties of the function $\varphi_o(x)$ we can use ideas related to mathematical programs with equilibrium constraints.

To proceed use some abbreviations and assumptions:

$$\Lambda(\overline{x}, \overline{y}) = \{\lambda : \mathscr{L}(\overline{x}, \overline{y}, \lambda) = 0,\ \lambda \geq 0,\ g(\overline{x}, \overline{y}) \leq 0,\ \lambda^\top g(\overline{x}, \overline{y}) = 0\},$$

$$\Lambda^{em}(\overline{x}, \overline{y}, \lambda, v) = \{(\beta, \gamma) :\ \text{conditions (3.79)–(3.81) are satisfied}\},$$

where

$$v + \nabla g(\overline{x}, \overline{y})^\top \beta + \nabla_{xy}\mathscr{L}(\overline{x}, \overline{y}, \lambda)^\top \gamma = 0, \tag{3.79}$$

$$\nabla_y g_{I_{0+}(\overline{x},\overline{y},\lambda)}(\overline{x}, \overline{y})\gamma = 0,\ \beta_{I_{-0}(\overline{x},\overline{y},\lambda)} = 0 \tag{3.80}$$

$$\forall\, i \in I_{00}(\overline{x}, \overline{y}, \lambda) : (\beta_i > 0 \wedge \nabla_y g_i(\overline{x}, \overline{y})\gamma > 0) \vee \beta_i \nabla_y g_i(\overline{x}, \overline{y})\gamma = 0 \tag{3.81}$$

and

$$\Lambda_y^{em}(\overline{x}, \overline{y}, \lambda, v) = \{(\beta, \gamma) :\ \text{conditions (3.80), (3.81), (3.82) are satisfied}\},$$

where

$$v + \nabla_y g(\overline{x}, \overline{y})^\top \beta + \nabla_y \mathscr{L}(\overline{x}, \overline{y}, \lambda)^\top \gamma = 0. \tag{3.82}$$

The following assumptions will be used:

(A_1^m) $(\beta, \gamma) \in \Lambda^{em}(\overline{x}, \overline{y}, \lambda, 0) \Longrightarrow \beta = 0,\ \gamma = 0.$
(A_2^m) $(\beta, \gamma) \in \Lambda_y^{em}(\overline{x}, \overline{y}, \lambda, 0) \Longrightarrow \nabla_x g(\overline{x}, \overline{y})^\top \beta + \nabla_x \mathcal{L}(\overline{x}, \overline{y}, \lambda)^\top \gamma = 0.$
(A_3^m) $(\beta, \gamma) \in \Lambda_y^{em}(\overline{x}, \overline{y}, \lambda, 0) \Longrightarrow \beta = 0,\ \gamma = 0.$

It is obvious that $(A_3^m) \Longrightarrow (A_1^m)$ and $(A_3^m) \Longrightarrow (A_2^m)$.

Theorem 3.44 (Dempe et al. [71]) *Let the lower level problem* (1.1) *be convex and let (MFCQ) be satisfied for the lower level problem at all points* $(x, y) \in \mathbf{gph}\,\Psi$. *Then, the following two assertions hold:*

1. *If the mapping*

$$x \mapsto S_0^h(x) := \{(y, \lambda) : \lambda \in \Lambda(x, y),\ g(x, y) \le 0,\ F(x, y) \le \varphi_o(x)\}$$

is inner semicontinuous at $(\overline{x}, \overline{y}, \lambda)$ *and if the assumption* (A_1^m) *holds at this point then*

$$\partial^M \varphi_o(\overline{x}) \subseteq \bigcup_{(\beta,\gamma) \in \Lambda_y^{em}(\overline{x}, \overline{y}, \lambda)} \left\{ \nabla_x F(\overline{x}, \overline{y}) + \beta^\top \nabla_x g(\overline{x}, \overline{y}) + \gamma^\top \nabla_x \mathcal{L}(\overline{x}, \overline{y}, \lambda) \right\}$$

with $\Lambda_y^{em}(\overline{x}, \overline{y}, \lambda) = \Lambda_y^{em}(\overline{x}, \overline{y}, \lambda, \nabla_y F(\overline{x}, \overline{y}))$.
Furthermore, if the assumption (A_2^{cm}) *is also satisfied at* $(\overline{x}, \overline{y}, \lambda)$, *the function* φ_o *is Lipschitz continuous around* $\overline{x}$.
2. *If the mapping* $x \mapsto S_0^h(x)$ *is inner semicompact at* $(\overline{x}, \overline{y}, \lambda)$ *and if the assumption* (A_1^m) *holds at all points* $(\overline{x}, \overline{y}, \lambda)$, $(\overline{y}, \lambda) \in S_0^h(\overline{x})$ *then*

$$\partial^M \varphi_o(\overline{x}) \subseteq \bigcup_{(\overline{y},\lambda) \in S_0^h(\overline{x})} \ \bigcup_{(\beta,\gamma) \in \Lambda_y^{em}(\overline{x}, \overline{y}, \lambda)} \left\{ \nabla_x F(\overline{x}, \overline{y}) + \beta^\top \nabla_x g(\overline{x}, \overline{y}) + \gamma^\top \nabla_x \mathcal{L}(\overline{x}, \overline{y}, \lambda) \right\}.$$

Again, if the assumption (A_2^{cm}) *is also satisfied at all points* $(\overline{x}, \overline{y}, \lambda)$ *with* $(\overline{y}, \lambda) \in S_0^h(\overline{x})$, *the function* φ_o *is Lipschitz continuous around* $\overline{x}$.

Proof The main ideas of the proof of the first assertion in Dempe et al. [71] follow: The task of computing a function value of $\varphi_o(x)$ is reduced to solving an MPEC:

$$\mu^c(x) = \min_y \{F(x, y) : g(x, y) \le 0,\ h(x, y) = 0,\ G(x, y) \ge 0,\ H(x, y) \ge 0,$$
$$G(x, y)^\top H(x, y) = 0\}$$

with the perturbed Lagrange multiplier set

$$\begin{aligned}
&\Lambda^{cm}(x, y, v) = \{(\alpha, \beta, \gamma, \zeta) : \alpha \ge 0,\ \alpha^\top g(x, y) = 0,\\
&\beta_i = 0,\ \forall\, i \in I_H(x, y) := \{j : H_j(x, y) = 0,\ G_j(x, y) > 0\},\\
&\zeta_i = 0,\ \forall\, i \in I_G(x, y) := \{j : H_j(x, y) > 0,\ G_j(x, y) = 0\},\\
&(\beta_i > 0 \wedge \zeta_i > 0) \vee (\beta_i \zeta_i = 0),\ \forall\, i \in I_{GH}(x, y) := \{j : H_j(x, y) = 0,\ G_j(x, y) = 0\},\\
&v + \nabla g(x, y)^\top \alpha + \nabla h(x, y)^\top \gamma + \nabla G(x, y)^\top \beta + \nabla H(x, y)^\top \zeta = 0\}.
\end{aligned}$$

Here, $y = (y, \lambda)$, $g(x, y, \lambda) \equiv 0$, $h(x, y, \lambda) = \mathscr{L}(x, y, \lambda)$, $G(x, y, \lambda) = \lambda$ and $H(x, y, \lambda) = -g(x, y)$ is used to obtain

$$\begin{aligned} \Lambda^{cm}(x, y, v) = \{(\beta, \gamma, \zeta) : \beta_i = 0, \ \forall\, i \in I_{+0}(x, y), \\ \zeta_i = 0, \ \forall\, i \in I_{0-}(x, y), \\ (\beta_i > 0 \wedge \zeta_i > 0) \vee (\beta_i \zeta_i = 0), \ \forall\, i \in I_{00}(x, y), \\ v_1 + \nabla_x \mathscr{L}(x, y, \lambda)^\top \gamma - \nabla_x g(x, y)^\top \beta = 0, \\ v_2 + \nabla_y \mathscr{L}(x, y, \lambda)^\top \gamma - \nabla_y g(x, y)^\top \beta = 0, \\ v_3 + \nabla_y g(x, y) \gamma + \zeta = 0\}. \end{aligned}$$

Then, setting $v_3 = 0$ in $v_3 + \nabla_y g(x, y)\gamma + \zeta = 0$ and $\zeta = -\nabla_y g(x, y)\gamma$, the equation

$$\Lambda^{cm}(x, y, \lambda, 0) = \{(\beta, \gamma, -\nabla_y g(x, y)\gamma) : (-\beta, \gamma) \in \Lambda^{em}(x, y, \lambda, 0)\}$$

is derived.

Now, assumption (A_1^m) implies the condition

$$(\beta, \gamma, \zeta) \in \Lambda^{cm}(x, y, \lambda, 0) \Rightarrow \beta = 0, \ \gamma = 0, \ \zeta = 0.$$

This regularity condition can be used to verify

$$\partial^M \varphi_o(x) \subseteq \bigcup_{(\beta,\gamma) \in \Lambda_y^{cm}(\overline{x}, \overline{y}, \lambda)} \left\{ \nabla_x F(\overline{x}, \overline{y}) - \nabla_x g(\overline{x}, \overline{y})^\top \beta + \nabla_x \mathscr{L}(\overline{x}, \overline{y}, \lambda)^\top \gamma \right\}$$

using Theorem 3.2 (i) in Dempe et al. [71]. This implies the first assertion.

Now, assumption (A_2^m) can be used to obtain

$$(\beta, \gamma, \zeta) \in \Lambda^{cm}(x, y, \lambda, 0) \Rightarrow -\nabla_x g(x, y)^\top \beta + \nabla_x \mathscr{L}(x, y, \lambda)\gamma = 0.$$

Using again Theorem 3.2 (i) in Dempe et al. [71] this implies local Lipschitz continuity of the function $\varphi_o(x)$.

The second assertion can similarly be shown. □

A related result for the pessimistic bilevel optimization problem (1.6) has been derived by Dempe et al. in [72].

3.8.2 An Approach Using Set-Valued Optimization

Dempe and Pilecka considered in [74] the formulation of necessary optimality conditions for the special case of the optimistic bilevel optimization problem

$$\begin{aligned} F(x, y) &\to \min_{x,y} \\ y &\in \Psi(x), \end{aligned} \tag{3.83}$$

where the mapping $\Psi(x) = \text{Argmin}_y\{f(x, y) : g(x, y) \leq 0\}$, cf. problem (1.1) with $T = \mathbb{R}^m$, is assumed to be upper semicontinuous, see Theorem 3.3. If the mapping $x \mapsto \Psi(x)$ is not locally Lipschitz continuous, (Lipschitz) continuity of the function $\varphi_o(\cdot)$ is in general violated (cf. Theorem 3.42). A (directional) convexificator can then be used to derive necessary optimality conditions.

Let $z : \mathbb{R}^n \to \overline{\mathbb{R}} := \mathbb{R} \cup \{\infty\}$ be a function, $x,\ d \in \mathbb{R}^n$. Then, the *lower* and the *upper Dini directional derivative* of the function z at x in direction d are defined as

$$D^+z(x, d) = \limsup_{t \downarrow 0} \frac{1}{t}[z(x + td) - z(x)],$$

$$D^-z(x, d) = \liminf_{t \downarrow 0} \frac{1}{t}[z(x + td) - z(x)].$$

Using these generalized directional derivatives the upper respectively lower convexificator can be defined.

Definition 3.11 (*Jeyakumar and Luc* [155]) The function z admits an *upper* (*lower*) *convexificator* $\partial^* z(x)$ at the point $x \in \mathbb{R}^n$ if the set $\partial^* z(x) \subset \mathbb{R}^n$ is closed and for each $d \in \mathbb{R}^n$ we have

$$D^-z(x, d) \leq \sup_{x^* \in \partial^* z(x)} d^\top x^*, \quad (D^+z(x, d) \geq \inf_{x^* \in \partial^* z(x)} d^\top x^*).$$

A function having both an upper and a lower convexificator is said the have a convexificator. The subdifferential of Clarke [42] is a convexificator and also the Michel-Penot and the symmetric subdifferentials, see Babbahadda and Gadhi [6]. In general, a convexificator needs not to be convex or bounded but most of the assertions have been verified under those assumptions.

Definition 3.12 (*Dutta and Chandra* [96]) The function z admits an *upper* (*lower*) *semiregular convexificator* $\partial^* z(x)$ at the point $x \in \mathbb{R}^n$ if the set $\partial^* z(x) \subset \mathbb{R}^n$ is a convexificator and for each $d \in \mathbb{R}^n$ we have

$$D^+z(x, d) \leq \sup_{x^* \in \partial^* z(x)} d^\top x^*, \quad (D^-z(x, d) \geq \inf_{x^* \in \partial^* z(x)} d^\top x^*).$$

The new notion in Dempe and Pilecka [74] is that of a directional convexificator which in general is an unbounded set.

Definition 3.13 (*Dempe and Pilecka* [74]) A vector $d \in \mathbb{R}^n$ is a *continuity direction* of the function $z : \mathbb{R}^n \to \overline{\mathbb{R}}$ at x if

$$\lim_{t \downarrow 0} z(x + td) = z(x).$$

Let $D_z(x)$ denote the set of all continuity directions of the function z at x. This set is in general neither convex or closed. But $0 \in D_z(x)$.

Definition 3.14 (*Dempe and Pilecka* [74]) The function $z : \mathbb{R}^n \to \overline{\mathbb{R}}$ admits a *directional convexificator* $\partial^*_D z(x)$ at x if the set $\partial^*_D z(x)$ is closed and for each $d \in D_z(x)$ we have

$$D^- z(x,d) \leq \sup_{x^* \in \partial^*_D z(x)} d^\top x^* \text{ and } D^+ z(x,d) \geq \inf_{x^* \in \partial^*_D z(x)} d^\top x^*.$$

Directional convexificators can be used to obtain convexificators:

Theorem 3.45 (Dempe and Pilecka [74]) *Let $z : \mathbb{R}^n \to \overline{\mathbb{R}}$ be a lower semicontinuous function which admits a bounded directional convexificator $\partial^*_D z(\overline{x})$ at $\overline{x}$ and assume $D_z(\overline{x})$ to be a closed and convex set. Then, the set*

$$K(\overline{x}) = \partial^*_D z(\overline{x}) + N_D(0)$$

is a convexificator, where $N_D(0) := \{v \in \mathbb{R}^n : v^\top d \leq 0 \,\forall\, d \in D_z(\overline{x})\}$ is the normal cone to $D_z(\overline{x})$ at $d = 0$.

Smaller (directional) convexificators make the formulation of more helpful necessary optimality conditions possible. Unfortunately, the aim of deriving a minimal (with respect to inclusion) convexificator is not an easy task. With respect to a directional convexificator the following can be shown, see Dempe and Pilecka [74]:

If $\widetilde{D} \subset D_z(\overline{x})$ is a closed and convex cone for which the directional convexificator $\partial^*_{\widetilde{D}} z(\overline{x})$ of the lower semicontinuous function $z : \mathbb{R}^n \to \overline{\mathbb{R}}$ at $\overline{x}$ exists and is bounded, the set $K(\overline{x}) = \partial^*_{\widetilde{D}} z(\overline{x}) + N_{\widetilde{D}}(0)$ is an upper convexificator.

If the cone $D_z(\overline{x})$ of continuity directions is itself convex and the lower semicontinuous function $z : \mathbb{R}^n \to \overline{\mathbb{R}}$ is convex with respect to all directions $d \in D_z(\overline{x})$, i.e.,

$$z(\lambda(\overline{x} + d_1) + (1-\lambda)(\overline{x} + d_2)) \leq \lambda z(\overline{x} + d_1) + (1-\lambda) z(\overline{x} + d_2)$$

for all $d_1, d_2 \in D_z(\overline{x})$ and $\lambda \in [0, 1]$ then, for all closed and convex cones $\widetilde{D} \subset D_z(\overline{x})$ for which the function z admits a bounded directional convexificator, the set

$$K(\overline{x}) = \partial^*_{\widetilde{D}} z(\overline{x}) + N_{\widetilde{D}}(0)$$

is a convexificator.

Convexity of the cone of continuity directions can be avoided as the following remark shows.

Remark 3.3 (*Dempe and Pilecka* [74]) Let $D_1 \subset D_z(x)$ and $D_2 \subset D_z(x)$ be closed convex cones with $D_1 \cup D_2 = D_z(x)$ and $z : \mathbb{R}^n \to \overline{\mathbb{R}}$. Suppose that the function z admits bounded directional convexificators $\partial^*_{D_1} z(x)$, $\partial^*_{D_2} z(x)$ at the point $x \in \mathbb{R}^n$. Then the set $\partial^*_D z(x) = \partial^*_{D_1} z(x) \cup \partial^*_{D_2} z(x)$ is a directional convexificator of the function z at x. Moreover, the set $K(x) = (\partial^*_{D_1} z(x) + N_{D_1}(0)) \cup (\partial^*_{D_2} z(x) + N_{D_2}(0))$ is a convexificator of z at the point x.

To derive a chain rule for a directional convexificator the following assumption is needed:

(A_0) Let the function $z : \mathbb{R}^n \to \mathbb{R}$ be lower semicontinuous and the set $D_z(x)$ denote all continuity directions of z at x. Then

$$\exists \delta > 0\ \forall d \in D_z(x)\ \forall \hat{x} \in (x, x+\delta d) : \lim_{t \downarrow 0} z(\hat{x}+td) = \lim_{t \downarrow 0} z(\hat{x}-td) = z(\hat{x}).$$

Theorem 3.46 *Let $z : \mathbb{R}^p \to \mathbb{R}^n$ be a vector valued function with the component functions $z_i : \mathbb{R}^p \to \mathbb{R}$, $i = 1, \ldots, n$. Assume that z_i, $i = 1, \ldots, n$ are lower semicontinuous and suppose that assumption (A_0) is satisfied for each z_i at $\overline{x}$. Let $g : \mathbb{R}^n \to \mathbb{R}$ be a continuous function. Assume that for each $i = 1, \ldots, n$ the function z_i admits a bounded directional convexificator $\partial^*_{D_i} z_i(\overline{x})$ at $\overline{x}$ with the continuity directions $D_i(\overline{x}) = D_{z_i}(\overline{x})$ while g admits a bounded convexificator $\partial^* g(z(\overline{x}))$ at $z(\overline{x})$. Additionally, suppose that $\partial^* g(\cdot)$ is upper semicontinuous at $z(\overline{x})$ and for each $i = 1, \ldots, n$ the directional convexificator $\partial^*_{D_i} z_i(\cdot)$ is upper semicontinuous at $\overline{x}$ on $U(\overline{x}) \cap (\{\overline{x}\} + \overline{D})$ where $U(\overline{x})$ denotes an open neighborhood of $\overline{x}$ and $\overline{D} = \bigcap_{i=1}^n D_i(\overline{x}) \neq \{\mathbf{0_p}\}$. Then the set*

$$\begin{aligned}\partial^*_{\overline{D}}(g \circ z)(\overline{x}) &= \partial^* g(z(\overline{x}))(\partial^*_{D_1} z_1(\overline{x}), \ldots, \partial^*_{D_n} z_n(\overline{x})) \\ &= \left\{ \sum_{i=1}^n a_i h_i : a \in \partial^* g(z(\overline{x})), h_i \in \partial^*_{D_i} z_i(\overline{x}) \right\}\end{aligned} \tag{3.84}$$

is a directional convexificator of $g \circ z$ at $\overline{x}$.

Proof Let $u \in \overline{D}$. Then, due to assumption A_0 there exists $\delta > 0$ such that z is continuous on $[\overline{x}, \overline{x} + \delta u]$. Thus, using the mean value theorem for convexificators (Jeyakumar and Luc [155]), we obtain

$$g(z(\bar{x} + tu)) - g(z(\bar{x})) \in \text{cl conv} \left\{ \langle a, z(\bar{x} + tu) - z(\bar{x}) \rangle : a \in \partial^* g(c^t) \right\}$$

for some $c^t \in (z(\bar{x}), z(\bar{x} + tu))$. In the same way we derive

$$z_i(\bar{x} + tu) - z_i(\bar{x}) \in \text{cl conv} \left\{ \langle h_i, tu \rangle : h_i \in \partial^*_{D_i} z_i(x_i^t) \right\}$$

for each $i \in \{1, \ldots, n\}$ where $x_i^t \in (\bar{x}, \bar{x} + tu)$, $t \in (0, \delta)$. Since the (directional) convexificators are assumed to be upper semicontinuous, for each $\varepsilon > 0$ there exists $t_0 > 0$ with

$$\begin{aligned}\partial^* g(c^t) &\subset \partial^* g(z(\bar{x})) + \varepsilon \mathbb{U}^n \\ \partial^*_{D_i} z_i(x_i^t) &\subset \partial^*_{D_i} z_i(\bar{x}) + \varepsilon \mathbb{U}^p\end{aligned}$$

for $t \in [0, t_0]$ and all $i = 1, \cdots, n$, where $\mathbb{U}^k$ denotes the open unit ball in $\mathbb{R}^k$. This implies that for each $t \leq t_0$ we have

$$\Big([g(z(\bar{x}+tu)) - g(z(\bar{x}))]/t\Big) \in \text{cl conv } \{\langle v, u\rangle : v \in S\}$$

with

$$S = \left\{ \sum_{i=1}^{n} (a_i + \varepsilon b_i)(h_i + \varepsilon d_i) : a \in \partial^* g(z(\bar{x})), b \in \mathbb{U}^n, h_i \in \partial^*_{D_i} z_i(\bar{x}), d_i \in \mathbb{U}^p \right\}.$$

Boundedness of the (directional) convexificators leads to

$$\begin{aligned}(g \circ z)^{-}(\bar{x}, u) \leq \sup\big\{\langle x^*, u\rangle : x^* \in \partial^* g(z(\bar{x}))(\partial^*_{D_1} z_1(\bar{x}), \ldots, \partial^*_{D_n} z_n(\bar{x}))\big\} \\ + (\varepsilon + \varepsilon^2) M.\end{aligned}$$

for sufficiently large constant $M < \infty$ which implies

$$(g \circ z)^{-}(\bar{x}, u) \leq \sup\big\{\langle x^*, u\rangle : x^* \in \partial^* g(z(\bar{x}))(\partial^*_{D_1} z_1(\bar{x}), \ldots, \partial^*_{D_n} z_n(\bar{x}))\big\}.$$

Since $u \in \overline{D}$ was chosen arbitrarily, we obtain that

$$\partial^*_{\overline{D}}(g \circ z)(\bar{x}) = \partial^* g(z(\bar{x}))(\partial^*_{D_1} z_1(\bar{x}), \ldots, \partial^*_{D_n} z_n(\bar{x}))$$

is an upper directional convexificator of $g \circ z$ at the point $\bar{x}$ with $\overline{D} = \bigcap_{i=1}^{n} D_i(\bar{x})$. Similar arguments can be used showing that $\partial^*_{\overline{D}}(g \circ z)(\bar{x})$ is a lower directional convexificator, too. □

This result can now be applied to max or penalty functions, see (Dempe and Pilecka [74]). Let $\overline{f}$ be the optimal objective function value of the optimization problem

$$\min\{f_0(x) : f(x) \leq 0\}$$

and $\overline{x}$ be a global optimal solution. Then, zero is the optimal function value of the problem to minimize the function

$$\alpha(x) = \max\{f_0(x) - \overline{f}, f(x)\}.$$

Hence, $D^{-}\alpha(\overline{x}, d) \geq 0$ for all directions d provided that the functions f_0, f admit a lower Dini derivative, see Demyanov [83]. This, obviously, is equivalent to $0 \in \text{conv}\, \partial^* \alpha(\bar{x}))$ if the function $\alpha(\cdot)$ has a convexificator at $\overline{x}$.

In the next theorem a necessary optimality condition using convexificators is given. For that we need the cl-*property* introduced by Dien [89, 90] for locally Lipschitz continuous set-valued mappings.

Definition 3.15 A set-valued mapping $H : \mathbb{R}^n \to 2^{\mathbb{R}^m}$ has the cl -property if the following holds true:

$$\forall x_n^* \to x^* \ \forall z_n^* \to z^* \ \forall x_n \to x : \ x_n^* \in \partial^* \mathscr{C}_H(z_n^*, \cdot)(x_n) \forall n$$
$$\text{implies } x^* \in \partial^* \mathscr{C}_H(z^*, \cdot)(x),$$

where

$$\mathscr{C}_H(z^*, \cdot) = \min_{z \in H(\cdot)} z^\top z^*$$

denotes the support function of the mapping H in direction z^*.

Remark 3.4 (Dempe and Gadhi [61]) The cl-property can be interpreted as a sequential upper semicontinuity of $\partial^* \mathscr{C}_H(z^*, \cdot)$ and can be established without difficulty in some cases (see Example 3.6).

Example 3.6 (*Dempe and Gadhi* [61]) Let $z^* \in \mathbb{B}^+_{\mathbb{R}^m} = \mathbb{B}_{\mathbb{R}^m} \cap \{z \in \mathbb{R}^m : z_i > 0 \ i = 1, \dots, m\}$ and $H(x) = f(x) + \mathbb{B}^+_{\mathbb{R}^m}$ where $f : \mathbb{B}^+_{\mathbb{R}^n} \to \mathbb{B}^+_{\mathbb{R}^m}$ is a locally Lipschitz continuous mapping.

Suppose that $x_n^* \in \partial^* \mathscr{C}_H(z_n^*, \cdot)(x_n)$ with $x_n^* \to x^*$, $z_n^* \to z^*$ and $x_n \to x$. Then we have $\mathscr{C}_H(z_n^*, x_n) = \langle z_n^*, f(x_n) \rangle$ for n sufficiently large as well as $x_n^* \in \partial^* \langle z_n^*, f \rangle (x_n)$ and it follows that $x^* \in \partial^* \langle z^*, f \rangle (x)$.

Theorem 3.47 (Dempe and Pilecka [74]) *Suppose that* $(\overline{x}, \overline{y})$ *is a local optimal solution of the bilevel programming problem* (1.4). *The following assumptions are satisfied:*

1. *Let the function* $F : \mathbb{R}^n \times \mathbb{R}^m \to \mathbb{R}$ *admit a bounded convexificator* $\partial^* F(\overline{x}, \overline{y})$ *at the point* $(\overline{x}, \overline{y})$ *which is upper semicontinuous at* $(\overline{x}, \overline{y})$ *and* $0 \notin \text{conv } (\partial^* F(\overline{x}, \cdot)(\overline{y}))$.
2. *There exists an open neighborhood* $U(\overline{x})$ *of* $\overline{x}$ *such that for each* $z^* \in Z^*_{\Psi(\overline{x})}$, $\|z^*\| \le 1$ *the support function*

$$\mathscr{C}_\Psi(z^*, \cdot) = \min_{z \in \Psi(\cdot)} z^\top z^*$$

is lower semicontinuous on $U(\overline{x})$, *continuous at* $\overline{x}$ *only in directions* $d \in \overline{D}_x \neq \{0\}$ *and the assumption* (A_0) *is satisfied for every*

$$z^* \in Z^*_{\Psi(\overline{x})} := \{z^* \in \mathbb{R}^m : \mathscr{C}_\Psi(z^*, \overline{x}) > -\infty\},$$

with $\|z^*\| \le 1$ *at* $\overline{x}$.

3. *The support function* $\mathscr{C}_\Psi(z^*, \cdot)$ *admits a bounded directional convexificator* $\partial^*_{\overline{D}_x} \mathscr{C}_\Psi(z^*, \cdot)(x)$ *on* $U(\overline{x}) \cap (\{\overline{x}\} + \overline{D}_x)$, *where*

$$\overline{D}_x = \bigcap_{z^* \in Z^*_{\Psi(\bar{x})}} D_{\mathscr{C}_\Psi(z^*, \cdot)}(\bar{x}) \neq \{0\}.$$

4. *The images of the solution set mapping Ψ of the lower level problem be non-empty, closed and convex for every fixed $x \in U(\overline{x})$ and Ψ possess the* cl *-property with respect to the convexificator and the directional convexificator.*
5. $G \equiv 0,\ T = \mathbb{R}^m,\ X = \mathbb{R}^n$ *and there exists $c < \infty$ such that the set $\{(x, y) : y \in \Psi(x),\ F(x, y) \le c\}$ is not empty and bounded.*
6. *The distance function*

$$\rho(y, \Psi(x)) = \min\{\|y - z\| : z \in \Psi(x)\} : \mathbb{R}^m \times 2^{\mathbb{R}^m} \to \mathbb{R}$$

is locally convex with respect to x in all its continuity directions at the point $\overline{x}$.
7. *For each sequence $(x_n, y_n) \to (\overline{x}, \overline{y})$ with $\rho(y_n, \Psi(x_n)) \to 0$ we have $\mathscr{C}_\Psi(z_n^*, x_n) \to \mathscr{C}_\Psi(z^*, \overline{x})$ for any $z_n^* \to z^*$.*

Then there exist a scalar α and a vector $z^* \in Z^*_{\Psi(\overline{x})}$ with $\|z^*\| = 1$ such that

$$\begin{cases} 0 \in \operatorname{conv} \partial^* F(\overline{x}, \overline{y}) + \alpha \cdot \operatorname{conv} \left\{ (\partial^*_{\overline{D}_x} \mathscr{C}_\Psi(z^*, \cdot)(\overline{x}) + N_{\overline{D}_x}(0)) \times \{-z^*\} \right\}, \\ \mathscr{C}_\Psi(z^*, \overline{x}) = \langle z^*, \overline{y} \rangle \end{cases}$$

with $\overline{D}_x = \bigcap_{z^* \in Z^*_{\Psi(\overline{x})}} D_{\mathscr{C}_\Psi(z^*, \cdot)}(\overline{x})$ being convex and closed.

Proof We will outline the proof from (Dempe and Pilecka [74]). Let $U(\overline{x})$ be a sufficiently small open neighborhood of $\overline{x}$ and define the following functions:

$$\Gamma_1(x, y) = F(x, y) - F(\overline{x}, \overline{y}) + \frac{1}{n},\ \ \Gamma_2(x, y) = \rho(y, \Psi(x)),$$

$$h_n(x, y) = \max\{\Gamma_1(x, y), \Gamma_2(x, y)\}.$$

Since $(\overline{x}, \overline{y})$ is a local optimal solution of problem (1.4) we have

$$h_n(\overline{x}, \overline{y}) \le \frac{1}{n} + \inf_{(x,y) \in U(\overline{x}) \times \mathbb{R}^m} h_n(x, y).$$

Using Ekeland's variational principle (Ekeland [299]) we obtain the existence of points $(x_k, y_k) \in U(\overline{x}) \times \mathbb{R}^m$ satisfying

$$\begin{cases} \|(x_n, y_n) - (\overline{x}, \overline{y})\| \le \dfrac{1}{\sqrt{n}} \\ h_n(x_n, y_n) \le h_n(x, y) + \dfrac{1}{\sqrt{n}} \|(x, y) - (x_n, y_n)\| \ \ \forall (x, y) \in U(\bar{x}) \times \mathbb{R}^m. \end{cases}$$

for sufficiently large n. Hence, (x_n, y_n) is a local minimum of the function $H(x, y) = h_n(x, y) + 1/\sqrt{n}\, \|(x, y) - (x_n, y_n)\|$ and consequently, the following necessary optimality condition is satisfied (see Jeyakumar and Luc [155]):

$$0 \in \operatorname{cl} \operatorname{conv} \partial^* H(x_n, y_n).$$

or, equivalently,

$$0 \in \operatorname{cl}\operatorname{conv} \partial^* h_n(x_n, y_n) + \frac{1}{\sqrt{n}} \mathbb{B}_{\mathbb{R}^n \times \mathbb{R}^m}.$$

Using the formulae for convexificators from Dempe and Pilecka [74] and the assumptions of the theorem, we derive

$$\partial^* h_n(x_n, y_n) = \operatorname{cl}\operatorname{conv}\left\{\partial^* \Gamma_i(x_n, y_n) : i \in I(x_n, y_n)\right\}$$

where $I(x_n, y_n) = \{i : h_n(x_n, y_n) = \Gamma_i(x_n, y_n)\}$ (we need the closure on the right hand side since the convexificator of the distance function may be unbounded).

Hence, we can find a $\lambda_n \in [0, 1]$ such that:

$$0 \in \lambda_n \cdot \operatorname{conv} \partial^* \Gamma_1(x_n, y_n) + (1 - \lambda_n) \cdot \operatorname{conv} \partial^* \Gamma_2(x_n, y_n) + \frac{1}{\sqrt{n}} \mathbb{B}_{\mathbb{R}^n \times \mathbb{R}^m}. \tag{3.85}$$

Now we can show $\max\{\Gamma_1(x_n, y_n), \Gamma_2(x_n, y_n)\} > 0$ and also $\Gamma_2(x_n, y_n) > 0$. Otherwise we would get a contradiction to local optimality of $(\overline{x}, \overline{y})$ or the assumptions of the theorem. Then, using the assumptions of the theorem again, we derive from (3.85) the existence of a sequence $z_n^* \in Z^*_{\Psi(\bar{x})}$ such that $\|z_n^*\| = 1$ and

$$\begin{cases} 0 \in \lambda_n \cdot \operatorname{conv} \partial^* F(x_n, y_n) \\ \quad + (1 - \lambda_n) \cdot \operatorname{conv}\left\{(\partial^*_{\overline{D}_x} \mathscr{C}_\Psi(z_n^*, \cdot)(x_n) + N_{\overline{D}_x}(0)) \times \{-z_n^*\}\right\} + \frac{1}{\sqrt{n}} \mathbb{B}_{\mathbb{R}^n \times \mathbb{R}^m}, \\ \rho(y_n, \Psi(x_n)) = \mathscr{C}_\Psi(z_n^*, x_n) - \langle z_n^*, y_n \rangle. \end{cases}$$

Tending n to infinity and using the cl -property, this leads to

$$\begin{aligned} &0 \in \lambda \cdot \operatorname{conv} \partial^* F(\overline{x}, \overline{y}) \\ &\quad + (1 - \lambda) \cdot \operatorname{conv}\left\{(\partial^*_{\overline{D}_x} \mathscr{C}_\Psi(z^*, \cdot)(\bar{x}) + N_{\overline{D}_x}(0)) \times \{-z^*\}\right\}. \end{aligned}$$

Since $\lambda \in (0, 1)$ we find $\alpha > 0$ such that

$$0 \in \operatorname{conv} \partial^* F(\overline{x}, \overline{y}) + \alpha \cdot \operatorname{conv}\left\{(\partial^*_{\overline{D}_x} \mathscr{C}_\Psi(z^*, \cdot)(\bar{x}) + N_{\overline{D}_x}(0)) \times \{-z^*\}\right\}.$$

The second equation in the assertion follows again using Ekeland's variational principle and the assumptions of the theorem. □

The following example shows the usefulness of the introduced necessary optimality conditions in terms of a bilevel programming problem with upper semicontinuous solution set mapping of the lower level problem.

Example 3.7 (*Dempe and Pilecka* [74]) Consider the following optimistic bilevel optimization problem:

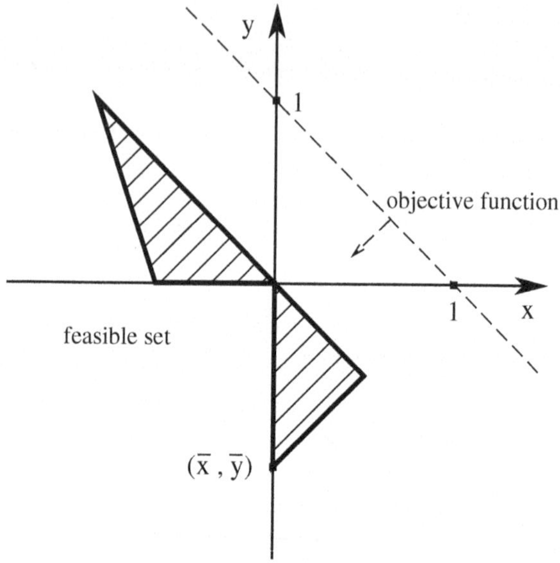

Fig. 3.8 The *dashed area* illustrates the feasible set of the bilevel programming problem (3.86) and the *dashed line* is the level set of the objective function of the *upper level* problem. The uniquely determined global optimal solution of this problem is given by the point $(\bar{x}, \bar{y}) = (0, -1)$ [there is also a local optimal solution at the point $(-\frac{2}{3}, 0)$]

$$
\begin{aligned}
x + y &\longrightarrow \min_{x,y} \\
y \in \Psi(x) = \operatorname*{arg\,min}_y \Big\{ &\max\{0, xy\} : (x - y - 1)^3 \leq 0, \\
&x + y \leq 0,\ -3x - y - 2 \leq 0 \Big\}.
\end{aligned}
\tag{3.86}
$$

The corresponding feasible set can be illustrated as in the Fig. 3.8. The support function of the solution set mapping Ψ of the lower level problem as a function of x for a fixed $z^* \geq 0$ is given by:

$$
\mathscr{C}_\Psi(z^*, x) = \begin{cases} z^*(-3x - 2) & \text{if } x \in [-1, -\frac{2}{3}), \\ 0 & \text{if } x \in [-\frac{2}{3}, 0), \\ z^*(x - 1) & \text{if } x \in [0, \frac{1}{2}], \end{cases}
$$

and for $z^* < 0$ we have $\mathscr{C}_\Psi(z^*, x) = -z^* x$.
The continuity directions of the support function at the point $\bar{x} = 0$ with respect to x are $\overline{D}_x = [0, +\infty)$.
Let us now consider $z^* = 1$, then $\mathscr{C}_\Psi$ is lower semicontinuous as a function of x and Dini directional derivatives at $\bar{x} = 0$ are:

$$
\mathscr{C}_\Psi^+(1, \cdot)(\bar{x}, v) = \mathscr{C}_\Psi^-(1, \cdot)(\bar{x}, v) = \begin{cases} v & \text{if } v \geq 0, \\ \infty & \text{if } v < 0. \end{cases}
$$

Obviously, this support function admits the directional convexificator $\partial^*_{\overline{D}_x} \mathscr{C}_\Psi(1,\cdot)(\bar{x}) = \{1\}$ at $\bar{x}$. Together with the normal cone to the set $\overline{D}_x$ at the point $d = 0$, we obtain the following convexificator of the function $\mathscr{C}_\Psi(z^*,\cdot)$ as a function of x at the point $\bar{x}$ with $z^* = 1$:

$$\partial^*\mathscr{C}_\Psi(1,\cdot)(\bar{x}) = \{1\} + (-\infty, 0] = (-\infty, 1].$$

Definitely, this set is a convexificator of the function $\mathscr{C}_\Psi(z^*,\cdot)$ at the point $\bar{x}$ since we have:

$$\mathscr{C}_\Psi^-(1,\cdot)(\bar{x}, v) = \begin{cases} v & \le \sup_{x^*\in(-\infty,1]}\langle x^*, v\rangle = v \quad \text{if } v \ge 0 \\ \infty & \le \sup_{x^*\in(-\infty,1]}\langle x^*, v\rangle = \infty \quad \text{if } v < 0, \end{cases}$$

$$\mathscr{C}_\Psi^+(1,\cdot)(\bar{x}, v) = \begin{cases} v & \ge \inf_{x^*\in(-\infty,1]}\langle x^*, v\rangle = -\infty \quad \text{if } v \ge 0 \\ \infty & \ge \inf_{x^*\in(-\infty,1]}\langle x^*, v\rangle = v \quad \text{if } v < 0. \end{cases}$$

Now we can state that the necessary optimality conditions are satisfied with $\alpha = 1$ because we obtain:

$$0 \in (1,1) + \alpha \cdot ((-\infty, 1] \times \{-1\}),$$

$$\mathscr{C}_\Psi(1,0) = -1 = 1 \cdot (-1).$$

□

3.8.3 Optimality Conditions Using Convexificators

A nonsmooth version of the *Mangasarian-Fromovitz constraint qualification* is used in Kohli [190]:

Definition 3.16 Let $\overline{y} \in Y(\overline{x}) = \{y : g(x,y) \le 0\}$, where the functions g_i are assumed to be Lipschitz continuous. The point $(\overline{x}, \overline{y})$ is said to be *lower-level regular* if

$$\sum_{i\in I(\overline{x},\overline{y})} \lambda_i v_i = 0,\ \lambda_i \ge 0,\ \forall\, i \Rightarrow \lambda_i = 0,\ \forall\, i \in I(\overline{x},\overline{y}),$$

whenever $(u_i, v_i) \in \partial^{Cl} g_i(\overline{x},\overline{y}),\ i \in I(\overline{x},\overline{y})$. A similar condition with respect to the upper level constraints is called *upper-level regularity*.

The *Abadie constraint qualification* using the concept of convexificators and the subdifferential of Clarke applied to problem (3.7) reads as

$$\left(\bigcup_{i\in I(\overline{x},\overline{y})} \operatorname{conv}\partial^* g_i(\overline{x},\overline{y}) \cup \bigcup_{j\in J(\overline{x})} \operatorname{conv}\partial^* G_j(\overline{x}) \times \{0\} \cup \partial^{Cl}(f(\overline{x},\overline{y}) - \varphi(\overline{x}) \times \{0\})\right)^- \subseteq C_{\mathscr{M}}(\overline{x},\overline{y}),$$

where $Z^-(z) = N_Z(z)$ denotes the normal cone to a set Z at z, $C_{\mathcal{M}}(\overline{x}, \overline{y})$ is the tangent (Bouligand) cone to $\mathcal{M}$ at $(\overline{x}, \overline{y})$ and

$$\mathcal{M} = \{(x, y) : G(x) \leq 0,\ f(x, y) \leq \varphi(x),\ g(x, y) \leq 0\}$$

denotes the feasible set of problem (3.7) with $T = \mathbb{R}^m$, $X = \mathbb{R}^n$, $I(\overline{x}, \overline{y})$ and $J(\overline{x})$ denote the index sets of active constraints in the lower respectively the upper level problem. The Abadie constraint qualification is a weaker constraint qualification than the Mangasarian-Fromovitz constraint qualification and can be satisfied for bilevel optimization problems.

Theorem 3.48 (Kohli [190]) *Let $(\overline{x}, \overline{y})$ be a local optimal solution of the optimistic bilevel optimization problem* (1.1), (1.4) *with $T = \mathbb{R}^m$, $X = \mathbb{R}^n$. Assume that the functions F, f, g_i are locally Lipschitz continuous and F has a bounded upper semiregular convexificator $\partial^* F(\overline{x}, \overline{y})$. Let the functions g_i and G_j have upper convexificators at $(\overline{x}, \overline{y})$. Assume that the Abadie constraint qualification is satisfied. Let the mapping Ψ be inner semicompact at $\overline{x}$ and let all points $(\overline{x}, y) \in$ **gph** Ψ be lower-level and upper-level regular. Then, there exist $\mu \geq 0, \lambda \geq 0, \tau_i \geq 0, \lambda_i \geq 0$ and $y^* \in \Psi(\overline{x})$ such that the following conditions are satisfied:*

$$(0, 0) \in \mathrm{cl}\left[\mathrm{conv}\, \partial^* F(\overline{x}, \overline{y}) - \left\{\lambda(\partial^{Cl} f(\overline{x}, \overline{y}) - \partial_x^{Cl} f(\overline{x}, y^*) \times \{0\} - \sum_{i=1}^{p} \lambda_i \partial_x^{Cl} g_i(\overline{x}, y^*)) + \sum_{i=1}^{p} \mu_i \mathrm{conv}\, \partial^* g_i(\overline{x}, \overline{y}) + \sum_{j=1}^{q} \tau_j \mathrm{conv}\, \partial^* G_j(\overline{x}) \times \{0\}, \right\}\right] \tag{3.87}$$

$$0 \in \partial_y^{Cl} f(\overline{x}, y^*) + \sum_{i=1}^{p} \lambda_i \partial_y^{Cl} g_i(\overline{x}, y^*), \tag{3.88}$$

$$\lambda_i g_i(\overline{x}, y^*) = 0,\ i = 1, \ldots, p. \tag{3.89}$$

Proof If $(\overline{x}, \overline{y})$ is a local optimal solution of problem (1.1), (1.4) then it is also a local optimal solution of problem (3.7). Using the definition of the Bouligand cone and local optimality we obtain (see Demyanov [83])

$$D^+ F((\overline{x}, \overline{y}), (d_x, d_y)) \geq 0\, \forall\, (d_x, d_y)^\top \in C_{\mathcal{M}}(\overline{x}, \overline{y}).$$

By upper semiregularity of the convexificator this implies

$$\max_{\eta \in \partial^* F(\overline{x}, \overline{y})} \eta^\top (d_x, d_y)^\top \geq 0\, \forall\, (d_x, d_y)^\top \in C_{\mathcal{M}}(\overline{x}, \overline{y}),$$

which by the Abadie constraint qualification leads to

$$\max_{\eta \in \partial^* F(\overline{x}, \overline{y})} \eta^\top (d_x, d_y)^\top \geq 0\, \forall\, (d_x, d_y)^\top \in (\mathcal{A})^-, \tag{3.90}$$

$$\mathscr{A} = \bigcup_{i \in I(\overline{x},\overline{y})} \text{conv}\, \partial^* g_i(\overline{x},\overline{y}) \cup \bigcup_{j \in J(\overline{x})} \text{conv}\, \partial^* G_j(\overline{x}) \times \{0\} \cup \partial^{Cl}(f(\overline{x},\overline{y}) - \varphi(\overline{x}) \times \{0\}).$$

By Theorem 16.3 in Dem'yanov and Rubinov [84] this implies that

$$(0,0) \in \text{cl conv}\left(\partial^* F(\overline{x},\overline{y}) - \mathscr{A}\right).$$

Thus, there exists a sequence $\{(a_n, b_n)\}_{n=1}^{\infty}$ converging to zero with

$$(a_n, b_n) \in \text{conv}\left(\partial^* F(\overline{x},\overline{y}) - \mathscr{A}\right) = \text{conv}\, \partial^* F(\overline{x},\overline{y}) + \text{conv}\,(-\mathscr{A}),$$

see Li and Zhang [202]. Using the convex hull property $\text{conv}\,(-S) = -\text{conv}\, S$ we get $(a_n, b_n) \in \text{conv}\, \partial^* F(\overline{x},\overline{y}) - \text{conv}\, \mathscr{A}$. Using the formula for $\mathscr{A}$ this implies

$$(0,0) \in \text{cl}\left[\text{conv}\, \partial^* F(\overline{x},\overline{y}) - \left\{\lambda(\partial^{Cl} f(\overline{x},\overline{y}) - \partial_x^{Cl} \varphi(\overline{x}) \times \{0\}) + \sum_{i=1}^{p} \mu_i \text{conv}\, \partial^* g_i(\overline{x},\overline{y}) + \sum_{j=1}^{q} \tau_j \text{conv}\, \partial^* G_j(\overline{x}) \times \{0\}\right\}\right].$$

The proof then follows from the formula for the subdifferential in the sense of Clarke for the optimal value function (see Mordukhovich et al. [243]). □

Chapter 4
Convex Bilevel Programs

4.1 Optimality Conditions for a Simple Convex Bilevel Program

4.1.1 A Necessary but Not Sufficient Condition

A special case of a bilevel optimization problem arises if a convex function $F : \mathbb{R}^n \to \mathbb{R}$ is minimized on the set of optimal solutions Ψ of a convex optimization problem

$$\min\{f(x) : x \in P\}, \tag{4.1}$$

where $f : \mathbb{R}^n \to \mathbb{R}$ is a convex, at least twice continuously differentiable function and P a closed convex set. Then,

$$\Psi := \operatorname*{Argmin}_x \{f(x) : x \in P\}$$

is a convex set and the bilevel optimization problem

$$\min\{F(x) : x \in \Psi\} \tag{4.2}$$

is a convex optimization problem.

This problem is a generalization of a convex optimization problem: let $Q = \{x : g(x) \leq 0,\ Ax = b\}$ with a convex function $g : \mathbb{R}^n \to \mathbb{R}^p$, A be a matrix of appropriate dimension and $b \in \mathbb{R}^m$. Consider the convex optimization problem

$$\min\{F(x) : x \in Q\}$$

with a convex function F. Then, problem (4.2) arises if

$$f(x) = \sum_{i=1}^{p} (\max\{0, g_i(x)\})^2 + \|Ax - b\|^2$$

is a penalty function for the set Q and $P = \mathbb{R}^n$.

S. Dempe et al., *Bilevel Programming Problems*, Energy Systems,
DOI 10.1007/978-3-662-45827-3_4

In general, problem (4.2) is a convex optimization problem and we expect that necessary optimality conditions are also sufficient. Let v be the optimal objective function value of problem (4.1). Then, problem (4.2) can be replaced by

$$\min\{F(x) : f(x) \leq v,\ x \in P\} \tag{4.3}$$

and, using the same ideas as in the proof to Theorem 3.30, we find, that the standard regularity conditions as (MFCQ) are violated for this problem.

A necessary and sufficient optimality condition for problem (4.1) is

$$0 \in \nabla f(x) + N_P(x),$$

where $N_P(x)$ is the normal cone from convex analysis to the set P at x. Thus,

$$\Psi = \{x \in P : (x, -\nabla f(x)) \in \mathbf{gph}\, N_P\}$$

and problem (4.2) can equivalently be replaced with

$$\min\{F(x) : x \in P,\ (x, -\nabla f(x)) \in \mathbf{gph}\, N_P\}. \tag{4.4}$$

Using results from variational analysis (Mordukhovich [241, 242], Rockafellar and Wets [274]) the following necessary optimality condition is obtained:

Theorem 4.1 (Dempe et al. [54]) *Let $\overline{x}$ be an optimal solution of problem* (4.2)*, the function F be convex and differentiable, the function f is assumed to be convex and at least twice continuously differentiable, P is a closed convex set. Suppose that the (basic) constraint qualification*

$$\left.\begin{array}{l} (v, w) \in N^M_{\mathbf{gph}\, N_P}(\overline{x}, -\nabla f(\overline{x})) \\ 0 \in w - \nabla^2 f(\overline{x}) w + N_P(\overline{x}) \end{array}\right\} \Rightarrow w = 0, v = 0.$$

Then, there is $(\overline{v}, \overline{w}) \in N^M_{\mathbf{gph}\, N_P}(\overline{x}, -\nabla f(\overline{x}))$ such that

$$0 \in \nabla F(\overline{x}) + \overline{w} - \nabla^2 f(\overline{x})\overline{w} + N_P(\overline{x}).$$

Unfortunately, this optimality condition is not sufficient which is surprising since (4.2) is a convex optimization problem.

Example 4.1 (*Dempe et al.* [54]) Consider the problem (4.2) with the function $F : \mathbb{R} \to \mathbb{R}$ given by $F(x) = x^2$ and the lower-level objective $f : \mathbb{R} \to \mathbb{R}$ given as follows: $f(x) = x^3$ when $x \geq 0$ and $f(x) = 0$, $x \leq 0$. The lower-level constraint set is $P = [-1, +1]$. Observe that $\Psi = [-1, 0]$. Thus, $x = 0$ is the only solution to the problem (4.2). However, the optimality condition given in Theorem 4.1 is satisfied at the point $x = -1$ which obviously is not a solution of the problem (4.2). This fact can be seen by noting that $(-1, 0) \in \mathbf{gph}\, N_P$ and also observing that

$\nabla F(-1) = -2$, $\nabla^2 f(-1) = 0$ and $(4, 0) \in N^M_{\mathbf{gph}\, N_P}(-1, 0)$. Now the optimality condition is satisfied by choosing the element -2 from $N_P(-1) = (-\infty, 0]$. □

4.1.2 Necessary Tools from Cone-Convex Optimization

The following material taken from Dempe et al. [54] is used to formulate another problem which is equivalent to (4.2).

Consider a class of *cone-convex programs* given as:

$$\min \theta(x) \quad \text{subject to} \quad g(x) \in -D, \text{ and } x \in C, \tag{4.5}$$

where $\theta : \mathbb{R}^n \to \overline{\mathbb{R}} := (-\infty, \infty]$ is a proper, convex, lower semicontinuous (l.s.c.) function with values in the extended real line $\overline{\mathbb{R}}$, $g : \mathbb{R}^n \to \mathbb{R}^m$ is a continuous $D-$convex mapping with D is a closed convex cone in $\mathbb{R}^m$ and $C \subset \mathbb{R}^n$ is closed and convex.

Here, a function $g : \mathbb{R}^n \to \mathbb{R}^m$ is *D-convex* to a closed and convex cone D provided that

$$f(\lambda x + (1-\lambda)y) - \lambda f(x) - (1-\lambda) f(y) \in -D\ \forall\, x, y \text{ and } \forall\, \lambda \in [0, 1].$$

For a set $C \subset \mathbb{R}^n$, the *indicator function* δ_C is defined as $\delta_C(x) = 0$ if $x \in C$ and $\delta_C(x) = +\infty$ if $x \notin C$. Let us recall that if C is nonempty, closed and convex, then δ_C is a proper l.s.c. convex function.

Let $A = \{x \in C : g(x) \in -D\}$. Further, let D^+ be the *positive dual cone* of D, i.e.,

$$D^+ := \{s^* \in \mathbb{R}^m : \langle s^*, s\rangle \geq 0,\ \forall s \in D\}.$$

Assume that $\operatorname{dom} \theta \cap A \neq \emptyset$, where $\operatorname{dom} \theta := \{x \in \mathbb{R}^n \mid \theta(x) < \infty\}$

Considering further an extended-real-valued function $\xi : \mathbb{R}^n \to \overline{\mathbb{R}}$, we always assume that it is proper, i.e., $\xi(x) \not\equiv \infty$ on $\mathbb{R}^n$. The *conjugate function* $\xi^* : \mathbb{R}^n \to \overline{\mathbb{R}}$ to ξ is defined by

$$\begin{aligned} \xi^*(x^*) &:= \sup\{\langle x^*, x\rangle - \xi(x) \mid x \in \mathbb{R}^n\} \\ &= \sup\left\{\langle x^*, x\rangle - \xi(x) \middle| \, x \in \operatorname{dom} \xi\right\}. \end{aligned} \tag{4.6}$$

We say that problem (4.5) satisfies the *Farkas-Minkowski constraint qualification*

(FM) if the cone

$$K := \bigcup_{\lambda \in D^+} \text{epi } (\lambda^\top g)^* + \text{epi } \delta_C^* \tag{4.7}$$

is closed in the space $\mathbb{R}^n \times \mathbb{R}$.

It is important to note that the set $\bigcup_{\lambda \in D^+} \text{epi } (\lambda g)^*$ is a closed convex cone. This was shown in Jeyakumar et al. [156].

We say that problem (4.5) satisfies the *(CC) constraint qualification*

(CC) if the set

$$\text{epi } \theta^* + K \tag{4.8}$$

is closed in the space $\mathbb{R}^n \times \mathbb{R}$ where K is given in (4.7).

Remark 4.1 It is worth noting that if θ is continuous at one point in $A \subset \mathbb{R}^n$ then (see Dinh et al. [91])

$$\text{epi } (\theta + \delta_A)^* = \text{cl}\left\{\text{epi } \theta^* + \text{epi } \delta_A^*\right\} = \text{epi } \theta^* + \text{epi } \delta_A^* = \text{epi } \theta^* + \text{cl } K.$$

So, if (FM) holds (i.e. K is closed) then (CC) holds.

Theorem 4.2 (Dinh et al. [92]) *Let the qualification condition* (*CC*) *hold for the convex program* (4.5). *Then* $\overline{x} \in A \cap \text{dom } \theta$ *is a* (*global*) *solution to* (4.5) *if and only if there is* $\lambda \in D^+$ *such that*

$$0 \in \partial\theta(\overline{x}) + \partial(\lambda^\top g)(\overline{x}) + N_C(\overline{x}) \tag{4.9}$$

$$\lambda g(\overline{x}) = 0. \tag{4.10}$$

We are now going to apply this result to the equivalent reformulation (4.3) of problem (4.2).

Theorem 4.3 (Dempe et al. [54]) *For the problem* (4.3), *assume that*

$$\{\text{cone }\{(0, 1)\} \cup \text{cone } \left[(0, v) + \text{epi } f^*\right]\} + \text{epi } \delta_P^*$$

is closed. Then $\overline{x} \in P$ *is a* (*global*) *solution to* (4.3) *if and only if there is* $\lambda \in \mathbb{R}_+$ *such that*

$$0 \in \partial F(\overline{x}) + \lambda\partial f(\overline{x}) + N_P(\overline{x}) \tag{4.11}$$

$$\lambda(f(\overline{x}) - v) = 0. \tag{4.12}$$

Proof First observe that problem (4.3) is of the type (4.5) with $D = D^+ = \mathbb{R}_+$ and $C = P$. Second, for each $u^* \in \mathbb{R}^n$, and $\mu \in \mathbb{R}_+$

$$(\mu(f(\cdot) - v))^*(u^*) = \mu v + (\mu f)^*(u^*).$$

Then it follows that

$$\text{epi } (\mu(f(.) - v))^* = (0, \mu v) + \text{epi } (\mu f)^*.$$

Now observe that when $\mu = 0$ we have

$$\text{epi } (\mu f)^* = \text{cone}\,\{(0, 1)\},$$

and when $\mu > 0$ we have

$$\text{epi } (\mu f)^* = \mu \text{epi } f^*.$$

Thus, we obtain

$$\text{epi } (\mu(f(.) - v))^* = \text{cone}\,\{(0, 1)\} \cup \left\{ \bigcup_{\mu > 0} \mu[(0, v) + \text{epi } f^*] \right\}.$$

Noting that $\text{cone}\,\{(0, 1)\} \cup \{(0, 0)\} = \text{cone}\,\{(0, 1)\}$ we have

$$\text{epi } (\mu(f(.) - v))^* = \text{cone}\,\{(0, 1)\} \cup \text{cone}\,[(0, v) + \text{epi } f^*].$$

Now from the hypothesis of the theorem it is clear that the problem (4.3) satisfies (FM) and hence, it satisfies (CC) since F is continuous (see Remark 4.1).

It now follows from Theorem 4.2 that there is $\lambda \in \mathbb{R}_+$ such that

$$\begin{aligned} &0 \in \partial F(\overline{x}) + \lambda \partial [f(.) - v](\overline{x}) + N_P(\overline{x}) \\ &\lambda(f(\overline{x}) - v) = 0. \end{aligned} \quad (4.13)$$

Since $\partial(f(.) - v)(\overline{x}) = \partial f(\overline{x})$, the conclusion follows. □

The following example is used in Dempe et al. [54] to illustrate this result.

Example 4.2 Let us consider the bilevel problem of the model (4.5) where $F(x) = x^2 + 1$, $P = [-1, 1]$, and $f(x) = \max\{0, x\}$.

It is easy to see that $\text{epi } \delta_P^* = \text{epi } |.|$, $\Psi = [-1, 0]$, and $v = 0$. The optimization problem reformulated from this bilevel problem is

$$\min F(x) := x^2 + 1 \text{ subject to } f(x) = \max\{0, x\} \leq 0, \ x \in [-1, 1]. \quad (4.14)$$

Note that for each $u \in \mathbb{R}$,

$$f^*(u^*) = \begin{cases} +\infty & \text{if } u^* < 0 \text{ or } u^* > 1 \\ 0 & \text{if } u^* \in [0, 1]. \end{cases}$$

We have

$$\text{epi } f^* = \{(u^*, r) \mid u \in [0, 1], r \geq 0\} = [0, 1] \times \mathbb{R}_+,$$

and

$$\text{cone}\,\{\text{epi } f^*\} + \text{epi } \delta_P^* = \mathbb{R}_+^2 \cup \{(u, r) \mid u \leq 0, \ r \geq -u\}$$

is a closed subset of $\mathbb{R}^2$. This shows that for the problem (4.14), (FM) holds since cone $\{(0, 1)\} \subset$ epi f^*. Since F is continuous, (CC) holds as well (note that the Slater's condition fails to hold for (4.14)). It is easy to see that $\overline{x} = 0$ is a solution of the bilevel problem. Since $N_P(0) = \{0\}$, $\partial F(0) = \{0\}$, $\partial f(0) = [0, 1]$, and (4.11)–(4.12) are satisfied with $\lambda = 0$. □

4.1.3 A Solution Algorithm

Consider problem (4.2), let the convex functions F, f be differentiable and $\Pi_P : \mathbb{R}^n \to \mathbb{R}$ denote the projection on the convex and closed set P. For $\sigma > 0$ let

$$w_\sigma(x) = \sigma F(x) + f(x).$$

In Solodov [292] a gradient type descent method solving problem (4.2) is suggested:

Algorithm: Initialization: Choose parameters $\overline{\alpha} > 0$, $\theta \in (0, 1)$ and $\eta \in (0, 1)$. Take $x^0 \in P$, $\sigma_0 > 0$ and set $k := 1$.

Step: Given x^k compute $x^{k+1} = z^k(\alpha_k)$ with $\alpha_k = \eta^{m_k}\overline{\alpha}$, where

$$z^k(\alpha) = \Pi_P(x^k - \alpha \nabla w_{\sigma_k}(x^k))$$

and m_k is the smallest nonnegative integer m such that

$$w_{\sigma_k}(z^k(\eta^m\overline{\alpha})) \leq w_{\sigma_k}(x^k) + \theta \nabla w_{\sigma_k}(x^k)(z^k(\eta^m\overline{\alpha}) - x^k). \tag{4.15}$$

Choose $0 < \sigma_{k+1} \leq \sigma_k$, set $k := k + 1$ and repeat the step.

The projection on a convex set is uniquely determined and, if $x^k = \Pi_P(x^k - \alpha \nabla w_\sigma(x^k))$ for some $\alpha > 0$, then x^k is a minimum of the function $w_\sigma(x)$ on P. In this case, $m_k = 0$, $x^{k+1} = z^k(\overline{\alpha}) = x^k$ and the algorithm stops.

Let the functions F and f be bounded on P:

$$-\infty < \overline{F} = \inf\{F(x) : x \in P\}$$

and

$$-\infty < \overline{f} = \inf\{f(x) : x \in P\}.$$

Theorem 4.4 (Solodov [292]) *Let the set P be closed and convex and the functions F, f be convex, differentiable with Lipschitz continuous gradients L_k locally around some point $x^k \in P$. Then, the step-size procedure in the above algorithm stops with some finite number m_k such that*

$$\alpha_k = \eta^{m_k}\overline{\alpha} \geq \min\left\{\overline{\alpha}; \frac{2(1-\theta)}{(1-\sigma_k)L_k}\right\}.$$

Proof Using the necessary optimality conditions for convex optimization applied to the projection problem in the above algorithm we obtain

$$(x^k - \alpha \nabla w_{\sigma_k}(x^k) - z^k(\alpha))^\top (x^k - z^k(\alpha)) \le 0$$

for $\alpha > 0$ implying

$$\|z^k(\alpha) - x^k\|^2 \le \alpha \nabla w_{\sigma_k}(x^k)(x^k - z^k(\alpha)). \tag{4.16}$$

$\nabla w_{\sigma_k}(\cdot)$ is locally Lipschitz continuous around x^k with Lipschitz modulus $(1 + \sigma_k)L_k$ and, hence, assuming that $\overline{\alpha}$ is small enough, $z^k(\alpha)$ belongs to the respective neighborhood for all $\alpha \le \overline{\alpha}$. Then, using a Taylor series for $w_{\sigma^k}(\cdot)$ we derive

$$\begin{aligned} w_{\sigma_k}(z^k(\alpha)) &\le w_{\sigma_k}(x^k) + \nabla w_{\sigma_k}(x^k)(z^k(\alpha) - x^k) + \frac{(1+\sigma_k)L_k}{2}\|z^k(\alpha) - x^k\|^2 \\ &\le w_{\sigma_k}(x^k) + \frac{1 - L_k(1+\sigma_k)\alpha}{2} \nabla w_{\sigma_k}(x^k)(z^k(\alpha) - x^k) \end{aligned}$$

using (4.16). This implies (4.15) for $\frac{1-L_k(1+\sigma_k)\alpha}{2} \le \theta$ and the assertion follows due to $\alpha_k \le \overline{\alpha}$ by construction. □

Theorem 4.5 (Solodov [292]) *Let $P \subseteq \mathbb{R}^n$ be closed and convex, $F, f : \mathbb{R}^n \to \mathbb{R}$ be convex and differentiable with locally Lipschitz continuous gradients and assume that both functions are bounded from below on P by $\overline{F}$ respectively $\overline{f}$. Assume that the solution set S of problem* (4.2) *is not empty and bounded. Assume further that*

$$\lim_{k\to\infty} \sigma_k = 0, \sum_{k=0}^{\infty} \sigma_k = \infty.$$

Then

$$\lim_{k\to\infty} d(x^k, S) = 0,$$

where $\{x^k\}_{k=1}^{\infty}$ is the sequence computed by the above algorithm and $d(x^k, S) := \inf\{\|x^k - z\| : z \in S\}$ *is the distance function for the set S.*

Proof We will give only some main steps of the proof here, the complete proof can be found in the paper Solodov [292].

By (4.15) we derive

$$\begin{aligned} \theta \nabla w_{\sigma_k}(x^k)(x^k - x^{k+1}) &\le w_{\sigma_k}(x^k) - w_{\sigma_k}(x^{k+1}) \\ &= \sigma_k(F(x^k) - \overline{F}) - \sigma_k(F(x^{k+1}) - \overline{F}) + (f(x^k) - \overline{f}) \\ &\quad - (f(x^{k+1}) - \overline{f}). \end{aligned}$$

Summing up these inequalities the inequality

$$\begin{aligned}
&\theta \sum_{k=0}^{\overline{k}} \nabla w_{\sigma_k}(x^k)(x^k - x^{k+1}) \\
&\quad \le \sigma_0(F(x^0) - \overline{F}) + \sum_{k=0}^{\overline{k}-1} (\sigma_{k+1} - \sigma_k)(F(x^{k+1}) - \overline{F}) \\
&\quad\quad - \sigma_{\overline{k}}(F(x^{\overline{k}+1}) - \overline{F}) + (f(x^0) - \overline{f}) - (f(x^{\overline{k}+1}) - \overline{f}) \\
&\quad \le \sigma_0(F(x^0) - \overline{F}) + (f(x^0) - \overline{f})
\end{aligned}$$

is obtained since $F(x^k) \ge \overline{F}$, $f(x^k) \ge \overline{f}$ by $x^k \in P$ and $0 < \sigma_{k+1} \le \sigma_k$ for all k. Hence,

$$\lim_{k\to\infty} \nabla w_{\sigma_k}(x^k)(x^k - x^{k+1}) = 0. \tag{4.17}$$

Take any sequence $\{x^k\}_{k=0}^{\infty}$ computed by the algorithm and assume that this set is bounded, i.e. it has accumulation points $\overline{x}$. We will first show that $\overline{x} \in \Psi$. Let L be a uniform Lipschitz constant for the functions F, f. Then, by Theorem 4.4, and $\sigma_k \le \sigma_0$ we obtain

$$\alpha_k \ge \min\left\{\overline{\alpha}; \frac{2(1-\theta)}{(1-\sigma_0)L}\right\}\ \forall k. \tag{4.18}$$

Hence, using (4.16) and (4.17) $(x^k - x^{k+1}) \to 0$ for $k \to \infty$ follows, i.e.

$$x^k - \Pi_P(x^k - \alpha_k(\sigma_k \nabla F(x^k) + \nabla f(x^k))) \to 0 \text{ for } k \to \infty. \tag{4.19}$$

Due to continuity of the projection operator, convergence of σ_k to zero and boundedness of α_k from zero (by $\beta > 0$) we obtain

$$\overline{x} = \Pi_P(\overline{x} - \widehat{\alpha}\nabla f(\overline{x}))$$

which implies $\overline{x} \in \Psi$.

Take $\widehat{x} \in S$. Then, by convexity of F, f and $S \subseteq \Psi$ implying $f(\widehat{x}) \le f(x^k)$ we derive

$$\nabla w_{\sigma_k}(x^k)(\widehat{x} - x^k) \le \sigma_k(F(\widehat{x}) - F(x^k)). \tag{4.20}$$

Further

$$\|x^{k+1} - \widehat{x}\|^2 = \|x^k - \widehat{x}\|^2 - \|x^{k+1} - x^k\|^2 + 2(x^{k+1} - x^k)^\top (x^{k+1} - \widehat{x})$$

by direct calculation.

We have

$$
\begin{aligned}
&(x^{k+1}-x^k)^\top(x^{k+1}-\widehat{x})\\
&=(x^{k+1}-x^k+\alpha_k\nabla w_{\sigma_k}(x^k))^\top(x^{k+1}-\widehat{x})-\alpha_k\nabla w_{\sigma_k}(x^k)(x^{k+1}-\widehat{x})\\
&\le -\alpha_k\nabla w_{\sigma_k}(x^k)(x^{k+1}-\widehat{x})\\
&=\alpha_k\nabla w_{\sigma_k}(x^k)(x^k-x^{k+1})+\alpha_k\nabla w_{\sigma_k}(x^k)(\widehat{x}-x^k)\\
&\le \overline{\alpha}\nabla w_{\sigma_k}(x^k)(x^k-x^{k+1})+\alpha_k\sigma_k(F(\widehat{x})-F(x^k)),
\end{aligned}
$$

where the first inequality comes from projection and the last one is (4.20). Combining the last two inequalities gives

$$
\|x^{k+1}-\widehat{x}\|^2\le\|x^k-\widehat{x}\|^2+2\overline{\alpha}\nabla w_{\sigma_k}(x^k)(x^k-x^{k+1})+2\alpha_k\sigma_k(F(\widehat{x})-F(x^k)).
$$

If $F(x^k)\ge F(\widehat{x})$ for all k (for the opposite case the interested reader is referred to the original paper Solodov [292]) and using (4.17) this implies that the sequence $\{\|x^k-\widehat{x}\|^2\}_{k=1}^\infty$ converges to zero. Thus, $\{x^k\}_{k=1}^\infty$ is bounded. Using this it can be shown that $\liminf_{k\to\infty}F(x^k)=F(\widehat{x})$. Repeating the above with an accumulation point $\widetilde{x}$ of the bounded sequence $\{x^k\}_{k=1}^\infty$ such that $F(\widetilde{x})=F(\widehat{x})$ we obtain that the sequence $\{\|x^k-\widetilde{x}\|^2\}_{k=1}^\infty$ converges to zero. This then implies that the whole sequence $\{x^k\}_{k=1}^\infty$ converges to $\widetilde{x}\in S$. □

4.2 A Penalty Function Approach to Solution of a Bilevel Variational Inequality

In this section, an approach to the solution of a mathematical program with variational inequality or nonlinear complementarity constraints is presented. It consists of a variational re-formulation of the optimization criterion and looking for a solution of the thus obtained variational inequality among the points satisfying the initial variational constraints. The main part of this section is an extension of the previous work of Kalashnikov and Kalashnikova [172].

4.2.1 Introduction

The problem of solving a mathematical program with variational inequalities or complementarity conditions as constraints arises quite frequently in the analysis of physical and socio-economic systems. According to a remark in the recent paper by Harker and Choi [137], these problems are usually solved using heuristic methods. The authors of the this paper [137] present an exterior-point penalty method based on Smith's optimization formulation of the finite-dimensional variational inequality problem (Smith [290]). Outrata in his paper [261] also studies this type of optimization problems.

In Sect. 4.2.2, we examine conditions under which the set of the feasible points is non-empty, and compare these conditions with those established previously in Rockafellar [272]. Section 4.2.3 describes a penalty function method to solve the bilevel problem after having reduced it to a single variational inequality with a penalty parameter.

4.2.2 An Existence Theorem

Let X be a nonempty closed convex subset of $\mathbb{R}^n$ and h a continuous mapping from X into $\mathbb{R}^n$. Suppose that h is *pseudo-monotone* with respect to X, i.e.,

$$(x-y)^\top h(y) \geq 0 \ \text{ implies } \ (x-y)^\top h(x) \geq 0 \ \ \forall x, y \in X, \tag{4.21}$$

and that there exists a vector $x^0 \in X$ such that

$$h(x^0) \in \text{int } (0^+X)^+, \tag{4.22}$$

where int X denotes the interior of the set X. Here 0^+X is the recession cone of the set X, i.e., the subset of all directions $s \in \mathbb{R}^n$ such that $X + s \subset X$, at last, C^+ is the dual cone of $C \subseteq \mathbb{R}^n$, i.e.

$$C^+ = \{y \in \mathbb{R}^n : y^\top x \geq 0 \quad \forall x \in C\}. \tag{4.23}$$

Hence, condition (4.22) implies that the vector $h(x^0)$ lies in the interior of the dual to the recession cone of the set X.

Under these assumptions, the following result is obtained:

Proposition 4.1 *Assume that the mapping h is continuous, pseudo-monotone over X and inclusion* (4.22) *holds. Then the variational inequality problem (VI):*

$$\textit{Find a vector } z \in X \textit{ such that } (x-z)^\top h(z) \geq 0 \quad \forall x \in X, \tag{4.24}$$

has a nonempty, compact, convex solution set.

Proof It is well-known (Karamardian [184]) that pseudo-monotonicity (4.21) and continuity of the mapping h imply convexity of the solution set

$$Z = \{z \in X : (x-z)^\top h(z) \geq 0 \quad \forall x \in X\}, \tag{4.25}$$

of problem (4.24). Now we show the existence of at least one solution to problem (4.24). In order to do that, we use the following result from Eaves [97]: if there exists a nonempty bounded subset D of X such that for every $x \in X \setminus D$ there is a $y \in D$ with

$$(x - y)^\top h(x) > 0, \tag{4.26}$$

then problem (4.24) has a solution. Moreover, the solution set (4.25) is bounded because $Z \subset D$. Now, we select the set D as follows:

$$D = \{ x \in X\colon \ (x - x^0)^\top h(x^0) \le 0 \} \tag{4.27}$$

for an arbitrary fixed point $x^0 \in X$. The set D is clearly nonempty, since it contains the point x^0. Now we show that D is bounded, even if X is unbounded. Suppose on the contrary that a sequence $\{x^k\}_{k=1}^{\infty} \subseteq D$ exists with $\lim_{k\to\infty} \|x^k - x^0\| = \infty$. Without loss of generality, assume that $x^k \ne x^0, \ k = 1, 2, \ldots,$ and consider the inequality

$$\frac{(x^k - x^0)^\top h(x^0)}{\|x^k - x^0\|} \le 0, \qquad k = 1, 2, \ldots, \tag{4.28}$$

which follows from definition (4.27) of the set D. Again not affecting generality, assume that the sequence $(x^k - x^0)/\|x^k - x^0\|_{k=1}^{\infty}$ converges to a vector $s \in \mathbb{R}^n, \ \|s\| = 1$. This implies $s \in 0^+X$ (see Rockafellar [272, Theorem 8.2]). From (4.28), we deduce the limit relationship

$$s^\top h(x^0) \ \le \ 0. \tag{4.29}$$

Since $0^+X \ne \{0\}$ (as X is unbounded and convex), we conclude that 0 is a boundary point of the dual cone $(0^+X)^+$, hence $h(x^0) \ne 0$. Now it is easy to see that inequality (4.29) contradicts assumption (4.22). Indeed, the inclusion $h(x^0) \in \text{int } (0^+X)^+$ implies that $s^\top h(x^0) > 0$ for any $s \in 0^+X, s \ne 0$. This contradiction proves the boundedness of the set D, and therewith the statement of Proposition 4.1. In effect, for a given $x \in X \backslash D$, one can pick $y = x^0 \in D$ with the inequality $(x - y)^\top h(y) > 0$ being valid. The latter inequality, together with the pseudo-monotonicity of the mapping h, yields the required condition (4.24) and thus completes the proof. □

Remark 4.2 The assertion of Proposition 4.1 was obtained earlier in Harker and Pang [138] under the same assumptions except for the inclusion (4.22), which is obviously invariant with respect to an arbitrary translation of the set X followed by the corresponding transformation of the mapping h. Instead of (4.22), Harker and Pang [138] used another assumption $h(x^0) \in \text{int}(X^+)$ which is clearly non-translation-invariant. Moreover, it is easy to verify that for any convex and closed subset $X \subseteq \mathbb{R}^n$ one has the inclusion

$$X^+ \subseteq (0^+X)^+, \tag{4.30}$$

which clearly means that condition (4.22) is weaker than that in the paper [138] by Harker and Pang.

Now suppose additionally that the solution set Ψ of problem (4.24) contains more than one element, and consider the following variational inequality problem:

$$\text{Find a vector } z^* \in \Psi \text{ such that } (z - z^*)^\top H(z^*) \geq 0 \qquad \text{for all} \quad z \in \Psi. \tag{4.31}$$

Here, the mapping $H : X \to \mathbb{R}^n$ is continuous and strictly monotone over X, i.e.,

$$(x - y)^\top [H(x) - H(y)] > 0 \qquad \forall x, y \in X, x \neq y. \tag{4.32}$$

In this case, the compactness and convexity of the set Ψ guarantees (cf. Eaves [97]) the existence of a unique (due to strict monotonicity of H) solution z of problem (4.31). We refer to the combined problem (4.24), (4.25) and (4.31) as the *lexicographical variational inequality* (*LVI*). This problem is similar to problem (4.2). In the next subsection, we present a penalty function algorithm solving the LVI *without* an explicit construction of the lower level solution set Ψ.

4.2.3 The Penalty Function Method

Let us fix a positive parameter value $\varepsilon > 0$ and consider the following *parametric* variational inequality problem:

$$\text{Find a vector } x_\varepsilon \in X \text{ such that } (x - x_\varepsilon)^\top [h(x_\varepsilon) + \varepsilon H(x_\varepsilon)] \geq 0 \; \forall \, x \in X. \tag{4.33}$$

If we assume that the mapping h is monotone over X, i.e.

$$(x - y)^\top [h(x) - h(y)] \geq 0 \qquad \forall x, y \in X, \tag{4.34}$$

and keep intact all the assumptions from the previous section regarding h, H and Ψ, then the following result is obtained. Note the similarity to the algorithm in Sect. 4.1.3.

Proposition 4.2 *Let the mapping h in the lower level problem be continuous and monotone over X, condition* (4.22) *be valid, whereas the (upper level) mapping H be continuous and strictly monotone on X. Then, for each sufficiently small value $\varepsilon > 0$, problem* (4.33) *has a unique solution x_ε. Moreover, x_ε converges to the solution z^* of LVI* (4.24), (4.25) *and* (4.31) *when $\varepsilon \to 0$.*

Proof Since h is monotone and H is strictly monotone, the mapping $\Phi_\varepsilon = G + \varepsilon F$ is strictly monotone on X for any $\varepsilon > 0$. It is also clear that, if some $x^0 \in X$ satisfies (4.22), then the following inclusion holds

$$\Phi_\varepsilon(x^0) = h(x^0) + \varepsilon H(x^0) \in \text{int } (0^+X)^+, \tag{4.35}$$

whenever $\varepsilon > 0$ is small enough. Hence, Proposition 4.1 implies the validity of the first assertion of Proposition 4.2, namely, for every $\varepsilon > 0$ satisfying (4.35), the variational inequality (4.33) has a unique solution x_ε.

From continuity of H and h it follows that each (finite) limit point $\overline{x}$ of the generalized sequence $Q = \{x_\varepsilon\}_{\varepsilon>0}$ (the net of solutions to problem (4.33)) solves the variational inequality (4.24). That is, $\overline{x} \in \Psi$. Now we prove that the point $\overline{x}$ solves problem (4.31), too. In order to do that, we use the following relationships valid for any $z \in \Psi$ due to (4.25), (4.33) and (4.34):

$$(z - x_\varepsilon)^\top [h(z) - h(x_\varepsilon)] \geq 0, \tag{4.36}$$

$$(z - x_\varepsilon)^\top h(z) \leq 0, \tag{4.37}$$

$$(z - x_\varepsilon)^\top h(x_\varepsilon) \geq -\varepsilon (z - x_\varepsilon)^\top H(x_\varepsilon). \tag{4.38}$$

Subtracting (4.38) from (4.37) and making use of (4.36), we obtain the following inequalities

$$0 \leq (z - x_\varepsilon)^\top [h(z) - h(x_\varepsilon)] \leq \varepsilon (z - x_\varepsilon)^\top H(x_\varepsilon). \tag{4.39}$$

From (4.39) we have $(z - x_\varepsilon)^\top H(x_\varepsilon) \geq 0$ for all $\varepsilon > 0$ and $z \in \Psi$. Since H is continuous, the following limit relationship holds: $(z - \overline{x})^\top H(\overline{x}) \geq 0$ for each $z \in \Psi$, which means that $\overline{x}$ solves (4.31).

Thus, we have proved that every limit point of the generalized sequence Q (the net) solves LVI (4.24), (4.25) and (4.31) and (since the mapping H is strictly monotone) Q can have at most one limit point, i.e. the sequence Q converges to this solution provided an accumulation point exists. To complete the proof it suffices to show that the net Q is bounded. In order to do that, consider a sequence $\{x_{\varepsilon_k}\}_{k=1}^\infty$ of solutions to the parametric problem (4.33) with $\|x_{\varepsilon_k}\| \to \infty$ where $\varepsilon_k \to 0$ as $k \to \infty$. Let x^0 be the vector from condition (4.22). Without loss of generality, suppose that $x_{\varepsilon_k} \neq x^0$ for each k, and

$$\lim_{k\to\infty} \frac{(x_{\varepsilon_k} - x^0)}{\|x_{\varepsilon_k} - x^0\|} = s \in \mathbb{R}^n,\ \|s\| = 1.$$

Due to $\|x_{\varepsilon_k} - x^0\| \to \infty$ this implies $s \in 0^+X$ (see Rockafellar [272]). Since x_{ε_k} is the solution to the parametric variational inequality (4.33) with the strictly monotone mapping $\Phi_\varepsilon = h + \varepsilon H$, and $x^0 \neq x_{\varepsilon_k}$, we obtain the following inequalities for all $k = 1, 2, \ldots$:

$$\left(x_{\varepsilon_k} - x^0\right)^\top \left[h\left(x_{\varepsilon_k}\right) + \varepsilon_k H\left(x_{\varepsilon_k}\right)\right] \leq 0, \tag{4.40}$$

and

$$\left(x_{\varepsilon_k} - x^0\right)^\top \left[h\left(x^0\right) + \varepsilon_k H\left(x^0\right)\right] < 0. \tag{4.41}$$

Dividing inequality (4.41) by $\|x_{\varepsilon_k} - x^0\| \neq 0$ we derive

$$\frac{\left(x_{\varepsilon_k} - x^0\right)^\top}{\|x_{\varepsilon_k} - x^0\|} \cdot \left[h\left(x^0\right) + \varepsilon_k H\left(x^0\right)\right] \leq 0, \quad k = 1, 2, \ldots. \tag{4.42}$$

For k tending to infinity this yields $s^\top h\left(x^0\right) \le 0$. Since $s \ne 0$, the latter inequality contradicts assumption (4.22). This contradiction demonstrates that the net Q is indeed bounded thus completing the proof of Proposition 4.2. □

4.2.4 An Example

Let $\Omega \subseteq \mathbb{R}^m$, $\Lambda \subseteq \mathbb{R}^n$ be subsets of two finite-dimensional Euclidean spaces and $f : \Omega \times \Lambda \to \mathbb{R}$, $\overline{h} : \Omega \times \Lambda \to \mathbb{R}^n$ continuous mappings. Consider the following mathematical program with variational inequality (or, equilibrium) constraint (MPEC):

$$\min_{(u,v)\in\Omega\times\Lambda} f(u,v), \tag{4.43}$$

s.t.

$$\overline{h}(u,v)^\top (w-v) \ge 0, \qquad \forall w \in \Lambda. \tag{4.44}$$

If the function f is continuously differentiable, denote by $H = H(u,v)$ the gradient mapping of f, i.e., $H(u,v) = \nabla_{(u,v)} f(u,v)$. Then problem (4.43)–(4.44) is obviously tantamount to the following *bilevel variational inequality* (*BVI*):

$$\begin{aligned} &\text{Find } z^* = (u^*, v^*) \in \Omega \times \Lambda \text{ such that } \left(z - z^*\right)^\top H\left(z^*\right) \ge 0, \\ &\forall z = (u,v) \in \Omega \times \Lambda, \end{aligned} \tag{4.45}$$

subject to $v^* \in \Lambda$ being a solution of the lower level variational inequality (4.44) for $u = u^*$, that is,

$$\overline{h}\left(u^*, v^*\right)^\top (v - v^*) \ge 0, \qquad \forall v \in \Lambda. \tag{4.46}$$

It is clear that the bilevel variational inequality (BVI) (4.45)–(4.46) is equivalent to the lexicographical variational inequality (LVI) (4.24), (4.25) and (4.31), in which the gradient mapping ∇f is used as H, while $h(u,v) = [0; \overline{h}(u,v)]$ for all $(u,v) \in \Omega \times \Lambda$. It is interesting to notice that in this case, the penalty function approach also may be useful to solve the bilevel variational inequality.

Example 4.3 As an example, consider the mathematical program with equilibrium constraints (MPEC) where

$$f(u,v) = (u-v-1)^2 + (v-2)^2; \quad \overline{h}(u,v) = uv; \quad \Omega = \Lambda = \mathbb{R}^1_+. \tag{4.47}$$

Then it can be readily verified that its optimal solution is unique: $z^* = (u^*, v^*) = (1;0)$, and the penalized mapping is described as follows:

$$\Phi_\varepsilon(u,v) = \left[\varepsilon(2u - 2v - 2); uv + \varepsilon(-2u + 4v - 2)\right]. \tag{4.48}$$

Solving the variational inequality:

$$\text{Find } (u_\varepsilon, v_\varepsilon) \in \mathbb{R}^2_+ \text{ such that } \left[(u, v) - (u_\varepsilon, v_\varepsilon)\right]^\top \Phi_\varepsilon(u_\varepsilon, v_\varepsilon) \geq 0 \quad \forall (u, v) \in \mathbb{R}^2_+, \tag{4.49}$$

we obtain

$$u_\varepsilon = v_\varepsilon + 1; \quad v_\varepsilon = -\frac{1}{2} - \varepsilon + \sqrt{\left(\frac{1}{2} + \varepsilon\right)^2 + 4\varepsilon}. \tag{4.50}$$

It is easy to check that $(u_\varepsilon, v_\varepsilon) \to z^*$ when $\varepsilon \to 0$. □

Unfortunately, this is not always the case, since the mapping $h(u, v) = [0; \overline{h}(u, v)]$ is in general not monotone with respect to (u, v), even if $\overline{h} = \overline{h}(u, v)$ is monotone with respect to v for each fixed u.

Chapter 5
Mixed-Integer Bilevel Programming Problems

5.1 Location of Integrality Conditions in the Upper or Lower Level Problems

Bilevel optimization problems may involve decisions in both discrete and continuous variables. For example, a chemical engineering design problem may involve discrete decisions regarding the existence of chemical process units besides to decisions in continuous variables, such as temperature or pressure values. Problems of this kind, dealing with both discrete and continuous decision variables, are referred to as mixed-integer bilevel optimization problems (MIBLP).

A particular case of the mixed-integer bilevel program is presented by the real-world problem of minimizing the cash-out penalty costs of a natural gas shipping company (Dempe et al. [65]). This problem arises when a (gas) shipper draws a contract with a pipeline company to deliver a certain amount of gas at several delivering meters. What is actually shipped may be higher or lower than the amount that had been originally agreed upon (this phenomenon is called an imbalance). When such an imbalance occurs, the pipeline penalizes the shipper by imposing a cash-out penalty policy. As this penalty is a function of the operating daily imbalances, an important problem for the shippers is how to carry out their daily imbalances so as to minimize the incurred penalty. On the other hand, the pipeline (the follower) tries to minimize the absolute values of the cash-outs, which produce the optimal response function taken into account by the leader in order to find the optimal imbalance operating strategy. Here, integer variables are involved at the lower level problem. Various solution algorithms for the natural gas cash-out problem are described e.g. in Dempe et al. [65] or Kalashnikov et al. [117].

In general, mixed-integer BLPs can be classified into four classes (see Gümüş and Floudas [134] and also Vicente et al. [304]):

(I) **Integer upper, continuous lower**: If the sets of inner (lower level) integer and outer (upper level) continuous variables are empty, and on the contrary, the sets of outer integer and inner continuous variables are nonempty, then the MIBLP is of Type I.

S. Dempe et al., *Bilevel Programming Problems*, Energy Systems,
DOI 10.1007/978-3-662-45827-3_5

(II) **Purely integer**: If the sets of inner and outer integer variables are nonempty, and the sets of inner and outer continuous variables are empty, then the problem is a purely integer BLP.

(III) **Continuous upper, integer lower**: When the sets of inner continuous and outer integer variables are empty, and vice versa, the sets of inner integer and outer continuous variables are nonempty, then the problem is a MIBLP of Type III.

(IV) **Mixed-integer upper and lower**: If the sets of both inner and outer continuous and integer variables are nonempty, then the problem is a MIBLP of Type IV.

Advances in the solution of the mixed-integer bilevel optimization problems (MIBLP) of all four types can greatly expand the scope of decision making instances that can be modeled and solved within a bilevel optimization framework. However, little attention has been paid in the literature to both the solution and the application of BLP governing discrete variables. This is mainly because these problems pose major algorithmic challenges in the development of efficient solution strategies.

In the literature, methods developed for the solution of the MIBLP have so far addressed a very restricted class of problems. For instance, for the solution of the purely integer (Type II) linear BLP, a branch-and-bound type algorithm has been proposed by Moore and Bard [236], whereas Nishizaki et al. [255] applied a kind of genetic algorithm to the same problem. For the solution of the mixed-integer BLP of Type I, another branch-and-bound approach has been developed by Wen and Yang [314]. Cutting plane and parametric solution techniques have been elaborated by Dempe [52] to solve MIBLP with only one upper level (outer) variable involved in the lower level objective function. Bard [10] obtained upper bounds for the objective functions at both levels. Thus, he generated a non-decreasing sequence of lower bounds for the objective function at the upper level, which, under certain conditions, converges to the solution of the general BLPP with continuously differentiable functions. Methods based upon decomposition technique have been proposed by Saharidis and Ierapetritou [277] and Zhang and Wu [329]. Further solution algorithms can be found in Domínguez and Pistikopoulos [94] and Mitsos [233].

Mixed-integer *nonlinear* bilevel optimization problems have received even less attention in the literature. The developed methods include an algorithm making use of parametric analysis to solve separable monotone nonlinear MIBLP proposed by Jan and Chern [154], a stochastic simulated annealing method presented by Sahin and Ciric [278], a global optimization approach based on parametric optimization technique published by Faísca et al. [103]. Additionally, Gümüş and Floudas [134] and Floudas et al. [117] developed several algorithms dealing with global optimization of mixed-integer bilevel optimization problems of both deterministic and stochastic nature. The sensitivity analysis for MIBLPP was also considered by Wendel [315]. Xu and Wang [320] formulate a branch-and-bound algorithm solving mixed-integer linear bilevel optimization problems.

In Dempe and Kalashnikov [62] and Dempe et al. [63], we already started considering and solving mixed-integer linear BLP of Type I. In particular cases, a BLP can be reduced to solving a multiobjective optimization problem, which is efficiently processed by Liu and Wang in [205]. Bilevel optimization problems with discrete variables are also examined by Hu et al. in [148].

The existence of an algorithm which solves the linear mixed-integer bilevel optimization problem of type I has been shown by Köppe et al. in [192] under the assumption that all data are integral and the lower level problem has parameters in the right-hand side only. This algorithm runs in polynomial time when the number of variables in the lower level is fixed. If the bilevel problem has feasible solutions, this algorithm also decides if the infimal objective function value is attained. In the same paper, the existence of a polynomial time algorithm has been shown for problems of type II if the number of all variables in fixed.

If discrete bilevel optimization problems are investigated at least two new aspects appear. Firstly, the feasible set mapping $x \mapsto Y(x)$ is not upper semicontinuous in general. To see this, consider an integer optimization in the lower level where the right-hand side of the constraints depend on the parameter: $Ay \leq x$. Then, if this parameter x converges to $\overline{x}$ a new integer vector can become feasible. This violates upper semicontinuity of the feasible set mapping Y. In Remark 3.2 this property was used to show lower semicontinuity of the function $\varphi(\cdot)$ which in turn was an essential assumption for existence of an optimal solution of the optimistic bilevel optimization problem. Hence, an optimal solution of the optimistic bilevel optimization problem need not to exist in the discrete case. To circumvent this difficulty, we need to define a weak optimal solution in Sect. 5.3.

The second surprising fact is that global optimal solutions of continuous relaxations of the bilevel optimization problem need not to be global optimal solutions of the discrete problem even if they are feasible. This is shown in the next example:

Example 5.1 (*Moore and Bard* [253]) Consider the problem

$$\min_x \{-10y - x : y \in \Psi_D(x),\ x \text{ integer}\},$$

where the lower level problem is

$$\begin{aligned}\Psi_D(x) = \operatorname*{Argmin}_y \{y : & \ 20y - 25x \leq 30, 2y + x \leq 10 \\ & - y + 2x \leq 15, 10y + 2x \geq 15, y \text{ integer}\}.\end{aligned}$$

The feasible set of this problem as well as its continuous relaxation are shown in Fig. 5.1.

The thick line in Fig. 5.1 is the graph of the solution set mapping of the continuous relaxation of the lower level problem. The unique global optimal solution of minimizing the objective function of the upper level problem on this set is the point $(\widehat{x}, \widehat{y}) = (1, 8)^\top$. This point is feasible for the discrete bilevel optimization problem but it is not a global optimal solution of this problem. The unique global optimal solution of the discrete bilevel problem is found at $(x^*, y^*) = (2, 2)^\top$. The upper level objective function value of $(\widehat{x}, \widehat{y})$ is $F(\widehat{x}, \widehat{y}) = -18$ and that of (x^*, y^*) is $F(x^*, y^*) = -22$. □

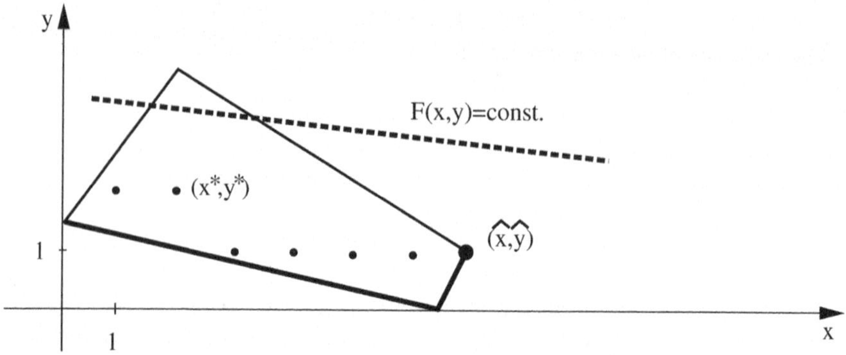

Fig. 5.1 Feasible set in Example 5.1. *Thick line* is the feasible set of the continuous relaxation

After defining the weak optimal solution in Sect. 5.3 a first algorithm for computing such a solution is formulated.

Optimality conditions for discrete bilevel optimization can be obtained using the notion of a radial-directional derivative or a radial subdifferential. This is done in Sect. 5.4. Later, Sect. 5.5 is devoted to the formulation of a solution algorithm using an upper estimation of the optimal value function of the lower level problem as it has also been done in Sect. 2.3.2. This idea will again be used in Sect. 5.6.4.

5.2 Knapsack Constraints

In Example 1.4 a linear bilevel optimization problem with a continuous knapsack problem in the lower level has been considered. Now, let the parameterized zero-one knapsack problem be the lower level problem:

$$\Psi(x) := \operatorname*{Argmax}_y \{f(x,y) := c^\top y : a^\top y \le x,\ y_i \in \{0,1\},\ i = 1,\dots,n\} \tag{5.1}$$

and consider the bilevel problem

$$\text{“}\max_x\text{”}\{F(x,y) := d^\top y + tx : y \in \Psi(x),\ \underline{x} \le x \le \overline{x}\}, \tag{5.2}$$

where $a_i > 0$, $c_i > 0$ are integer coefficients, $i = 1,\dots,n$, $0 \le \underline{x} < \overline{x} \le \sum_{i=1}^n a_i$ have integer values and $d \in \mathbb{R}^n$. This problem has been investigated by Dempe and Richter in [75] and by Brotcorne et al. in [33]. Note that we consider here the maximization problems since we will apply dynamic programming to solve them which is an efficient algorithm for the solution of the $\mathcal{NP}$-hard Boolean knapsack problem. The Boolean knapsack problem can be solved by dynamic programming

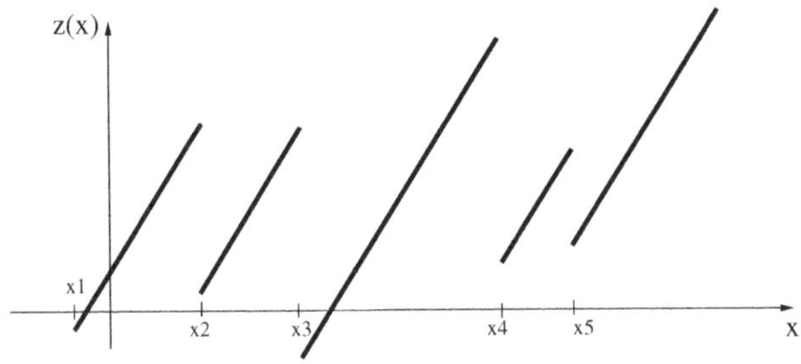

Fig. 5.2 The mapping $x \to \bigcup_{y\in\Psi(x)}\{z(x)\}$ with $z(x) = d^\top y + tx$

with a running time of $O(nx)$ (Nemhauser and Wolsey [251]), since $x \le \overline{x}$ this leads to an $O(n\overline{x})$ running time algorithm for problem (5.1).

The point-to-set mapping $x \to \bigcup_{y\in\Psi(x)}\{d^\top y + tx\}$ is discontinuous, as can be seen in see Fig. 5.2.

In Fig. 5.2, $x1, \ldots, x5 \in \left\{\sum_{i\in I} a_i : I \subseteq \{1,\ldots,n\}\right\}$, since the feasible set of problem (5.1) changes at those points, and the slopes of the lines equals t. If $\Psi(x)$ does not reduce to a singleton then, the optimistic function value $\varphi_o(x)$ is attained at the top of the lines, the pessimistic one $\varphi_p(x)$ at the bottom. If $\Psi(x)$ is a singleton for increasing x and $z(\cdot)$ is discontinuous there, the function value of $z(x)$ is attained at the line starting at point x in Fig. 5.2. This implies the existence of both an optimistic and a pessimistic solution of problem (5.2) for $t \le 0$.

Theorem 5.1 (Dempe and Richter [75]) *Let $\overline{x} \le \sum_{i=1}^n a_i$. Then, optimistic and pessimistic solutions exist for problem* (5.2) *if $t \le 0$. If $t > 0$ then, problem* (5.2) *has no (optimistic or pessimistic) solution or $(\overline{x}, \overline{y})$ with $\overline{y} \in \Psi(\overline{x})$ is optimistic or pessimistic optimal.*

Proof Note that $\{y : a^\top y \le x,\ y_i \in \{0,1\},\ i = 1,\ldots,n,\ y \in \Psi(x) \text{ for some } x\}$ is a finite set and the *region of stability* $\mathscr{R}(y)$ is an interval for each such y.

1. First, let $t \le 0$. Let $\alpha = \sup\{d^\top y + tx : \underline{x} \le x \le \overline{x},\ y \in \Psi(y)\}$. $\overline{z}(x) = d^\top\overline{y} + tx$ with $\overline{y} \in \Psi(x)$ is piecewise affine linear and there exist points $x^i \in [\underline{x}, \overline{x}]$ with $\underline{x} = x^1 < x^2 < \cdots < x^p = \overline{x}$ such that $\overline{z}(x)$ is affine linear and decreasing on $[x^i, x^{i+1})$, $i = 1,\ldots,p-1$. $\mathscr{R}(\overline{y}) = [x^i, x^{i+1})$ or $\mathscr{R}(\overline{y}) = [x^i, x^{i+1}]$ for some Boolean vector $y \in \{0,1\}^n$. Hence, $\alpha = z(x^i)$. If $|\Psi(x^i)| = 1$, then, $\varphi_o(z^i) = \varphi_p(x^i) = z(x^i)$ and the solution exists. If the optimal solution of problem (5.1) is not unique for $x = x^i$, then $\alpha = \max\{d^\top y^k + tx^i : y^k \in \Psi(x^i)\} = \varphi_0(x^i)$ and the optimistic solution exists.
2. $\varphi_p(\cdot)$ is piecewise affine linear and decreasing on the intervals $[x^i, x^{i+1})$ where it is continuous. For each interval $[x^i, x^{i+1})$ there is y^i with $\varphi_p(x) = d^\top y^i + tx$ and $\sup\{\varphi_p(x) : x \in [x^i, x^{i+1})\} = \varphi_p(x^i)$. Hence, the solution exists in this case, too.

3. Now, assume $t > 0$. Then, all the functions $x \mapsto d^\top \overline{y} + tx$ with $x \in \mathscr{R}(y)$ are increasing. If $\Psi(x)$ is a singleton for all x, either $\varphi_o(x) = \varphi_p(x)$ is increasing or there exist points $\widehat{x} \in (\underline{x}, \overline{x})$ with $\lim_{x \uparrow \widehat{x}} \varphi_o(x) > \varphi_o(\widehat{x})$. In the first case, this is the optimal solution. In the second case either $(\overline{y}, \overline{x})$ is a solution or the problem does not have one.
4. At the end, let $\Psi(x)$ contain more than one point for certain x and consider the function $\varphi_o(\cdot)$ or $\varphi_p(\cdot)$. If this function is increasing on $[\underline{x}, \overline{x}]$, the point $(\overline{x}, \overline{y})$ is optimistic or pessimistic optimal. Otherwise, consider a point $\widehat{x}$ with $\lim_{x \uparrow \widehat{x}} \varphi_o(x) \geq \varphi_o(\widehat{x})$. If $\varphi_o(\cdot)$ is not increasing around $\widehat{x}$, $\Psi(\widehat{x}) \setminus \Psi(x) \neq \emptyset$ for $x < \widehat{x}$ close to $\widehat{x}$ and we have a strict inequality $\lim_{x \uparrow \widehat{x}} \varphi_o(x) > \varphi_o(\widehat{x})$. Then, we can repeat part 3. The same is true for $\varphi_p(\cdot)$. □

Weak optimistic and pessimistic solutions exist for all t, see Sect. 5.3.

An example in Brotcorne et al. [33] can be used to show that an optimal solution for the bilevel optimization problem with Boolean knapsack constraints need not to exist.

Example 5.2 (*Brotcorne et al.* [33]) Consider the problem

$$\max_{x,y}\{5y_1 + y_2 + y_3 + y_4 + x : 1 \leq x \leq 3,\ y \in \Psi(x)\}$$

with

$$\Psi(x) = \operatorname*{Argmax}_y \{F(x, y) = 4y_1 + 5y_2 + 10y_3 + 15y_4 :$$
$$y_1 + 2y_2 + 3y_3 + 4y_4 \leq x,\ y_i \in \{0, 1\},\quad i = 1, 2, 3, 4\}.$$

For $x \in [1, 2)$ the optimal solution of the lower level problem is $y = (1,\ 0,\ 0,\ 0)^\top$, and $y = (0,\ 1,\ 0,\ 0)^\top$ is the optimal solution for $x \in [2, 4)$. Hence, the leader's objective function value is

$$F(x, y) = \begin{cases} 5 + x, & x \in [1, 2), \\ 1 + x, & x \in [2, 3), \\ 6, & x = 3. \end{cases}$$

The upper level objective function is bounded from above by 7, but the value of 7 is not attained. Hence, the problem does not have an optimal solution. □

For solving the problem (5.1), (5.2) in the optimistic case (maximization of the objective function in problem (5.2) with respect to both x, y), and in the pessimistic case a dynamic programming algorithm is applied. For this, the upper level problem is replaced with

$$\max_{x,y}\{F(x,y) := d^\top y + tx : y \in \Psi(x),\ \underline{x} \le x \le \overline{x},\ \text{integer}\} \tag{5.3}$$

or

$$\max_{x}\min_{y}\{F(x,y) := d^\top y + tx : y \in \Psi(x),\ \underline{x} \le x \le \overline{x},\ \text{integer}\}, \tag{5.4}$$

respectively. Let

$$\varphi_k(x) = \max_y\{\sum_{i=1}^k c_i y_i : \sum_{i=1}^k a_i y_i \le x,\ y_i \in \{0,1\},\quad i = 1,\dots,k\}$$

and

$$\widetilde{F}_k(x) = \max_x\{\sum_{i=1}^k d_i y_i : y \in \Psi(x)\}.$$

Algorithm: (Solution of the bilevel optimization problem with 0–1 knapsack problem in the lower level)

1. For all x, set

$$\varphi_k(x) = \begin{cases} 0, & \text{if} \quad x \le a_1 - 1, \\ c_1, & \text{if} \quad a_1 \le x \le \overline{x}, \end{cases} \qquad \widetilde{F}_k(x) = \begin{cases} 0, & \text{if} \quad x \le a_1 - 1, \\ d_1, & \text{if} \quad a_1 \le x \le \overline{x}. \end{cases}$$

2. For $k = 2$ to n do
   for $x = 0$ to $\overline{x}$ do
   if $x < a_k$ then

$$\varphi_k(x) = \varphi_{k-1}(x), \qquad \widetilde{F}_k(x) = \widetilde{F}_{k-1}(x).$$

   otherwise

$$\varphi_k(x) = \max\{\varphi_{k-1}(x), \varphi_{k-1}(x - a_k) + c_k\}.$$

   If $\varphi_{k-1}(x) \ne \varphi_{k-1}(x - a_k) + c_k$ set

$$\widetilde{F}_k(x) = \begin{cases} \widetilde{F}_{k-1}(x), & \text{if} \quad \varphi_k(x) = \varphi_{k-1}(x), \\ \widetilde{F}_{k-1}(x - a_k) + d_k, & \text{if} \quad \varphi_k(x) = \varphi_{k-1}(x - a_k) + c_k. \end{cases}$$

   If $\varphi_{k-1}(x) = \varphi_{k-1}(x - a_k) + c_k$ set

$$\widetilde{F}_k(x) = \max\{\widetilde{F}_{k-1}(x), \widetilde{F}_{k-1}(x - a_k) + d_k\}$$

   in the optimistic and

$$\widetilde{F}_k(x) = \min\{\widetilde{F}_{k-1}(x), \widetilde{F}_{k-1}(x - a_k) + d_k\}$$

```
in the pessimistic case.
end if,  end for.
```

Theorem 5.2 (Brotcorns et al. [33]) *If* $t \le 0$, *the point* (x^*, y^*) *with*

$$\widetilde{F}_n(x^*) + tx^* = \max\{\widetilde{F}_n(x) + tx : \underline{x} \le x \le \overline{x},\ integer\}$$

is an optimal solution for the optimistic or the pessimistic bilevel optimization problem, respectively.

Since a_i are integer coefficients and (by $t \le 0$) the value of x is as small as possible, it is also integral and we can restrict us to the fully discrete problem.

Example 5.3 Consider the problem

$$\max_{x,y}\{3y_1 + 5y_2 + y_3 + 9y_4 - 2x : 0 \le x \le 6,\ y \in \Psi(x)\},$$

where

$$\Psi(x) = \underset{y}{\text{Argmax}}\ \{y_1+y_2+y_3+y_4 : 5y_1+3y_2+2y_3+y_4 \le x,\ y_i \in \{0,1\},\ \ i = 1,2,3,4\}.$$

Then, Table 5.1 is computed using the above algorithm.

Using Theorem 5.2, the optimal objective function value for the upper level problem is

$$\max\{0-0, 9-2, 9-4, 10-6, 14-8, 14-10, 15-12\} = 7.$$

Hence, $x^* = 1$, and $y^* = (0,\ 0,\ 0,\ 1)^\top$ and the optimal function values of the lower and upper level problems are 1 and 7. □

Table 5.1 Realization of the dynamic programming algorithm

x	k							
	y_1		y_2		y_3		y_4	
	$\widetilde{F}_1$	φ_1	$\widetilde{F}_2$	φ_2	$\widetilde{F}_3$	φ_3	$\widetilde{F}_4$	φ_4
0	0	0	0	0	0	0	0	0
1	0	0	0	0	0	0	1	9
2	0	0	0	0	1	1	1	9
3	0	0	1	5	1	5	2	10
4	0	0	1	5	1	5	2	14
5	1	3	1	5	2	6	2	14
6	1	3	1	5	2	6	3	15

5.3 Weak Solution

In this section, we investigate the bilevel optimization problem (1.1), (1.2) with $G(x) \equiv 0$, some nonempty, compact set $X \subseteq \mathbb{R}^n$ and a nonempty and discrete set $T \subseteq \mathbb{R}^k$. I.e. there exists a number $\omega > 0$ with $\|y^1 - y^2\| > \omega$ for all $y^1, y^2 \in T$. Constraint $G(x) \leq 0$ can be added to the upper level. The results presented in this chapter are taken from Fanghänel and Dempe [106], see also Fanghänel [105].

Let

$$\overline{T} := \{y \in T : \exists x \in X \text{ with } y \in \Psi(x)\}. \tag{5.5}$$

Throughout the section we need the following general assumptions:

(A1) There exists at least one optimal solution for the follower's problem for all parameter selections: $\Psi(x) \neq \emptyset$ for all $x \in X$.

(A2) The set $\overline{T}$ is finite: $\text{card}(\overline{T}) < \infty$.

5.3.1 Regions of Stability

Regions of stability have been used in Sect. 2.3 for the formulation of a descent algorithm to solve a linear bilevel optimization problem. They are sets of all parameters for which a fixed feasible solution of the lower level problem is (global) optimal. Here, they are formulated as follows:

Definition 5.1 The set

$$\mathscr{R}(y) := \{x \in X : \ y \in \Psi(x)\}$$

is called *region of stability* for $y \in T$.

Consequently, $\mathscr{R}(y) \neq \emptyset$ if and only if $y \in \overline{T}$. Assumption (A1) induces

$$X = \bigcup_{y \in \overline{T}} \mathscr{R}(y). \tag{5.6}$$

The regions of stability have an equivalent formulation given in

Lemma 5.1 *For all $y \in \overline{T}$ it holds*

$$\mathscr{R}(y) = \{x \in X : g(x, y) \leq 0, \ (g(x, \overline{y}) \not\leq 0 \, or \, f(x, y) \leq f(x, \overline{y})) \ \forall \overline{y} \in \overline{T}\}.$$

Proof Consider a point (x, y) with $y \in \overline{T}$ and $x \in \mathscr{R}(y)$. Then, by Definition 5.1, $y \in \Psi(x)$ and hence, $g(x, y) \leq 0$. Further, no feasible point $\overline{y} \in \overline{T}$ may have a better function value than y, i.e. we have

$$\mathscr{R}(y) \subseteq \{x \in X : g(x, y) \leq 0, \ (g(x, \overline{y}) \not\leq 0 \text{ or } f(x, y) \leq f(x, \overline{y})) \ \forall \overline{y} \in \overline{T}\}.$$

Now, consider a point $(x, y) \in X \times \overline{T}$ satisfying

$$g(x, y) \leq 0, \ (g(x, \overline{y}) \not\leq 0 \text{ or } f(x, y) \leq f(x, \overline{y})) \, \forall \overline{y} \in \overline{T}. \tag{5.7}$$

We have to show $x \in \mathscr{R}(y)$. Due to assumption (A1), there exists $\hat{y} \in \overline{T}$ with $\hat{y} \in \Psi(x)$. Thus, the point $\hat{y} \in \overline{T}$ is feasible and $f(x, \hat{y}) \leq f(x, y)$. Furthermore, due to the properties of (x, y) in Eq. (5.7), we have $f(x, \hat{y}) \geq f(x, y)$. Consequently, $f(x, \hat{y}) = f(x, y)$ and $g(x, y) \leq 0$, $g(x, \hat{y}) \leq 0$. Thus, if $\hat{y}$ is global optimal, the point y is global optimal, too. Hence, $y \in \Psi(x)$ and thus $x \in \mathscr{R}(y)$. Consequently, the lemma is true. □

Obviously, by the definition of the set $\overline{T}$, the region of stability can also be written in the form

$$\mathscr{R}(y) = \{x \in X : g(x, y) \leq 0, \ (g(x, \overline{y}) \not\leq 0 \text{ or } f(x, y) \leq f(x, \overline{y})) \, \forall \overline{y} \in T\}, \tag{5.8}$$

where the set $\overline{T}$ is replaced by T.

Unfortunately, the regions of stability are in general neither open nor closed nor connected, they can be empty and can overlap. This is shown in the following three examples.

Example 5.4 Let $T = \{0, 1, 2\}$, $X = [0; 3]$, $g(x, y) = y - x \leq 0$ and $f(x, y) = (y - x)^2$. Then, $f(x, y) \leq f(x, \overline{y})$ if and only if $y^2 - 2xy \leq \overline{y}^2 - 2x\overline{y}$, i.e. if $2x(\overline{y} - y) \leq \overline{y}^2 - y^2$ holds. Thus, using formula (5.8) we obtain

$$\begin{aligned}
\mathscr{R}(0) &= \{x \in [0; 3] : x \leq 0.5 \text{ or } x < 1, \ x \leq 1 \quad \text{or } x < 2\} = [0; 1),\\
\mathscr{R}(1) &= \{x \in [1; 3] : x \geq 0.5 \text{ or } x < 0, \ x \leq 1.5 \text{ or } x < 2\} = [1; 2),\\
\mathscr{R}(2) &= \{x \in [2; 3] : x \geq 1 \quad \text{or } x < 0, \ x \geq 1.5 \text{ or } x < 1\} = [2; 3].
\end{aligned}$$
□

Example 5.5 Let $T = \{-0.5, 0, 1\}$, $X = [-1; 1]$, $g(x, y) = y^2 - x^2 \leq 0$ and $f(x, y) = -y^2$. Then,

$$\mathscr{R}(-0.5) = (-1; -0.5] \cup [0.5; 1), \ \mathscr{R}(0) = (-0.5, 0.5) \text{ and } \mathscr{R}(1) = \{-1\} \cup \{1\}.$$

In this example the set $\mathscr{R}(0)$ is open, $\mathscr{R}(1)$ is closed and $\mathscr{R}(-0.5)$ is neither open nor closed. Furthermore, $\mathscr{R}(1)$ and $\mathscr{R}(-0.5)$ are not connected. □

Example 5.6 Let $T = \{-1, 0, 1\}$, $X = [-2; 1]$, $g(x, y) = x + y \leq 0$ and $f(x, y) = -y^2$. Then, we obtain by (5.8)

$$\mathscr{R}(-1) = [-2; 1], \quad \mathscr{R}(0) = \emptyset, \quad \mathscr{R}(1) = [-2; -1].$$

Thus, $\mathscr{R}(0) = \emptyset$, i.e. $0 \notin \overline{T}$ and $\operatorname{int} \mathscr{R}(-1) \cap \operatorname{int} \mathscr{R}(1) = (-2; -1) \neq \emptyset$. □

5.3.2 Properties of the Solution Sets

The regions of stability describe an inverse operation to the solution set mapping:

$$\Psi(x) = \{y \in \overline{T} : x \in \mathscr{R}(y)\}. \tag{5.9}$$

Thus, knowing all regions of stability, it is easy to determine $\Psi(x)$ for all $x \in X$.

Example 5.7 Consider Example 5.6 again. In this example the regions of stability are $\mathscr{R}(-1) = [-2; 1]$, $\mathscr{R}(0) = \emptyset$ and $\mathscr{R}(1) = [-2; -1]$. Thus, $\Psi(x) = \{1, -1\}$ for all $x \in [-2; -1]$ and $\Psi(x) = \{-1\}$ for all $x \in (-1; 1]$. □

An often satisfied and useful property in bilevel optimization with continuous lower level problems is the upper semicontinuity of the solution set mapping $\Psi(\cdot)$, see Definition 3.1. This is used for the investigation of the existence of an optimal solution of the bilevel optimization problem (cf. Dempe [52]). Since the set $\overline{T}$ has finitely many elements in the discrete case, upper semicontinuity can be verified using the following lemma.

Lemma 5.2 *A point-to-set mapping $\Gamma : X \rightarrow 2^{\overline{T}}$ is upper semicontinuous at $x^0 \in X$ if and only if there exists some $\delta > 0$ with*

$$\Gamma(x^0) \supseteq \Gamma(x) \textit{ for all } x \in X \textit{ with } \|x - x^0\| < \delta.$$

Proof Arguing by contradiction, assume that there does not exist $\delta > 0$ with $\Gamma(x^0) \supseteq \Gamma(x)$ for all $x \in X$ with $\|x - x^0\| < \delta$. Then there is a sequence $\{x^k\}_{k=1}^{\infty} \subseteq X$ with $\lim_{k\to\infty} x^k = x^0$ and $\Gamma(x^0) \not\supseteq \Gamma(x^k)$ for all k. This implies the existence of a sequence $\{y^k\}_{k=1}^{\infty} \subseteq \overline{T}$ with $y^k \in \Gamma(x^k)\backslash\Gamma(x^0)$ for all k. Due to $\text{card}(\overline{T}) < \infty$, we can assume w.l.o.g. that all y^k coincide, i.e. $y = y^k$ for all k. This implies the existence of an open set $\Omega \supseteq \Gamma(x^0)$ with $y \notin \Omega$. Thus, $\Gamma(\cdot)$ is not upper semicontinuous at $x^0 \in T$.

Let an arbitrary $\delta > 0$ be given and assume $\Gamma(x^0) \supseteq \Gamma(x)$ for all $x \in X$ with $\|x - x^0\| < \delta$. Then, for each open set Ω with $\Omega \supseteq \Gamma(x^0)$ it holds $\Omega \supseteq \Gamma(x^0) \supseteq \Gamma(x)\ \forall\, x \in X$ with $\|x - x^0\| < \delta$, i.e. $\Gamma(\cdot)$ is upper semicontinuous. □

As a result of this property, we obtain that the solution set mapping $\Psi(\cdot)$ is in general not upper semicontinuous in parametric discrete optimization.

Example 5.8 Consider Example 5.4 again, i.e. let $T = \{0, 1, 2\}$, $X = [0; 3]$, $g(x, y) = y - x \leq 0$ and $f(x, y) = (y - x)^2$. Then it holds

$$\mathscr{R}(0) = [0; 1), \quad \mathscr{R}(1) = [1; 2) \quad \text{and} \quad \mathscr{R}(2) = [2; 3].$$

Thus, it is

$$\Psi(1) = \{1\} \quad \text{but} \quad \Psi(x) = \{0\} \quad \text{for all} \quad x \in [0; 1).$$

There does not exist $\delta > 0$ with $\Psi(x) \subseteq \Psi(1)$ for all $x \in X$ with $\|x - 1\| < \delta$. Consequently, the solution set mapping is not upper semicontinuous at $x^0 = 1$. □

5.3.3 Extended Solution Sets

Since the solution set mapping $\Psi(\cdot)$ is not upper semicontinuous in general, the existence of an optimal solution of the optimistic bilevel optimization problem cannot be guaranteed. To obtain sufficient conditions for the existence of an optimal solution, we introduce the so-called *extended solution set mapping* $\overline{\Psi}(\cdot)$.

Definition 5.2 The set

$$\overline{\Psi}(x) := \{y \in \overline{T} : x \in \mathrm{cl}\,\mathscr{R}(y)\}$$

is the extended solution set at the point $x \in X$. The mapping $\overline{\Psi} : X \to 2^{\overline{T}}$ is called extended solution set mapping.

Example 5.9 Let $T = \{0, 1, 2\}$, $X = [0; 3]$, $g(x, y) = y - x \leq 0$ and $f(x, y) = (y - x)^2$. Then, remembering Examples 5.4 and 5.8, we know

$$\mathscr{R}(0) = [0; 1), \quad \mathscr{R}(1) = [1; 2) \quad \text{and} \quad \mathscr{R}(2) = [2; 3].$$

Thus, it holds

$$\mathrm{cl}\,\mathscr{R}(0) = [0; 1], \quad \mathrm{cl}\,\mathscr{R}(1) = [1; 2] \quad \text{and} \quad \mathrm{cl}\,\mathscr{R}(2) = [2; 3].$$

Consequently, we obtain $\overline{\Psi}(1) = \{0, 1\} \neq \Psi(1) = \{1\}$. □

The point-to-set mapping $\overline{\Psi}(\cdot)$ is an extension of the solution set mapping:

$$\Psi(x) \subseteq \overline{\Psi}(x) \quad \text{for all} \quad x \in X. \tag{5.10}$$

If the regions of stability are closed for all $y \in \overline{T}$, then $\Psi(x) = \overline{\Psi}(x)$ for all $x \in X$. The most important property of $\overline{\Psi}(\cdot)$ is its upper semicontinuity.

Theorem 5.3 *$\overline{\Psi}(\cdot)$ is the smallest upper semicontinuous point-to-set mapping with $\Psi(x) \subseteq \overline{\Psi}(x)$ for all $x \in X$.*

Proof First we show the upper semicontinuity of $\overline{\Psi}(\cdot)$. Assume on the contrary that $\overline{\Psi}(\cdot)$ is not upper semicontinuous at $x^0 \in X$. Then, by Lemma 5.2, there exists some sequence $\{x^k\}_{k=1}^{\infty} \subseteq X$ with $\lim_{k\to\infty} x^k = x^0$ and $\overline{\Psi}(x^k) \not\subseteq \overline{\Psi}(x^0)$ for all k. Because of $\mathrm{card}(\overline{T}) < \infty$ and $\overline{\Psi}(x^k) \subseteq \overline{T}$ we can assume w.l.o.g. that there exists some $y \in \overline{T}$ with

$$y \in \overline{\Psi}(x^k)\ \forall k \quad \text{and} \quad y \notin \overline{\Psi}(x^0).$$

Thus, we obtain

$$x^k \in \operatorname{cl}\mathscr{R}(y)\ \forall k \quad \text{and} \quad x^0 \notin \operatorname{cl}\mathscr{R}(y).$$

But this is a contradiction to $\operatorname{cl}\mathscr{R}(y)$ being closed. Consequently, $\overline{\Psi}(\cdot)$ is upper semicontinuous.

Assume now that there exists another upper semicontinuous point-to-set mapping $\Gamma(\cdot)$ with

$$\Psi(x^0) \subseteq \Gamma(x^0) \subseteq \overline{\Psi}(x^0)$$

for some $x^0 \in X$ and $\Psi(x) \subseteq \Gamma(x)$ for all $x \in X$. Take an arbitrary point $y \in \overline{\Psi}(x^0)$. Then, $x^0 \in \operatorname{cl}\mathscr{R}(y)$ and there exists a sequence $\{x^k\}_{k=1}^{\infty} \subseteq \mathscr{R}(y)$ with $\lim_{k\to\infty} x^k = x^0$. Consequently, $y \in \Psi(x^k) \subseteq \Gamma(x^k)$ for all k. Using the upper semicontinuity of $\Gamma(\cdot)$ we obtain $y \in \Gamma(x^0)$. Hence, $\Gamma(x^0) = \overline{\Psi}(x^0)$ and the theorem is true. □

5.3.4 Solution Functions

In general, $\Psi(x)$ does not reduce to a singleton (see Example 5.7). This has been used in Sect. 1.2 to formulate the optimistic and the pessimistic bilevel optimization problems. The optimistic solution function $\varphi_o : X \to \mathbb{R}$ defined in (1.3) and the pessimistic solution function $\varphi_p(\cdot) : X \to \mathbb{R}$ defined in Eq. (1.5) are in general not continuous. Thus, in the optimistic and the pessimistic bilevel optimization problems

$$\min\{\varphi_o(x) : x \in X\}$$

and

$$\min\{\varphi_p(x) : x \in X\},$$

discontinuous functions are minimized over the compact set X.

Let us introduce the following sets:

Definition 5.3 For a given point $y \in T$, let

$$O(y) := \{x \in \mathscr{R}(y) : \ \varphi_o(x) = F(x, y)\} \quad \text{and}$$
$$P(y) := \{x \in \mathscr{R}(y) : \ \varphi_p(x) = F(x, y)\}$$

denote the set of all $x \in X$, for which y can be chosen using the optimistic and pessimistic solution approaches, respectively.

These sets are in general neither open nor closed, they can be empty or not connected.

Example 5.10 Consider again the problem in Example 5.6, i.e. let $T = \{-1, 0, 1\}$, $X = [-2; 1]$, $g(x, y) = x + y \leq 0$ and $f(x, y) = -y^2$. Then,

$$\mathscr{R}(1) = [-2, -1], \quad \mathscr{R}(0) = \emptyset \quad \text{and} \quad \mathscr{R}(-1) = [-2, 1].$$

Thus, $\Psi(x) = \{1, -1\}$ for all $x \in [-2, -1]$ and $\Psi(x) = \{-1\}$ for all $x \in (-1, 1]$. Now let $F(x, y) = xy$. Then, it holds

$$\varphi_o(x) = \begin{cases} x & \text{for } x \in [-2, -1] \\ -x & \text{for } x \in (-1, 1] \end{cases} \quad \text{and} \quad \varphi_p(x) = -x \quad \text{for all} \quad x \in X.$$

Consequently,

$$\begin{aligned} O(-1) &= (-1, 1], \quad O(0) = \emptyset \quad \text{and} \quad O(1) = [-2; -1], \\ P(-1) &= [-2, 1], \quad P(0) = \emptyset \quad \text{and} \quad P(1) = \emptyset. \end{aligned}$$

□

If some lower semicontinuous function is minimized over a compact set local and global minima always exist (cf. Rockafellar and Wets [274]). Unfortunately, the optimistic and the pessimistic solution functions are in general not lower semicontinuous.

Example 5.11 Let $f(x, y) = (y - x)^2$, $g(x, y) = y - x$, $F(x, y) = 1 + y - x$, $T = \{0, 1, 2\}$ and $X = [0; 2]$. These are the data of Example 5.4 with modified set X. Analogously to Example 5.4 it holds $\mathscr{R}(0) = [0; 1)$, $\mathscr{R}(1) = [1; 2)$ and $\mathscr{R}(2) = \{2\}$. Thus, $\text{card}(\Psi(x)) = 1$ for all $x \in X$, i.e. the optimistic and the pessimistic solution functions coincide. It holds

$$\varphi_o(x) = \varphi_p(x) = \begin{cases} 1 - x, & x \in [0; 1) \\ 2 - x, & x \in [1; 2) \\ 1, & x = 2 \end{cases}.$$

Hence, the solution functions are not lower semicontinuous at the points $x = 1$ and $x = 2$. In Fig. 5.3 we see that no local minimum exists. □

Thus, in general there do not exist local minima for the optimistic and pessimistic solution functions, respectively. Therefore, the goal of computing an optimal solution of the bilevel optimization problem needs to be replaced by the search for a "nearly" optimal solution. For that reason we introduce the so-called weak solution functions.

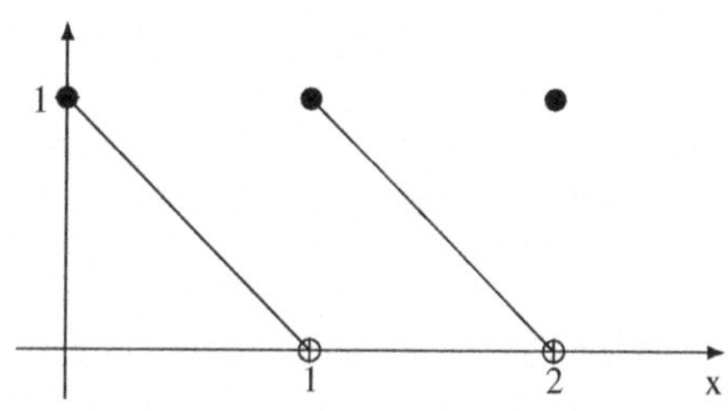

Fig. 5.3 Local optima do not exist in general

5.3.5 Weak Solution Functions

If the function $\varphi_o : X \to \mathbb{R}$ is not lower semicontinuous at some point $x^0 \in X$ it holds $\varphi_o(x^0) > \liminf_{x\to x^0,\, x\in X} \varphi_o(x)$. This motivates us to introduce the following functions:

Definition 5.4 (*Schmidt* [281]) For $x^0 \in X$, let

$$\overline{\varphi}_o(x^0) := \liminf_{x\to x^0,\, x\in X} \varphi_o(x) \tag{5.11}$$

and

$$\overline{\varphi}_p(x^0) := \liminf_{x\to x^0,\, x\in X} \varphi_p(x). \tag{5.12}$$

Then $\overline{\varphi}_o(\cdot)$ is called *weak optimistic* and $\overline{\varphi}_p(\cdot)$ is the *weak pessimistic solution function*.

Definition 5.4 obviously implies

$$\overline{\varphi}_o(x) \leq \varphi_o(x) \quad \text{and} \quad \overline{\varphi}_p(x) \leq \varphi_p(x) \quad \text{for all} \quad x \in X. \tag{5.13}$$

In the next theorem we prove that both the weak optimistic and the weak pessimistic solution functions are lower semicontinuous implying the existence of local and global solutions of the optimistic and the pessimistic optimization problem.

Theorem 5.4 (Fanghänel and Dempe [106]) *The functions $\overline{\varphi}_o(\cdot)$ and $\overline{\varphi}_p(\cdot)$ are the largest lower semicontinuous functions with $\overline{\varphi}_o(x) \leq \varphi_o(x)$ and $\overline{\varphi}_p(x) \leq \varphi_p(x)$ for all $x \in X$.*

Proof We prove the theorem only for the weak optimistic solution function. For the weak pessimistic solution function the proof is analogous.
Let an arbitrary point $x^0 \in X$ and an arbitrary sequence $\{x^k\}_{k=1}^{\infty} \subseteq X$ with $\lim_{k\to\infty} x^k = x^0$ be given. Due to the definition of $\overline{\varphi}_o(\cdot)$, there exist points $z^k \in X$ for all k with $\|z^k - x^k\| < 1/k$ and $|\varphi_o(z^k) - \overline{\varphi}_o(x^k)| < 1/k$. Thus, it holds

$$0 \leq \lim_{k\to\infty} \|z^k - x^0\| \leq \lim_{k\to\infty} (\|z^k - x^k\| + \|x^k - x^0\|) = 0,$$

i.e. the sequence $\{z^k\} \subseteq X$ converges to x^0. Further, it is

$$\begin{aligned}\overline{\varphi}_o(x^0) &= \liminf_{x\to x^0,\, x\in X} \varphi_o(x) \leq \liminf_{k\to\infty} \varphi_o(z^k)\\ &= \liminf_{k\to\infty} \left[(\varphi_o(z^k) - \overline{\varphi}_o(x^k)) + \overline{\varphi}_o(x^k)\right]\\ &\leq \liminf_{k\to\infty} (1/k + \overline{\varphi}_o(x^k)) = \liminf_{k\to\infty} \overline{\varphi}_o(x^k).\end{aligned}$$

Since $\{x^k\}$ was chosen arbitrarily, $\overline{\varphi}_o(\cdot)$ is lower semicontinuous at x^0. Since $x^0 \in X$ was chosen arbitrarily $\overline{\varphi}_o(\cdot)$ is lower semicontinuous.

It remains to show that there is no lower semicontinuous function between $\varphi_o(\cdot)$ and $\overline{\varphi}_o(\cdot)$. Arguing by contradiction, assume that there exists a lower semicontinuous function $\pi : T \to \mathbb{R}$ with $\pi(x) \leq \varphi_o(x)$ for all $x \in X$ and $\overline{\varphi}_o(x^0) \leq \pi(x^0) \leq \varphi_o(x^0)$ for some $x^0 \in X$. Let $\{x^k\} \subseteq X$ be an arbitrary sequence with $\lim_{k\to\infty} x^k = x^0$ and $\overline{\varphi}_o(x^0) = \lim_{k\to\infty} \varphi_o(x^k)$. Then, due to lower semicontinuity of $\pi(\cdot)$ and $\pi(x) \leq \varphi_o(x)$ for all $x \in X$, it holds

$$\overline{\varphi}_o(x^0) = \lim_{k\to\infty} \varphi_o(x^k) \geq \lim_{k\to\infty} \pi(x^k) \geq \pi(x^0).$$

Thus, it is $\overline{\varphi}_o(x) \geq \pi(x)$ for all $x \in X$. □

Next, we intend to find alternative definitions of the weak optimistic and weak pessimistic solution functions, resp., which can easier be used in solution algorithms. Therefore, we introduce the following sets:

$$\widehat{\Psi}_o(x) := \{y \in \overline{T} : x \in \text{cl}\, O(y)\} \tag{5.14}$$

$$\widehat{\Psi}_p(x) := \{y \in \overline{T} : x \in \text{cl}\, P(y)\} \tag{5.15}$$

Due to the fact, that $O(y) \subseteq \mathscr{R}(y)$ and $P(y) \subseteq \mathscr{R}(y)$, these sets are subsets of $\overline{\Psi}(x)$ for all $x \in X$:

$$\widehat{\Psi}_o(x) \subseteq \overline{\Psi}(x) \quad \text{and} \quad \widehat{\Psi}_p(x) \subseteq \overline{\Psi}(x) \quad \text{for all} \quad x \in X. \tag{5.16}$$

These sets define two point-to-set mappings $\widehat{\Psi}_o : X \to 2^{\overline{T}}$ and $\widehat{\Psi}_p : X \to 2^{\overline{T}}$.

Theorem 5.5 (Fanghänel and Dempe [106]) *The point-to-set mappings $\widehat{\Psi}_o : X \to 2^{\overline{T}}$ and $\widehat{\Psi}_p : X \to 2^{\overline{T}}$ are upper semicontinuous.*

Proof We prove the theorem for the point-to-set mapping $\widehat{\Psi}_o : X \to 2^{\overline{T}}$. For $\widehat{\Psi}_p : X \to 2^{\overline{T}}$ the proof is analogous.

Suppose that the conclusion of the theorem does not hold. Then, due to Lemma 5.2, there exists $x^0 \in X$ and a sequence $\{x^k\}_{k=1}^{\infty} \subseteq X$ with $\lim_{k\to\infty} x^k = x^0$ and $\widehat{\Psi}_o(x^k) \not\subseteq \widehat{\Psi}_o(x^0)$ for all k. Due to $\text{card}(\overline{T}) < \infty$, we can assume that there exists a point $y \in \overline{T}$ with $y \in \widehat{\Psi}_o(x^k)$ for all k and $y \notin \widehat{\Psi}_o(x^0)$. Hence, $x^k \in \text{cl}\, O(y)$ for all k. Since the set $\text{cl}\, O(y)$ is closed, this implies $x^0 \in \text{cl}\, O(y)$. But this is a contradiction to $y \notin \widehat{\Psi}_o(x^0)$. This proves the theorem. □

Now we use the mappings $\widehat{\Psi}_o(\cdot)$ and $\widehat{\Psi}_p(\cdot)$ to rewrite the functions $\overline{\varphi}_o : X \to \mathbb{R}$ and $\overline{\varphi}_p : X \to \mathbb{R}$.

Theorem 5.6 (Fanghänel and Dempe [106]) *For all $x \in X$ it is*

$$\overline{\varphi}_o(x) = \min_{y \in \widehat{\Psi}_o(x)} F(x, y) \quad \textit{and} \quad \overline{\varphi}_p(x) = \min_{y \in \widehat{\Psi}_p(x)} F(x, y).$$

Proof We prove the theorem for the weak optimistic solution function. For the weak pessimistic solution function the proof is analogous.

Let $x \in X$. Then, since (5.11), there exists a sequence $\{x^k\}_{k=1}^{\infty} \subseteq X$ with $\lim_{k\to\infty} x^k = x$ and $\overline{\varphi}_o(x) = \lim_{k\to\infty} \varphi_o(x^k)$. Further, because of $\text{card}(\overline{T}) < \infty$, there exists w.l.o.g. some $y \in \overline{T}$ with $x^k \in O(y)$ for all k. Hence, $x \in \text{cl}\, O(y)$, i.e. $y \in \widehat{\Psi}_o(x)$. Thus, it holds

$$\overline{\varphi}_o(x) = \lim_{k\to\infty} \varphi_o(x^k) = \lim_{k\to\infty} F(x^k, y) = F(x, y)$$

for this $y \in \widehat{\Psi}_o(x)$. This implies

$$\overline{\varphi}_o(x) \geq \min_{y \in \widehat{\Psi}_o(x)} F(x, y).$$

Take an arbitrary $y \in \widehat{\Psi}_o(x)$. Then, $x \in \text{cl}\, O(y)$, i.e. there exists some sequence $\{x^k\}_{k=1}^{\infty} \subseteq O(y)$ with $\lim_{k\to\infty} x^k = x$. This implies $\varphi_o(x^k) = F(x^k, y)$ for all k. Thus, it holds

$$\overline{\varphi}_o(x) \leq \liminf_{k\to\infty} \varphi_o(x^k) = \liminf_{k\to\infty} F(x^k, y) = F(x, y).$$

Consequently, we obtain

$$\overline{\varphi}_o(x) \leq \min_{y \in \widehat{\Psi}_o(x)} F(x, y).$$

□

Example 5.12 Let $f(x, y) = -(y_1 + y_2)$, $g(x, y) = y_1 + 2y_2 - x$, $F(x, y) = y_1 - y_2 - x + 2$, $T = \{0, 1\}^2$ and $X = [1; 3]$. Using formula (5.8) we obtain the following regions of stability

$$\mathscr{R}(y^1) = \emptyset,\ \mathscr{R}(y^2) = [1; 3),\ \mathscr{R}(y^3) = [2; 3) \quad \text{und} \quad \mathscr{R}(y^4) = \{3\}$$

with

$$y^1 = \begin{pmatrix} 0 \\ 0 \end{pmatrix},\ y^2 = \begin{pmatrix} 1 \\ 0 \end{pmatrix},\ y^3 = \begin{pmatrix} 0 \\ 1 \end{pmatrix} \quad \text{and} \quad y^4 = \begin{pmatrix} 1 \\ 1 \end{pmatrix}.$$

Next we consider the functions $F(\cdot, y)$ over the sets $\mathscr{R}(y)$.

Using the Fig. 5.4 we easily obtain

$$\varphi_o(x) = \begin{cases} 3 - x, & x \in [1; 2) \\ 1 - x, & x \in [2; 3) \\ -1, & x = 3 \end{cases} \quad \text{and} \quad \varphi_p(x) = \begin{cases} 3 - x, & x \in [1; 3) \\ -1, & x = 3 \end{cases}.$$

Thus, the optimistic solution function has no local minimum. Further, for all $x \in X$ we have

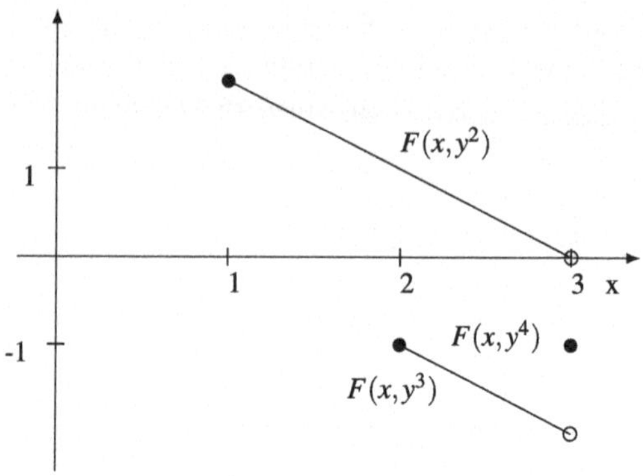

Fig. 5.4 Objective function value over regions of stability

$$\widehat{\Psi}_o(x) = \begin{cases} \{y^2\},\ x \in [1;2) \\ \{y^2, y^3\},\ x = 2 \\ \{y^3\},\ x \in (2;3) \\ \{y^3, y^4\},\ x = 3 \end{cases} \quad \text{and} \quad \widehat{\Psi}_p(x) = \begin{cases} \{y^2\},\ x \in [1;3) \\ \{y^2, y^4\},\ x = 3. \end{cases}$$

Hence,

$$\overline{\varphi}_o(x) = \begin{cases} 3 - x,\ x \in [1;2) \\ 1 - x,\ x \in [2;3] \end{cases} \quad \text{and} \quad \overline{\varphi}_p(x) = \varphi_p(x).$$

The weak solution functions $\overline{\varphi}_o(\cdot)$ and $\overline{\varphi}_p(\cdot)$ have at the point $x = 3$ a local and global minimum, resp. □

5.3.6 Optimality Conditions

Focus in this subsection is on the formulation of necessary optimality conditions for minimizing the optimistic and the pessimistic solution functions on the set X.

In the following let $\operatorname{locmin}\{\varphi(x) : x \in X\}$ denote the set of all local minima of the function $\varphi(\cdot)$ over X. Using this, we denote by

$\operatorname{locmin}\{\overline{\varphi}_o(x) : x \in X\}$ the set of all weak local optimistic solutions,
$\operatorname{locmin}\{\overline{\varphi}_p(x) : x \in X\}$ the set of all weak local pessimistic solutions,
$\operatorname{locmin}\{\varphi_o(x) : x \in X\}$ the set of all local optimistic solutions,
$\operatorname{locmin}\{\varphi_p(x) : x \in X\}$ the set of all local pessimistic solutions.

First we show, that the concepts of weak local optimistic and pessimistic solutions are weaker than the concepts of local optimistic and pessimistic solutions, respectively.

Lemma 5.3 *It holds*

1. $\operatorname{locmin}\{\varphi_o(x) : x \in X\} \subseteq \operatorname{locmin}\{\overline{\varphi}_o(x) : x \in X\}$,
2. $\operatorname{locmin}\{\varphi_p(x) : x \in X\} \subseteq \operatorname{locmin}\{\overline{\varphi}_p(x) : x \in X\}$.

Proof We prove the theorem for the optimistic solution approach. For the pessimistic solution approach, the proof is analogous.

Let $x^0 \in \operatorname{locmin}\{\varphi_o(x) : x \in X\}$. Then, there exists $\delta > 0$ with $\varphi_o(x^0) \leq \varphi_o(x)$ for all $x \in X$ satisfying $\|x - x^0\| < \delta$. Take $x \in X$ with $\|x - x^0\| < \delta$ and a sequence $\{x^k\}_{k=1}^{\infty} \subseteq X$ with $\lim_{k\to\infty} x^k = x$ and

$$\overline{\varphi}_o(x) = \lim_{k\to\infty} \varphi_o(x^k).$$

Then, there is an index k_0 such that $\|x^k - x^0\| < \delta$ and thus $\varphi_o(x^0) \leq \varphi_o(x^k)$ for all $k \geq k_0$. Consequently,

$$\overline{\varphi}_o(x) = \lim_{k\to\infty} \varphi_o(x^k) \geq \varphi_o(x^0) \geq \overline{\varphi}_o(x^0).$$

Since x was chosen arbitrarily, this implies $x^0 \in \operatorname{locmin}\{\overline{\varphi}_o(x) : x \in X\}$. □

Thus, each local optimistic or local pessimistic solution is also a weak local optimistic or weak local pessimistic solution, respectively. Before we can discuss optimality conditions for weak local optimistic and weak local pessimistic solutions, we introduce the following sets:

$$L_o(y) := \operatorname*{locmin}_x\{F(x, y) : x \in \operatorname{cl} O(y)\} \tag{5.17}$$

$$L_p(y) := \operatorname*{locmin}_x\{F(x, y) : x \in \operatorname{cl} P(y)\} \tag{5.18}$$

Theorem 5.7 (Fanghänel and Dempe [106]) *It holds*

1. $x^0 \in \operatorname{locmin}\{\overline{\varphi}_o(x) : x \in X\}$ *if and only if* $x^0 \in L_o(y)$ *for all* $y \in \widehat{\Psi}_o(x^0)$ *with* $\overline{\varphi}_o(x^0) = F(x^0, y)$.
2. $x^0 \in \operatorname{locmin}\{\overline{\varphi}_p(x) : x \in X\}$ *if and only if* $x^0 \in L_p(y)$ *for all* $y \in \widehat{\Psi}_p(x^0)$ *with* $\overline{\varphi}_p(x^0) = F(x^0, y)$.

Proof Again we prove the theorem only for the optimistic solution approach. For the pessimistic solution approach the proof is analogous.

Assume that $x^0 \notin L_o(y)$ for some $y \in \widehat{\Psi}_o(x^0)$ with $\overline{\varphi}_o(x^0) = F(x^0, y)$. Then, $y \in \widehat{\Psi}_o(x^0)$ implies $x^0 \in \operatorname{cl} O(y)$. Because of $x^0 \notin L_o(y)$, there exists a sequence $\{x^k\}_{k=1}^{\infty} \subseteq \operatorname{cl} O(y)$ with $\lim_{k\to\infty} x^k = x^0$ and

$$F(x^k, y) < F(x^0, y) \quad \text{for all} \quad k.$$

Further, Theorem 5.6 implies $\overline{\varphi}_o(x^k) \leq F(x^k, y)$ for all k because of $x^k \in \operatorname{cl} O(y)$, i.e. $y \in \widehat{\Psi}_o(x^k)$. Thus, it holds

$$\overline{\varphi}_o(x^k) \le F(x^k, y) < F(x^0, y) = \overline{\varphi}_o(x^0)$$

for all k. Consequently, $x^0 \notin \text{locmin}\{\overline{\varphi}_o(x) : x \in X\}$.

Assume that $x^0 \notin \text{locmin}\{\overline{\varphi}_o(x) : x \in X\}$ for some $x^0 \in X$. Then there exists a sequence $\{x^k\}_{k=1}^{\infty} \subseteq X$ with $\lim_{k\to\infty} x^k = x^0$ and

$$\overline{\varphi}_o(x^k) < \overline{\varphi}_o(x^0) \quad \text{for all} \quad k.$$

Because of $\text{card}(\overline{T}) < \infty$, there exists w.l.o.g. some $y \in \overline{T}$ with $y \in \widehat{\Psi}_o(x^k)$ and $\overline{\varphi}_o(x^k) = F(x^k, y)$ for all k (see Theorem 5.6). This implies $y \in \widehat{\Psi}_o(x^0)$ because of the upper semicontinuity of $\widehat{\Psi}_o(\cdot)$ (Theorem 5.5). Thus, it is

$$F(x^k, y) = \overline{\varphi}_o(x^k) < \overline{\varphi}_o(x^0) \le F(x^0, y). \tag{5.19}$$

For k converging to infinity this yields $\overline{\varphi}_o(x^0) = F(x^0, y)$. Since $y \in \widehat{\Psi}_o(x^k)$ implies $x^k \in \text{cl}\, O(y)$, using (5.19) we obtain $x^0 \notin L_o(y)$ for some $y \in \widehat{\Psi}_o(x^0)$ with $\overline{\varphi}_o(x^0) = F(x^0, y)$. □

Thus, using the sets $\widehat{\Psi}_o(x^0)$ and $L_o(x^0)$ or $\widehat{\Psi}_p(x^0)$ and $L_p(x^0)$, Theorem 5.7 provides us with a necessary and sufficient optimality condition. For many problems the computation of these sets is maybe complicated. Hence, we are interested in obtaining simpler optimality conditions. For that we introduce the set

$$\overline{L}(y) := \underset{x}{\text{locmin}}\{F(x, y) : x \in \text{cl}\,\mathscr{R}(y)\} \tag{5.20}$$

for $y \in \overline{T}$.

Lemma 5.4 *Let $x^0 \in X$. Then,*

$$\overline{\varphi}_o(x^0) \le F(x^0, y) \quad \textit{for all} \quad x \in \overline{\Psi}(x^0)$$

and the following two statements are equivalent:

(a) $x^0 \in L_o(y)$ *for all* $y \in \widehat{\Psi}_o(x^0)$ *with* $\overline{\varphi}_o(x^0) = F(x^0, y)$,
(b) $x^0 \in \overline{L}(y)$ *for all* $y \in \overline{\Psi}(x^0)$ *with* $\overline{\varphi}_o(x^0) = F(x^0, y)$.

Proof Let $y \in \overline{\Psi}(x^0)$. Then, $x^0 \in \text{cl}\,\mathscr{R}(y)$ by definition. Thus, there exists a sequence $\{x^k\}_{k=1}^{\infty} \subseteq \mathscr{R}(y)$ with $\lim_{k\to\infty} x^k = x^0$. Then it follows $y \in \Psi(x^k)$ and $\varphi_o(x^k) \le F(x^k, y)$ for all k. Tending $k \to \infty$, we obtain

$$\overline{\varphi}_o(x^0) \le \liminf_{k\to\infty} \varphi_o(x^k) \le \lim_{k\to\infty} F(x^k, y) = F(x^0, y).$$

Thus, the first statement of the lemma holds.

Next we investigate the equivalence statement. Assume statement (a) is not valid. Then there exists some $y \in \widehat{\Psi}_o(x^0)$ with $F(x^0, y) = \overline{\varphi}_o(x^0)$ and $x^0 \notin L_o(y)$. This implies the existence of a sequence $\{x^k\}_{k=1}^\infty \subseteq \operatorname{cl} O(x)$ with $\lim_{k\to\infty} x^k = x^0$ and $F(x^k, y) < F(x^0, y)$ for all k. Due to $\operatorname{cl} O(y) \subseteq \operatorname{cl} \mathscr{R}(y)$, we obtain $x^0 \notin \overline{L}(y)$. Moreover, $y \in \overline{\Psi}(x^0)$ by (5.16). Thus, statement (b) is not valid, too.

Assume statement (b) is not valid. Then there exists some $y \in \overline{\Psi}(x^0)$ with $\overline{\varphi}_o(x^0) = F(x^0, y)$ and $x^0 \notin \overline{L}(y)$. This implies the existence of a sequence $\{x^k\}_{k=1}^\infty \subseteq \operatorname{cl} \mathscr{R}(y)$ with $\lim_{k\to\infty} x^k = x^0$ and $F(x^k, y) < F(x^0, y)$ for all k. The property $\{x^k\}_{k=1}^\infty \subseteq \operatorname{cl} \mathscr{R}(y)$ is equivalent to $y \in \overline{\Psi}(x^k)$ for all k. Using the first statement of the lemma and because of $F(x^0, y) = \overline{\varphi}_o(x^0)$, we obtain

$$\overline{\varphi}_o(x^k) \le F(x^k, y) < F(x^0, y) = \overline{\varphi}_o(x^0) \quad \text{for all} \quad k.$$

Thus, $x^0 \notin \operatorname{locmin}\{\overline{\varphi}_o(x) : x \in X\}$. Then, because of Theorem 5.7, statement (a) is not valid, too. □

Using Lemma 5.4 we obtain the following obvious corollary of Theorem 5.7:

Corollary 5.1 $x^0 \in \operatorname{locmin}\{\overline{\varphi}_o(x) : x \in X\}$ *if and only if* $x^0 \in \overline{L}(y)$ *for all* $y \in \overline{\Psi}(x^0)$ *with* $\overline{\varphi}_o(x^0) = F(x^0, y)$.

Thus, in the optimistic case, we can check weak local optimality also by using the extended solution sets and the sets $\overline{L}(y)$.

Next, we want to investigate optimality conditions for the optimistic and pessimistic solution functions.

Theorem 5.8 (Fanghänel and Dempe [106]) *It holds*

1. $x^0 \in \operatorname{locmin}\{\varphi_o(x) : x \in X\}$ *if and only if the following two conditions are valid:*
   a. $\varphi_o(x^0) \le F(x^0, y)$ *for all* $y \in \widehat{\Psi}_o(x^0)$,
   b. $x^0 \in L_o(y)$ *for all* $y \in \widehat{\Psi}_o(x^0)$ *with* $\varphi_o(x^0) = F(x^0, y)$.
2. $x^0 \in \operatorname{locmin}\{\varphi_p(x) : x \in X\}$ *if and only if the following two conditions are valid:*
   a. $\varphi_p(x^0) \le F(x^0, y)$ *for all* $y \in \widehat{\Psi}_p(x^0)$,
   b. $x^0 \in L_p(y)$ *for all* $y \in \widehat{\Psi}_p(x^0)$ *with* $\varphi_p(x^0) = F(x^0, y)$.

Proof We prove the statements of the theorem for the pessimistic solution function. For the optimistic solution approach the proof is analogous.

Let $x^0 \in \operatorname{locmin}\{\varphi_p(x) : x \in X\}$. Assume that condition (a) is not valid, i.e., there exists some $y \in \widehat{\Psi}_p(x^0)$ with $F(x^0, y) < \varphi_p(x^0)$. Then, $y \in \widehat{\Psi}_p(x^0)$ implies $x^0 \in \operatorname{cl} P(y)$. Thus, there exists a sequence $\{x^k\}_{k=1}^\infty \subseteq P(y)$ with $\lim_{k\to\infty} x^k = x^0$. Hence, $F(x^k, y) = \varphi_p(x^k)$ for all k. Tending k to infinity, we obtain

$$\lim_{k\to\infty} \varphi_p(x^k) = \lim_{k\to\infty} F(x^k, y) = F(x^0, y) < \varphi_p(x^0).$$

But this is a contradiction to $x^0 \in \operatorname{locmin}\{\varphi_p(x) : x \in X\}$.

Assume condition (a) holds, but condition (b) is not valid. Then there exists some $y \in \widehat{\Psi}_p(x^0)$ with $F(x^0, y) = \varphi_p(x^0)$ and $x^0 \notin L_p(y)$. Further, condition (a) implies

$$\varphi_p(x^0) \leq \inf_{y \in \widehat{\Psi}_p(x^0)} F(x^0, y) = \overline{\varphi}_p(x^0),$$

i.e. $\varphi_p(x^0) = \overline{\varphi}_p(x^0) = F(x^0, y)$ [(see Eq. (5.13)]. Thus, using Theorem 5.7, we obtain $x^0 \notin \text{locmin}\{\overline{\varphi}_p(x) : x \in X\}$. Consequently, $x^0 \notin \text{locmin}\{\varphi_p(x) : x \in X\}$ because of Lemma 5.3. Hence, $x^0 \in \text{locmin}\{\varphi_p(x) : x \in X\}$ implies the validity of the conditions (a) and (b).

Assume $x^0 \in X$ is not a local pessimistic solution. Then there exists a sequence $\{x^k\}_{k=1}^{\infty} \subseteq X$ with $\lim_{k\to\infty} x^k = x^0$ and

$$\varphi_p(x^k) < \varphi_p(x^0) \quad \text{for all} \quad k. \tag{5.21}$$

Because of $\text{card}(\overline{T}) < \infty$, there exists w.l.o.g. some $y \in \Psi(x^k)$ with $\varphi_p(x^k) = F(x^k, y)$ for all k, i.e. $x^k \in P(y)$ for all k. This implies $x^0 \in \text{cl}\, P(y)$ and thus $y \in \widehat{\Psi}_p(x^0)$. Then, using the inequality (5.21), we obtain

$$F(x^k, y0) = \varphi_p(x^k) < \varphi_p(x^0) \quad \text{for all} \quad k.$$

With $k \to \infty$ this yields $F(x^0, y) \leq \varphi_p(x^0)$. Thus, if condition (a) is fulfilled, it holds $F(x^0, y) = \varphi_p(x^0)$. But then we obtain

$$F(x^k, y) = \varphi_p(x^k) < \varphi_p(x^0) = F(x^0, y) \quad \text{for all} \quad k,$$

i.e. condition (b) is not fulfilled. □

Corollary 5.2 *We have*

1. $x^0 \in \text{locmin}\{\varphi_o(x) : x \in X\}$ *if and only if*

$$\varphi_o(x^0) = \overline{\varphi}_o(x^0) \quad \textit{and} \quad x^0 \in \text{locmin}\{\overline{\varphi}_o(x) : x \in X\}.$$

2. $x^0 \in \text{locmin}\{\varphi_p(x) : x \in X\}$ *if and only if both*

$$\varphi_p(x^0) = \overline{\varphi}_p(x^0) \quad \textit{and} \quad x^0 \in \text{locmin}\{\overline{\varphi}_p(x) : x \in X\}.$$

Proof We prove the statements of the corollary for the pessimistic solution function. For the optimistic solution approach the proof is analogous.

Assume $x^0 \in X$ is a local pessimistic solution. Then, x^0 is also a weak local pessimistic solution because of Lemma 5.3. Further it is $\varphi_p(x^0) \leq F(x^0, y)$ for all $y \in \widehat{\Psi}_p(x^0)$ (Theorem 5.8). Thus, applying Theorem 5.6, we obtain $\varphi_p(x^0) \leq \overline{\varphi}_p(x^0)$. Consequently, equality holds because of (5.13).

Assume that both $\varphi_p(x^0) = \overline{\varphi}_p(x^0)$ and $x^0 \in \operatorname{locmin}\{\overline{\varphi}_p(x) : x \in X\}$ are satisfied. Then,

$$\overline{\varphi}_p(x^0) = \min_{y \in \widehat{\Psi}_p(x^0)} F(x^0, y) = \varphi_p(x^0).$$

Thus, we obtain $\varphi_p(x^0) \leq F(x^0, y)$ for all $y \in \widehat{\Psi}_p(x^0)$. Consequently, condition (a) of Theorem 5.8 is fulfilled. Furthermore, because of Theorem 5.7 it is $x^0 \in L_p(y)$ for all $y \in \widehat{\Psi}_p(x^0)$ with $F(x^0, y) = \overline{\varphi}_p(x^0)$. With $\varphi_p(x^0) = \overline{\varphi}_p(x^0)$ this yields condition (b) of Theorem 5.8. Hence, x^0 is a local pessimistic solution. □

Example 5.13 Let $f(x, y) = -(y_1 + y_2)$, $g(x, y) = y_1 + 2y_2 - x$, $T = \{0, 1\}^2$ and $X = [1; 3]$ as in Example 5.12 but $F(x, y) = (y_2 - y_1)x + y_1 - 3y_2$. Then the regions of stability

$$\mathscr{R}(y^1) = \emptyset, \quad \mathscr{R}(y^2) = [1; 3), \quad \mathscr{R}(y^3) = [2; 3) \quad \text{and} \quad \mathscr{R}(y^4) = \{3\}$$

for the feasible points

$$y^1 = \begin{pmatrix} 0 \\ 0 \end{pmatrix}, \quad y^2 = \begin{pmatrix} 1 \\ 0 \end{pmatrix}, \quad y^3 = \begin{pmatrix} 0 \\ 1 \end{pmatrix} \quad \text{and} \quad y^4 = \begin{pmatrix} 1 \\ 1 \end{pmatrix}$$

are unchanged. Because $F(\cdot, \cdot)$ was changed we obtain different solution functions. These solution functions are sketched in the Figs. 5.5 and 5.6.

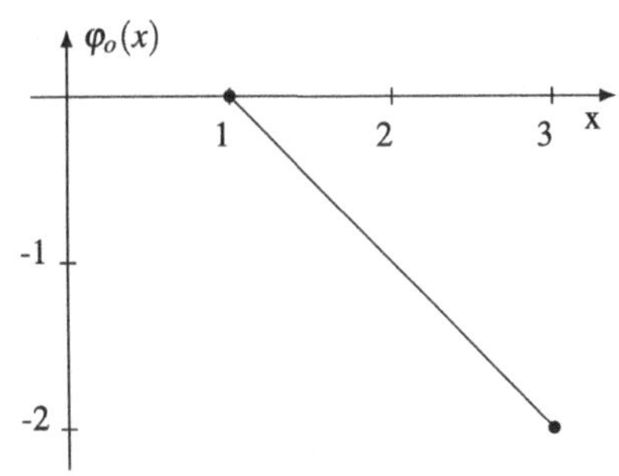

Fig. 5.5 The optimistic solution function in Example 5.13

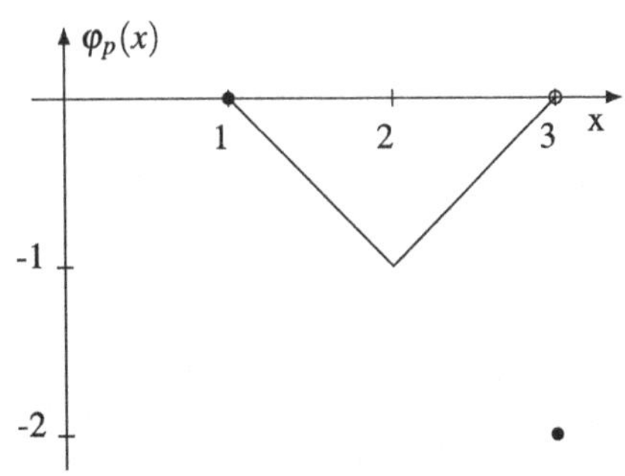

Fig. 5.6 The pessimistic solution function in Example 5.13

We investigate local optimality at the point $x^0 = 2$. Obviously it holds $\Psi(x^0) = \overline{\Psi}(x^0) = \{y^2, y^3\}$. The sets $\widehat{\Psi}_o(x^0)$ and $\widehat{\Psi}_p(x^0)$ are subsets of $\overline{\Psi}(x^0)$. Thus, because of $F(x^0, y^2) = -1 = F(x^0, y^3)$, we obtain

$$\widehat{\Psi}_o(x^0) = \widehat{\Psi}_p(x^0) = \{y^2, y^3\}$$

and

$$\varphi_o(x^0) = \varphi_p(x^0) = \overline{\varphi}_o(x^0) = \overline{\varphi}_p(x^0) = -1.$$

Consequently, at the point $x^0 = 2$ condition (a) of Theorem 5.8 is fulfilled for both the optimistic and the pessimistic solution approach. Now we check condition (b) of Theorem 5.8. It holds

$$O(y^2) = [1; 3), \quad O(y^3) = \{2\}, \quad P(y^2) = [1; 2] \text{ and } P(y^3) = [2; 3).$$

This yields

$$L_o(y^2) = \{3\}, \quad L_o(y^3) = \{2\}, \quad L_p(y^2) = \{2\} \text{ and } L_p(y^3) = \{2\}.$$

Thus, we have $\varphi_o(x^0) = F(x^0, y^2)$ but $x^0 \notin L_o(y^2)$, i.e. $x^0 = 2$ is not a (weak) local optimistic solution. For the pessimistic solution approach, it holds $x^0 \in L_p(x^2)$ and $x^0 \in L_p(x^3)$, i.e. $x^0 = 2$ is a (weak) local pessimistic solution. □

5.3.7 Computation of Optimal Solutions

Knowing optimality conditions we are looking for a way to compute a weak local optimistic and a weak local pessimistic solution.

Algorithm: `Computation of a weak optimistic solution.`

`Input:` $x^0 \in X$ `(starting point)`
0. `Set` $k := 0$.
1. `Compute` $\widehat{\Psi}_o(x^k)$, $\overline{\varphi}_o(x^k)$ `and` $M(x^k) := \{y \in \widehat{\Psi}_o(x^k) : \overline{\varphi}_o(x^k) = F(x^k, y)\}$.
2. `If there exists` $y^k \in M(x^k)$ `with` $x^k \notin L_o(y^k)$ `then choose` $x^{k+1} \in L_o(y^k)$, `set` $k := k + 1$ `and proceed with step 1;` `else stop.`

`Output:` $x^k \in \text{locmin}\{\overline{\varphi}_o(x) : x \in X\}$.

Theorem 5.9 (Fanghänel and Dempe [106]) *The above algorithm computes a weak local optimistic solution in a finite number of steps.*

Proof If the algorithm stops, the computed point is a weak local optimistic solution because of Theorem 5.7. Thus, we only have to prove that the algorithm stops after finitely many iterations. Assume this is not true. Then sequences $\{x^k\}_{k=1}^{\infty} \subseteq X$ and $\{y^k\}_{k=1}^{\infty} \subseteq \overline{T}$ are be computed with $y^k \in M(x^k)$, $x^k \notin L_o(y^k)$ and $x^{k+1} \in L_o(y^k)$

for all $k \geq 0$. Because of $x^{k+1} \in L_o(y^k)$ for all k, we obtain $y^k \in \widehat{\Psi}_o(x^{k+1})$, i.e. $\overline{\varphi}_o(x^{k+1}) \leq F(x^{k+1}, y^k)$ for all k. Thus, it holds

$$\overline{\varphi}_o(x^k) = F(x^k, y^k) > F(x^{k+1}, y^k) \geq \overline{\varphi}_o(x^{k+1}) \quad \text{for all} \quad k.$$

Hence, $\overline{\varphi}_o(x^k) > \overline{\varphi}_o(x^l)$ for all indices $k < l$. Because of $\operatorname{card}(\overline{T}) < \infty$, there exist two indices k, l with $k < l$ and $y^k = y^l$. Then it holds $y^k = y^l \in \widehat{\Psi}_o(x^l)$, i.e. $x^l \in \operatorname{cl} O(y^k)$. Thus, using $x^{k+1} \in L_o(y^k)$ and $x^l \notin L_o(y^l) = L_o(y^k)$, we obtain

$$F(x^{k+1}, y^k) < F(x^l, y^k) = F(x^l, y^l).$$

But this is a contradiction to

$$F(x^{k+1}, y^k) \geq \overline{\varphi}_o(x^{k+1}) \geq \overline{\varphi}_o(x^l) = F(x^l, y^l).$$

Thus, the algorithm stops after finitely many iterations. □

For weak local pessimistic solutions, there exists an analogous algorithm. Remark that the choice of the starting point is very important with respect to the quality of the computed solution. If we consider Example 5.13 we see that only with the choice of $x^0 = 3$ a global pessimistic solution is computed.

It is well-known that each global optimal solution of a mathematical optimization problem is also a local optimal solution. Thus, it holds

$$\begin{aligned} \operatorname{Argmin}\{\varphi_o(x) : x \in X\} &\subseteq \operatorname{locmin}\{\varphi_o(x) : x \in X\}, \\ \operatorname{Argmin}\{\varphi_p(x) : x \in X\} &\subseteq \operatorname{locmin}\{\varphi_p(x) : x \in X\}. \end{aligned}$$

But the existence of a local optimal solution is not a guarantee for the existence of a global optimal solution. Let

$$\overline{\varphi}_o^* := \min\{\overline{\varphi}_o(x) : \ x \in X\} \quad \text{and} \quad \overline{\varphi}_p^* := \min\{\overline{\varphi}_p(x) : \ x \in X\}. \tag{5.22}$$

Then, by lower semicontinuity of $\varphi_o(\cdot)$ and $\varphi_p(\cdot)$, there always exists $x^* \in X$ with $\overline{\varphi}_o(x^*) = \overline{\varphi}_o^*$ and $\overline{\varphi}_p(x^*) = \overline{\varphi}_p^*$, respectively. Let

$$\varphi_o^* := \inf\{\varphi_o(x) : \ x \in X\} \quad \text{and} \quad \varphi_p^* := \inf\{\varphi_p(x) : \ x \in X\}. \tag{5.23}$$

Then the following lemma holds.

Lemma 5.5 *It is* $\overline{\varphi}_o^* = \varphi_o^*$ *and* $\overline{\varphi}_p^* = \varphi_p^*$.

Proof Because of (5.13), we have $\overline{\varphi}_o^* \leq \varphi_o^*$. Further, there exists some $x^* \in X$ with $\overline{\varphi}_o(x^*) = \overline{\varphi}_o^*$. Due to Definition 5.4, there exists a sequence $\{x^k\}_{k=1}^{\infty} \subseteq X$ with $\overline{\varphi}_o(x^*) = \lim_{k\to\infty} \varphi_o(x^k)$. Thus, inequality

$$\overline{\varphi}_o^* = \overline{\varphi}_o(x^*) \geq \inf_{x \in X} \varphi_o(x) = \varphi_o^*$$

holds. For the pessimistic solution function, the proof is analogous. □

A point x^0 is a global optimistic (pessimistic) solution if and only if it is a weak global optimistic (pessimistic) solution with $\varphi_o(x^0) = \overline{\varphi}_o(x^0)$ ($\varphi_p(x^0) = \overline{\varphi}_p(x^0)$).

If there does not exist any global optimistic (pessimistic) solution, we are interested in finding some ε-optimal solution for the optimistic (pessimistic) case.

Definition 5.5 Let $\varepsilon > 0$. Some $x^0 \in X$ is called *ε-optimal solution* for the optimistic (pessimistic) case if it holds $\varphi_o(x^0) < \varphi_o^* + \varepsilon$ (resp. $\varphi_p(x^0) < \varphi_p^* + \varepsilon$).

For the computation of ε-optimal solutions we can use the following theorem.

Theorem 5.10 (Fanghänel and Dempe [106]) *Let $\varepsilon > 0$. Further let there be given some weak global optimistic (pessimistic) solution $x^0 \in X$ and some point $y^0 \in \widehat{\Psi}_o(x^0)$ (resp. $y^0 \in \widehat{\Psi}_p(x^0)$) with $F(x^0, y^0) = \overline{\varphi}_o(x^0)$ (resp. $F(x^0, y^0) = \overline{\varphi}_p(x^0)$). Then each point $x \in O(y^0)$ (resp. $x \in P(y^0)$) with $F(x, y^0) \leq F(x^0, y^0) + \varepsilon$ is an ε-optimal solution for the optimistic (pessimistic) case.*

Proof Let there be given some $x \in O(y^0)$ with $F(x, y^0) \leq F(x^0, y^0) + \varepsilon$. Then it holds

$$\varphi_o(x) = F(x, y^0) \leq F(x^0, y^0) + \varepsilon = \bar{\varphi}_o(x^0) + \varepsilon = \varphi_o^* + \varepsilon$$

because of the definition of $O(y^0)$, the assumptions of the theorem and Lemma 5.5. Thus, the conclusion of the theorem holds for the optimistic case.

For the pessimistic case, the proof is analogous. □

5.4 Optimality Conditions Using a Radial-Directional Derivative

Focus in the paper [104] of Fanghänel is on the application of the radial-directional derivative and the radial subdifferential for describing optimality conditions for bilevel optimization problems with discrete variables in the lower level problem. This is a generalization of the approach used in the monograph [52] of Dempe for linear bilevel optimization problems. It will be shown in this section that a similar approach can be used for discrete bilevel optimization problems. The following material in this Section is taken from [104] of Fanghänel.

5.4.1 A Special Mixed-Discrete Bilevel Problem

Consider in this section the bilevel optimization problem

$$\text{“}\min\text{”}\{F(x, y) : x \in \mathbb{R}^n, y \in \Psi_D(x)\} \tag{5.24}$$

with

$$\Psi_D(x) = \underset{y}{\text{Argmin}}\,\{f(x,y) : Ay \le b,\, y \in \mathbb{Z}^n\} \tag{5.25}$$

and $A \in \mathbb{R}^{m\times n}$, $b \in \mathbb{R}^m$, $f(x,y) = h(y) - y^\top x$ where $h : \mathbb{R}^n \to \mathbb{R}$ is a differentiable, *strongly convex* function with modulus θ (cf. Hiriart-Urruty and Lemarechal [145]) and $F(x,y)$ is continuous and continuously differentiable with respect to y. Recall that a function $h : \mathbb{R}^n \to \mathbb{R}$ is strongly convex if there exists $\theta > 0$ such that for all $z^0 \in \mathbb{R}^n$ the inequalities

$$h(z) \ge h(z^0) + \nabla h(z^0)(z - z^0) + \theta\|z - z^0\|^2\ \forall\, z \in \mathbb{R}^n$$

hold. Let

$$Y_D := \{y : Ay \le b,\ y \in \mathbb{Z}^n\}$$

denote the feasible set of the lower level problem in (5.24).

Thus, the problem under consideration is continuous in the upper level and discrete with somc special structure in the lower level, it is of type III.

The solution of the lower level problem is not unique, in general. This causes some uncertainty in the definition of the upper level objective function, see Sect. 1.2. Thus, instead of $F(x,y)$, we will minimize the following functions

$$\varphi_o(x) = \min_{y\in\Psi_D(x)} F(x,y), \tag{5.26}$$

$$\varphi_p(x) = \max_{y\in\Psi_D(x)} F(x,y). \tag{5.27}$$

The function $\varphi_o(x)$ is called *optimistic solution function* and $\varphi_p(x)$ *pessimistic solution function*. A local minimum of the optimististic/pessimistic solution function is called a local optimistic/pessimistic solution of (5.24).

In this section we will use the notation $\widehat{\varphi}(y)$ if the statement holds for both $\varphi_o(x)$ and $\varphi_p(x)$.

For our considerations, the so-called *regions of stability* are again very important. Let $y^0 \in Y_D$. Then the set

$$\begin{aligned}\mathscr{R}(y^0) &= \{x \in \mathbb{R}^n :\ f(x,y^0) \le f(x,y) \text{ for all } y \in Y_D\}\\ &= \{x \in \mathbb{R}^n :\ y^0 \in \Psi_D(x)\}\end{aligned}$$

is called *region of stability* for the point y^0.

To make the subject more clear consider the following example.

Example 5.14

$$\min\{\sin(xy) : x \in \mathbb{R},\ y \in \Psi_D(x)\},$$
$$\Psi_D(x) = \operatorname*{Argmin}_y \left\{\frac{1}{2}y^2 - xy : 0 \leq y \leq 5,\ y \in \mathbb{Z}\right\}$$

The regions of stability are

$$\mathscr{R}(0) = (-\infty, 0.5],\quad \mathscr{R}(1) = [0.5, 1.5],\quad \mathscr{R}(2) = [1.5, 2.5],\quad \mathscr{R}(3) = [2.5, 3.5],$$
$$\mathscr{R}(4) = [3.5, 4.5] \quad \text{and} \quad \mathscr{R}(5) = [4.5, \infty).$$

Using the definitions of the optimistic and pessimistic solution functions, we obtain

$$\varphi_o(x) = \begin{cases} 0 & x \leq 0.5 \\ \sin(x) & 0.5 < x < 1.5 \\ \sin(2x) & 1.5 \leq x \leq 2.5 \\ \sin(3x) & 2.5 < x \leq 3.5 \\ \sin(4x) & 3.5 < x \leq 4.5 \\ \sin(5x) & x > 4.5 \end{cases} \qquad \varphi_p(x) = \begin{cases} 0 & x < 0.5 \\ \sin(x) & 0.5 \leq x \leq 1.5 \\ \sin(2x) & 1.5 < x < 2.5 \\ \sin(3x) & 2.5 \leq x < 3.5 \\ \sin(4x) & 3.5 \leq x < 4.5 \\ \sin(5x) & x \geq 4.5 \end{cases}.$$

As it can be seen in Fig. 5.7, for $0 \leq x \leq 5$, the local optimal solutions of φ_o are

$$x \in [0, 0.5],\quad x = \frac{3\pi}{4},\quad x = 3.5,\quad x = \frac{11\pi}{8},\quad x = \frac{3\pi}{2}$$

and

$$x \in [0, 0.5),\quad x = \frac{3\pi}{4},\quad x = \frac{11\pi}{8},\quad x = \frac{3\pi}{2}$$

are the local optimal solutions of φ_p. □

In Example 5.14 the optimistic and the pessimistic solution functions are not continuous but rather selections of finitely many continuously differentiable functions.

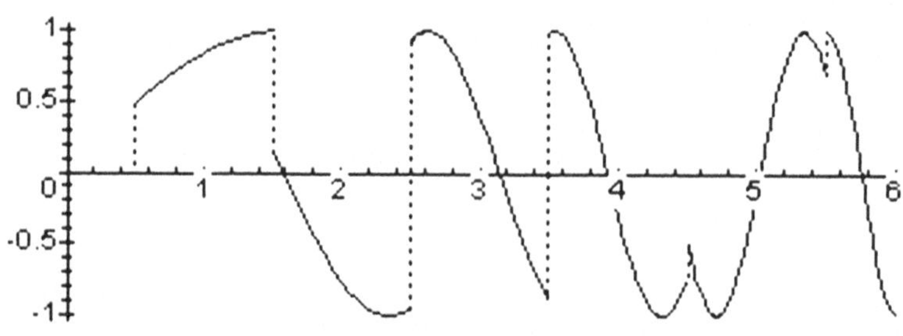

Fig. 5.7 Solution function $\widehat{\varphi}$ for Example 5.14

A special class of functions with that property is that of the so-called GPC^1-functions which have been investigated by Dempe [52] and in Dempe and Unger [78]. Optimistic and pessimistic solution functions for problem (5.24) are in general no GPC^1-functions but they have many properties in common with them. One of the most important property concerning this is that the solution functions are also radial-directionally continuous and radial-directionally differentiable. Using these concepts which where introduced by Recht in [266] we will obtain necessary and sufficient optimality criteria.

5.4.2 Some Remarks on the Sets $\Psi_D(x)$ and $\mathscr{R}(y)$

The set of optimal solutions of the lower level problem $\Psi_D(x)$ and the region of stability $\mathscr{R}(y)$ have the following properties:

Lemma 5.6 (Fanghänel [104]) *Consider problem* (5.24). *For each $y^0 \in Y_D$ the set $\mathscr{R}(y^0)$ is a closed convex set with $\nabla h(y^0)^\top$ in its interior.*

Proof Let $y^0 \in Y_D$. Then for all $x \in \mathscr{R}(y^0)$ it holds $f(x, y^0) \le f(x, y)$ for all $y \in Y_D$ and therefore

$$(y - y^0)^\top x \le h(y) - h(y^0) \quad \forall y \in Y_D.$$

Thus, $\mathscr{R}(y^0)$ corresponds to the intersection of (maybe infinitely many) halfspaces. This implies that $\mathscr{R}(y^0)$ is convex and closed.

Now we show that $\nabla h(y^0)^\top \in \operatorname{int} \mathscr{R}(y^0)$. Since $h : \mathbb{R}^n \to \mathbb{R}$ is strongly convex, there exists some $\theta > 0$ with $h(y) \ge h(y^0) + \nabla h(y^0)(y - y^0) + \theta \|y - y^0\|^2 \; \forall y \in \mathbb{R}^n$.

Consider $x = \nabla h(y^0)^\top + \alpha d$ with $d \in \mathbb{R}^n$, $\|d\| = 1$ and $\alpha \in [0, \theta]$. Then, for all $y \in Y_D$, $y \neq y^0$, the following sequence of inequalities is valid by $\|y - y^0\| \ge 1$ for $y \neq y^0$:

$$\begin{aligned} h(y) &\ge h(y^0) + \nabla h(y^0)(y - y^0) + \theta \|y - y^0\|^2 \\ &= h(y^0) + x^\top (y - y^0) - \alpha d^\top (y - y^0) + \theta \|y - y^0\|^2 \\ &\ge h(y^0) + x^\top (y - y^0) - \alpha \|y - y^0\| + \theta \|y - y^0\|^2 \\ &\ge h(y^0) + x^\top (y - y^0) + (\theta - \alpha) \|y - y^0\| \\ &\ge h(y^0) + x^\top (y - y^0). \end{aligned}$$

Thus, we obtain $(\nabla h(y^0)^\top + \alpha d) \in \mathscr{R}(y^0)$ for all $\alpha \in [0, \theta]$, i.e. the assumption holds. □

Lemma 5.7 (Fanghänel [104])

1. *For each $x \in \mathbb{R}^n$ the set $\Psi_D(x)$ has finite cardinality.*
2. *If $x^* \in \operatorname{int} \mathscr{R}(y^*)$, then $\Psi_D(x^*) = \{y^*\}$.*
3. *For $x^* \in \mathbb{R}^n$ there exists $\delta > 0$ such that $\Psi_D(x) \subseteq \Psi_D(x^*)$ for all $x \in U_\delta(x^*)$ with $U_\delta(x^*) := \{x : \|x - x^*\| < \delta\}$.*

Proof 1. If $Y_D = \emptyset$ the assumption obviously holds. Assume that $Y_D \neq \emptyset$ and take a point $y^* \in Y_D$. Let an arbitrary $x \in \mathbb{R}^n$ be given. Then for all $y \in \Psi_D(x)$ it holds

$$h(y) - x^\top y \leq h(y^*) - x^\top y^*$$

implying

$$h(y^*) + \nabla h(y^*)(y - y^*) + \theta \|y - y^*\|^2 \leq h(y^*) + x^\top (y - y^*)$$

for some $\theta > 0$ since h is strongly convex. Thus,

$$\begin{aligned} \theta \|y - y^*\|^2 &\leq (x - \nabla h(y^*)^\top)(y - y^*) \leq \|x - \nabla h(y^*)^\top\| \|y - y^*\|, \\ \|y - y^*\| &\leq \frac{1}{\theta} \|x - \nabla h(y^*)^\top\| \end{aligned}$$

for all $y \in \Psi_D(x)$. Therefore $\Psi_D(y) \subset \mathbb{Z}^n$ has finite cardinality.

2. The inclusion $x^* \in \operatorname{int} \mathscr{R}(y^*)$ implies $\{y^*\} \subseteq \Psi_D(x^*)$ by definition. To prove the opposite direction assume that there exists a point $y \in \Psi_D(x^*)$, $y \neq y^*$. Then,

$$\begin{aligned} h(y) - {x^*}^\top y &= h(y^*) - {x^*}^\top y^* \\ h(y) - h(y^*) &= {x^*}^\top (y - y^*) > \nabla h(y^*)(y - y^*) \end{aligned}$$

since h is strongly convex. Due to $x^* \in \operatorname{int} \mathscr{R}(y^*)$ there exists some $\varepsilon > 0$ such that $x := x^* + \varepsilon(x^* - \nabla h(y^*)^\top) \in \mathscr{R}(y^*)$. Now we obtain

$$\begin{aligned} f(x, y^*) = h(y^*) - x^\top y^* &= h(y) - x^\top y^* - {x^*}^\top (y - y^*) \\ &= f(x, y) + (x - x^*)^\top (y - y^*) \\ &= f(x, y) + \varepsilon({x^*}^\top - \nabla h(y^*))(y - y^*) > f(x, y) \end{aligned}$$

which is a contradiction to $x \in \mathscr{R}(y^*)$.

3. Assume that the assertion does not hold. Then there exist sequences

$$\{x^k\}_{k=1}^\infty \quad \text{with} \quad \lim_{k \to \infty} x^k = x^*,$$

and

$$\{y^k\}_{k=1}^\infty \quad \text{with} \quad y^k \in \Psi_D(x^k) \quad \text{but} \quad y^k \notin \Psi_D(x^*) \quad \text{for all} \quad k.$$

Thus, for fixed $y^* \in Y_D$, it holds

$$
\begin{aligned}
h(y^k) - {x^k}^\top y^k &\leq h(y^*) - {x^k}^\top y^* \\
h(y^*) + \nabla h(y^*)(y^k - y^*) + \theta \|y^k - y^*\|^2 &\leq h(y^*) + {x^k}^\top (y^k - y^*) \\
\|y^k - y^*\| &\leq \frac{\|x^k - \nabla h(y^*)^\top\|}{\theta}.
\end{aligned}
$$

This yields

$$
\|y^k - y^*\| \leq \underbrace{\frac{\|x^k - x^*\|}{\theta}}_{\to 0} + \frac{\|x^* - \nabla h(y^*)^\top\|}{\theta},
$$

i.e. $\{y^k\}$ is bounded and has finitely many elements. Therefore, we can assume that all y^k are equal, i.e. $\exists y \in Y_D$ with $y \in \Psi_D(x^k)\ \forall k$ but $y \notin \Psi_D(x^*)$. That means $x^k \in \mathscr{R}(y)\ \forall k$ but $x^* \notin \mathscr{R}(y)$. This is a contradiction to Lemma 5.6. □

5.4.3 Basic Properties of $\widehat{\varphi}(x)$

In this section we will see that, locally around $x^* \in \mathbb{R}^n$, the optimistic/pessimistic solution functions are a selection of finitely many continuously differentiable functions. Let $\varepsilon > 0$ and consider the support set

$$
Y_y(x^*) := \{x \in U_\varepsilon(x^*) \cap \mathscr{R}(y) : \ F(x, y) = \widehat{\varphi}(x)\}
$$

and its *Bouligand cone*

$$
C_{Y_y}(x^*) := \left\{ r : \exists \{x^s\} \subseteq Y_y(x^*)\ \exists \{t_s\} \subseteq \mathbb{R}_+ : \ x^s \to x^*, t_s \downarrow 0, \lim_{s \to \infty} \frac{x^s - x^*}{t_s} = r \right\}.
$$

That means, $Y_y(x^*)$ is the set of all $x \in U_\varepsilon(x^*)$ for which both $y \in \Psi_D(x)$ and $F(x, y) = \widehat{\varphi}(x)$ hold for a fixed point $y \in Y_D$. Properties of these sets are essential for the investigation of generalized PC^1–functions in the paper [78] of Dempe and Unger leading to optimality conditions for linear bilevel optimization problems in the monograph [52] by Dempe. The following two theorems show that the objective functions in the two auxiliary problems (5.26) and (5.27) have interesting properties. But they are not generalized PC^1–functions (as defined by Dempe in [52]), which is shown by an example thereafter.

Theorem 5.11 (Fanghänel [104]) *Consider problem* (5.24). *For the function $\widehat{\varphi}$ and each $x^* \in \mathbb{R}^n$ it holds:*

1. *There exists an open neighborhood $U_\varepsilon(x^*)$ of x^* and a finite number of points $y \in \Psi_D(x^*)$ with*

$$\widehat{\varphi}(x) \in \{F(x, y)\}_{y \in \Psi_D(x^*)} \quad \forall x \in U_\varepsilon(x^*).$$

2. $\operatorname{int} Y_y(x^*) = U_\varepsilon(x^*) \cap \operatorname{int} \mathscr{R}(y)$ *and* $Y_y(x^*) \subseteq \operatorname{cl} \operatorname{int} Y_y(x^*)$ *for* $x^*, y \in \mathbb{R}^n$.
3. $C_{Y_y}(x^*) \subseteq \operatorname{cl} \operatorname{int} C_{Y_y}(x^*)$ *for* $x^* \in \mathscr{R}(y)$.

Proof Let an arbitrary $x^* \in \mathbb{R}^n$ be given.

1. Because of Lemma 5.7, $\Psi_D(x^*)$ has finite cardinality and there exists some $\varepsilon > 0$ with $\Psi_D(x^*) \supseteq \Psi_D(x)$ for all $x \in U_\varepsilon(x^*)$. With $\widehat{\varphi}(x) \in \{F(x, y)\}_{x \in \Psi_D(x)}$ it follows $\widehat{\varphi}(x) \in \{F(x, y)\}_{y \in \Psi_D(x^*)} \quad \forall x \in U_\varepsilon(x^*)$.
2. Let $\bar{x} \in \operatorname{int} Y_{y^*}(x^*)$. Then there exists some $\delta > 0$ with $U_\delta(\bar{x}) \subseteq Y_{y^*}(x^*)$. Thus, $\bar{x} \in U_\varepsilon(x^*)$ and $U_\delta(\bar{x}) \subseteq \mathscr{R}(y^*)$, i.e. $\bar{x} \in U_\varepsilon(x^*) \cap \operatorname{int} \mathscr{R}(y^*)$.
   Let $\bar{x} \in U_\varepsilon(x^*) \cap \operatorname{int} \mathscr{R}(y^*)$. Then there exists some $\delta > 0$ with $U_\delta(\bar{x}) \subseteq U_\varepsilon(x^*)$ and $U_\delta(\bar{x}) \subseteq \operatorname{int} \mathscr{R}(y^*)$. From Lemma 5.7, it follows $\Psi_D(x) = \{y^*\} \, \forall x \in U_\delta(\bar{x})$. Thus, $\widehat{\varphi}(x) = F(x, y^*) \, \forall x \in U_\delta(\bar{x})$, i.e. $x \in Y_{y^*}(x^*) \, \forall x \in U_\delta(\bar{x})$. Therefore, $\bar{x} \in \operatorname{int} Y_{y^*}(x^*)$. This implies the first equation of part 2.
   Now let $\bar{x} \in Y_y(x^*)$. This means $\bar{x} \in \mathscr{R}(y)$, $\bar{x} \in U_\varepsilon(x^*)$ and $\widehat{\varphi}(\bar{x}) = F(\bar{x}, y)$. Since $\mathscr{R}(y)$ is convex with nonempty interior (cf. Lemma 5.6), there exists some sequence $\{x^k\} \subseteq \operatorname{int} \mathscr{R}(y)$ with $x^k \to \bar{x}$, $k \to \infty$. W.l.o.g. we can further assume that $x^k \in U_\varepsilon(x^*) \, \forall k$. Consequently, $x^k \in \operatorname{int} Y_y(x^*) \, \forall k$ and thus $\bar{x} \in \operatorname{cl} \operatorname{int} Y_y(x^*)$.
3. Let an arbitrary $r \in C_{Y_y}(x^*)$ be given. Then there exist sequences $\{x^s\} \subseteq Y_y(x^*)$ and $\{t^s\} \subseteq \mathbb{R}_+$ with $x^s \to x^*$, $t^s \downarrow 0$ and $\lim_{s\to\infty} \frac{x^s - x^*}{t^s} = r$. We can assume w.l.o.g. that $t_s < 1 \, \forall s$.
   Take any $\tilde{x} \in \operatorname{int} Y_y(x^*)$ and let $\widehat{x}^s := t^s \tilde{x} + (1 - t^s)x^* = x^* + t^s(\tilde{x} - x^*)$. Then, $\lim_{s\to\infty} \widehat{x}^s = x^*$ and $\frac{\widehat{x}^s - x^*}{t^s} = \tilde{x} - x^* =: \tilde{r} \, \forall s$. Since $\mathscr{R}(y)$ is convex it follows easily that $\widehat{x}^s \in \operatorname{int} Y_y(x^*) \, \forall s$ and $\tilde{r} \in \operatorname{int} C_{Y_y}(x^*)$.
   Now consider a point $z_\lambda^s := \lambda x^s + (1 - \lambda)\widehat{x}^s$ with $\lambda \in (0, 1)$. Since $\mathscr{R}(y)$ is convex it follows $z_\lambda^s \in \operatorname{int} Y_y(x^*) \, \forall \lambda \in (0, 1) \, \forall s$. Then, $z_\lambda^s \to x^*$ for $s \to \infty$ and $\lim_{s\to\infty} \frac{z_\lambda^s - x^*}{t^s} = \lambda r + (1 - \lambda)\tilde{r} =: r_\lambda \in C_{Y_y}(x^*) \, \forall \lambda \in (0, 1)$. Moreover, $r_\lambda \to r$ for $\lambda \to 1$.
   Now, from $z_\lambda^s \in \operatorname{int} Y_y(x^*)$ it follows easily that $z_\lambda^s - x^* \in \operatorname{int} C_{Y_y}(x^*)$ and thus $\frac{z_\lambda^s - x^*}{t^s} \in \operatorname{int} C_{Y_y}(x^*) \, \forall s \, \forall \lambda \in (0, 1)$.
   Hence $r_\lambda \in \operatorname{cl} \operatorname{int} C_{Y_y}(x^*) \, \forall \lambda \in (0, 1)$. This together with $r_\lambda \to r$ for $\lambda \to 1$ implies $r \in \operatorname{cl} \operatorname{cl} \operatorname{int} C_{Y_y}(x^*) = \operatorname{cl} \operatorname{int} C_{Y_y}(x^*)$. □

Theorem 5.12 (Fanghänel [104]) *We have* $\operatorname{int} C_{Y_{y^1}}(x^*) \cap \operatorname{int} C_{Y_{y^2}}(x^*) = \emptyset$ *if* $y^1 \neq y^2$.

Proof Let $r \in C_{Y_{y^1}}(x^*) \cap C_{Y_{y^2}}(x^*)$ be arbitrary. Due to $r \in C_{Y_{y^1}}(x^*)$ there exist sequences $\{x^s\} \subset Y_{y^1}(x^*)$, $x^s \to x^*$ and $\{t^s\}$, $t^s \downarrow 0$ with $r^s := \frac{x^s - x^*}{t^s} \to r$.

From $x^s \in Y_{y^1}(x^*) \, \forall s$ it follows $x^s \in \mathscr{R}(y^1) \, \forall s$, i.e. $h(y^1) - {x^s}^\top y^1 \le h(y^2) - {x^s}^\top y^2$. Since $y^1, y^2 \in \Psi_D(x^*)$ it holds $h(y^1) - {x^*}^\top y^1 = h(y^2) - {x^*}^\top y^2$. Hence,

$$x^{s\top}(y^1 - y^2) \geq h(y^1) - h(y^2) = x^{*\top}(y^1 - y^2)$$
$$(x^s - x^*)^\top(y^1 - y^2) \geq 0 \quad \forall s$$
$$r^{s\top}(y^1 - y^2) \geq 0 \quad \forall s.$$

With $r^s \to r$ this yields $r^\top(y^1 - y^2) \geq 0$.

From $r \in C_{Y_{y2}}(x^*)$ it follows analogously $(y^1 - y^2)^\top r \leq 0$. Therefore it holds

$$(y^2 - y^1)^\top r = 0 \quad \text{for all} \quad r \in C_{Y_{y1}}(x^*) \cap C_{Y_{y2}}(x^*).$$

Assume that there exists some $r \in \operatorname{int} C_{Y_{y1}}(x^*) \cap \operatorname{int} C_{Y_{y2}}(x^*)$. Then $\forall d \in \mathbb{R}^n$, $\|d\| = 1$ there exists a real number $\delta > 0$ with $r + \delta d \in C_{Y_{y1}}(x^*) \cap C_{Y_{y2}}(x^*)$, i.e.

$$(y^2 - y^1)^\top(r + \delta d) = 0$$
$$\delta(y^2 - y^1)^\top d = 0$$
$$(y^2 - y^1)^\top d = 0 \quad \forall d$$

and therefore $y^1 = y^2$. □

It is worth noting that $\widehat{\varphi}$ is not a GCP^1-function in general (cf. Dempe [52]; Dempe and Unger [78]). In order that $\widehat{\varphi}$ is a GPC^1-function one requires additionally assumption compared with the results in Theorems 5.11 and 5.12. Namely, there exists a number $\delta > 0$ such that, for all $r \in C_{Y_{y1}}(x^*) \cap C_{Y_{y2}}(x^*)$, $\|r\| = 1$, $y^1 \neq y^2$, a number $t_0 = t(r) \geq \delta$ can be found with $x^* + tr \in Y_{y^1}(x^*)$ or $x^* + tr \in Y_{y^2}(x^*)$ $\forall t \in (0, t_0)$ (see Dempe [52]).

We will see that the functions $\widehat{\varphi}$ usually do not have this property.

Example 5.15 Consider the lower level problem in (5.24) with the feasible set $Y_D = \{y^1 = (0,0,0)^\top, y^2 = (1,0,0)^\top, y^3 = (0,1,0)^\top\}$ and $f(x,y) = \frac{1}{2}y^\top y - y^\top x$. Then we obtain the following regions of stability:

$$\mathscr{R}(y^1) = \{y \in \mathbb{R}^3 : x_1 \leq 1/2, x_2 \leq 1/2\}$$
$$\mathscr{R}(y^2) = \{y \in \mathbb{R}^3 : x_1 \geq 1/2, x_2 \leq x_1\}$$
$$\mathscr{R}(y^3) = \{y \in \mathbb{R}^3 : x_2 \geq 1/2, x_2 \geq x_1\}$$

Let $F(x,y) = (1/2, -1, 0)^\top y$ be the objective function in the upper level problem. Then,

$$\varphi_o(x) = \begin{cases} -1 & x \in \mathscr{R}(y^3) \\ 0 & x \in \mathscr{R}(y^1) \backslash \mathscr{R}(y^3) \\ 1/2 & else \end{cases}.$$

Set $r = (0,0,1)^\top$ and $x^* = (1/2, 1/2, 0)^\top$.
The point $x^1(\varepsilon) := (1/2 - \varepsilon^2, 1/2 - \varepsilon^2, \varepsilon + \varepsilon^2)^\top \in Y_{y^1}$ $\forall \varepsilon > 0$. Then

$$\lim_{\varepsilon\to 0} x^1(\varepsilon) = (1/2, 1/2, 0)^\top = x^*,$$
$$\lim_{\varepsilon\to 0} \frac{x^1(\varepsilon) - x^*}{\varepsilon} = \lim_{\varepsilon\to 0} (-\varepsilon, -\varepsilon, 1+\varepsilon)^\top = r, \text{ i.e. } r \in C_{Y_{y^1}}(x^*).$$

Analogously, $x^2(\varepsilon) := (1/2 + \varepsilon^2, 1/2, \varepsilon + \varepsilon^2)^\top \in Y_{y^2}\ \forall \varepsilon > 0$. Then

$$\lim_{\varepsilon\to 0} x^2(\varepsilon) = (1/2, 1/2, 0)^\top = x^*,$$
$$\lim_{\varepsilon\to 0} \frac{x^2(\varepsilon) - x^*}{\varepsilon} = \lim_{\varepsilon\to 0} (\varepsilon, 0, 1+\varepsilon)^\top = r, \text{ i.e. } r \in C_{Y_{y^2}}(x^*).$$

Therefore, $r \in C_{Y_{y^1}}(x^*) \cap C_{Y_{y^2}}(x^*)$, $\|r\| = 1$, $y^1 \neq y^2$ but

$$\varphi_o(x^* + tr) = -1 < F(x^* + tr, y^i), \quad i = 1, 2, \quad \forall t > 0,$$

i.e. $x^* + tr \notin Y_{y^1}(x^*)$ and $x^* + tr \notin Y_{y^2}(x^*)$ $\forall t > 0$. □

5.4.4 The Radial-Directional Derivative

Definition 5.6 Let $U \subseteq \mathbb{R}^n$ be an open set, $x^0 \in U$ and $\widehat{\varphi} : U \to \mathbb{R}$. We say that $\widehat{\varphi}$ is *radial-continuous* at x^0 in direction $r \in \mathbb{R}^n$, if there exists a real number $\widehat{\varphi}(x^0; r)$ such that

$$\lim_{t\downarrow 0} \widehat{\varphi}(x^0 + tr) = \widehat{\varphi}(x^0; r).$$

If the radial limit $\widehat{\varphi}(x^0; r)$ exists for all $r \in \mathbb{R}^n$, $\widehat{\varphi}$ is called radial-continuous at y^0. $\widehat{\varphi}$ is *radial-directionally differentiable* at x^0, if there exists a positively homogeneous function $d\widehat{\varphi}_{x^0} : \mathbb{R}^n \to \mathbb{R}$ such that

$$\widehat{\varphi}(x^0 + tr) - \widehat{\varphi}(x^0; r) = t \cdot d\widehat{\varphi}_{x^0}(r) + o(x^0, tr)$$

with $\lim_{t\downarrow 0} o(x^0, tr)/t = 0$ holds for all $r \in \mathbb{R}^n$ and all $t > 0$. Obviously, $d\widehat{\varphi}_{x^0}$ is uniquely defined and is called the *radial-directional derivative* of $\widehat{\varphi}$ at x^0.

Theorem 5.13 (Fanghänel [104]) *Consider problem* (5.24). *Then, both the optimistic solution function* φ_o *and the pessimistic solution function* φ_p *are radial-continuous and radial-directionally differentiable.*

Proof Consider x^0 and some direction $r \in \mathbb{R}^n$, $\|r\| = 1$. Further let

$$I_r(x^0) := \{y \in \Psi_D(x^0) : \forall \varepsilon > 0\ \exists t \in (0, \varepsilon) \text{ with } x^0 + tr \in Y_y(x^0)\}.$$

Since $\Psi_D(x^0)$ has finite cardinality and the sets $\mathscr{R}(y)$ are convex,

$$\varphi_o(x^0 + tr) = G(x^0 + tr) := \min_{y \in I_r(x^0)} F(x^0 + tr, y)$$

for all sufficiently small reals $t > 0$. Since the function $G(\cdot)$ is the minimum function of finitely many continuously differentiable functions, it is continuous and quasidifferentiable (cf. Dem'yanov and Rubinov [84]) and thus directionally differentiable in $t = 0$. Therefore, the limits

$$\lim_{t \downarrow 0} G(x^0 + tr) = G(x^0) \text{ and } \lim_{t \downarrow 0} \frac{G(x^0 + tr) - G(x^0)}{t} = G'(x^0; r)$$

exist. Moreover, since $\forall y \in I_r(x^0)\, \exists \{t_k\} \downarrow 0 : x^0 + t_k r \in Y_y(x^0)$ and

$$\lim_{t \downarrow 0} G(x^0 + tr) = \lim_{k \to \infty} G(x^0 + t_k r) = \lim_{k \to \infty} F(x^0 + t_k r, y) = F(x^0, y),$$

we derive

$$\varphi_o(x^0; r) = \lim_{t \downarrow 0} G(x^0 + tr) = G(x^0) = F(x^0, y)\ \forall y \in I_r(x^0). \tag{5.28}$$

Concerning the radial-directional derivative, we obtain

$$\begin{aligned} d\varphi_{o x^0}(r) &= \lim_{t \downarrow 0} \frac{\varphi_o(x^0 + tr) - \varphi_o(x^0; r)}{t} = \lim_{t \downarrow 0} \frac{G(x^0 + tr) - G(x^0)}{t} \\ &= \nabla_x F(x^0, y) r \quad \forall y \in I_r(x^0) \end{aligned} \tag{5.29}$$

since g is continuously differentiable with respect to y.

For $\varphi_p(x)$ we can prove the assertions analogously. □

Example 5.16 Consider problem (5.24) with the lower level feasible set $Y_D = \{y^1 = (0, 0)^\top, y^2 = (0, 1)^\top, y^3 = (-1, 0)^\top\}$, $x \in \mathbb{R}^2$ and with the objective functions $f(x, y) = \frac{1}{2} y^\top y - x^\top y$ and

$$F(x, y) = y_1 + y_2 \cdot \begin{cases} x_1^3 \sin \frac{1}{x_1} & x_1 > 0 \\ 0 & x_1 \le 0. \end{cases}$$

Then the function $F(x, y)$ is continuously differentiable with respect to x. The regions of stability are

$$\begin{aligned} \mathscr{R}(y^1) &= \{x \in \mathbb{R}^2 : x_1 \ge -0.5,\ x_2 \le 0.5\} \\ \mathscr{R}(y^2) &= \{x \in \mathbb{R}^2 : x_1 + x_2 \ge 0,\ x_2 \ge 0.5\} \\ \mathscr{R}(y^3) &= \{x \in \mathbb{R}^2 : x_1 \le -0.5,\ x_1 + x_2 \le 0\}. \end{aligned}$$

Let $x^* = (0, \frac{1}{2})^\top$ and $r = (1,0)^\top$. Then $I_r(x^*) = \{y^1, y^2\}$ for both the optimistic and the pessimistic solution function. Thus, it holds

$$\varphi_o(x^*; r) = \varphi_p(x^*; r) = F(x^*, y^1) = F(x^*, y^2) = 0$$

and

$$d\varphi_{o_{x^*}}(r) = d\varphi_{p_{x^*}}(r) = \nabla_x F(x, y^i)(1,0)^\top, \quad i = 1, 2.$$

Further it holds $\varphi_o(x^*) = \varphi_p(x^*) = 0$. Remarkable in this example is the fact that $\forall \varepsilon > 0$ there exists some $t \in (0, \varepsilon)$ with either

$$\widehat{\varphi}(x * + tr) \neq F(x^* + tr, y^1) \quad \text{or} \quad \widehat{\varphi}(x^* + tr) \neq F(x^* + tr, y^2).$$

Now let $\bar{x} = (-\frac{1}{2}, \frac{1}{2})^\top$ and $r = (-1, 1)^\top$. Then, for the optimistic solution function it holds

$$I_r(\bar{x}) = \{y^3\} \quad \text{and} \quad \varphi_o(\bar{x}) = \varphi_o(\bar{x}; r) = -1$$

and for the pessimistic solution function it holds

$$I_r(\bar{x}) = \{y^2\} \quad \text{and} \quad \varphi_p(\bar{x}) = \varphi_p(\bar{x}; r) = 0.$$

Considering the direction $r = (0, 1)$ we obtain $I_r(\bar{x}) = \{y^2\}$ and $\widehat{\varphi}(\bar{x}; r) = 0$ for both the optimistic and the pessimistic case, but $\varphi_o(\bar{x}) = -1 \neq 0 = \varphi_p(\bar{x})$. □

Lemma 5.8 *For all $x^* \in \mathbb{R}^n$ and for all $r \in \mathbb{R}^n$ it holds:*

1. $\varphi_o(x^*) \leq \varphi_o(x^*; r)$
2. $\varphi_p(x^*) \geq \varphi_p(x^*; r)$

Proof Assume there exists some x^* and some r with $\varphi_o(x^*) > \varphi_o(x^*; r)$. Then from $I_r(x^*) \subseteq \Psi_D(x^*)$ and the proof of Theorem 5.13 it follows that there exists some $y \in \Psi_D(x^*)$ with $\varphi_o(x^*; r) = F(x^*, y)$. Hence, $\varphi_o(x^*) > F(x^*, y)$ for some $y \in \Psi_D(x^*)$. This is a contradiction to the definition of φ_o.

The proof for φ_p is similar. □

5.4.5 *Optimality Criteria Based on the Radial-Directional Derivative*

Let locmin $\widehat{\varphi}$ denote the set of all local minima of the function $\widehat{\varphi}(\cdot)$. The lower level problem of the bilevel optimization problem (5.24) has a fixed feasible set with finitely many elements and a parameter in the objective function. As it is shown in the next theorem, this implies that every pessimistic optimal solution is also optimistic optimal.

Theorem 5.14 (Fanghänel [104]) *For problem* (5.24) *it holds*

$$\text{locmin } \varphi_p \subseteq \text{locmin } \varphi_o\,.$$

Proof Arguing by contradiction, we assume that there is some x^* with $x^* \in \text{locmin } \varphi_p$ but $x^* \notin \text{locmin } \varphi_o$. Then there exists some sequence $\{x^k\}$ with $\lim_{k\to\infty} x^k = x^*$ and $\varphi_o(x^k) < \varphi_o(x^*)\ \forall k$. Since $\Psi_D(x^*)$ has finite cardinality and $\Psi_D(x^*) \supseteq \Psi_D(x)$ for all x in a neighborhood of x^*, we can assume w.l.o.g. that there exists some $y \in \Psi_D(x^*)$ with $x^k \in Y_y(x^*)\ \forall k$. Due to differentiability of $F(\cdot, y)$ with respect to x and $Y_y(x^*) \subseteq \text{cl int } Y_y(x^*)$, we can further assume that $x^k \in \text{int } Y_y(x^*)$ $\forall k$. Thus, it holds $\Psi_D(x^k) = \{y\}\ \forall k$, i.e. $\varphi_o(x^k) = \varphi_p(x^k) = F(x^k, y)\ \forall k$. Consequently,

$$\varphi_p(x^k) = \varphi_o(x^k) < \varphi_o(x^*) \le \varphi_p(x^*) \quad \forall k.$$

This is a contradiction to $x^* \in \text{locmin } \varphi_p$. □

Theorem 5.15 (Fanghänel [104]) *Consider some point* $x^0 \in \mathbb{R}^n$ *and let* $\widehat{\varphi} : \mathbb{R}^n \to \mathbb{R}$ *be the optimistic* (5.26) *or the pessimistic* (5.27) *solution function. Then* $x^0 \notin \text{locmin } \widehat{\varphi}$ *if there exists a direction* $r \in \mathbb{R}^n$, $\|r\| = 1$, *such that one of the following conditions is satisfied:*

1. $d\widehat{\varphi}_{y^0}(r) < 0$ *and* $\widehat{\varphi}(y^0; r) \le \widehat{\varphi}(y^0)$
2. $\widehat{\varphi}(y^0; r) < \widehat{\varphi}(y^0)$.

Proof Let the vector r^0 with $\|r^0\| = 1$ satisfy the first condition. That means

$$d\widehat{\varphi}_{x^0}(r^0) = \lim_{t\downarrow 0} t^{-1}(\widehat{\varphi}(x^0 + tr^0) - \widehat{\varphi}(x^0; r^0)) < 0.$$

Then there exists some $t_0 > 0$ such that $\widehat{\varphi}(x^0 + tr^0) < \widehat{\varphi}(x^0; r^0)\ \forall t \in (0, t_0)$. Because of $\widehat{\varphi}(x^0; r^0) \le \widehat{\varphi}(x^0)$ we have $\widehat{\varphi}(x^0 + tr^0) < \widehat{\varphi}(x^0)$ for all these t, too. Thus, x^0 cannot be a local minimum of $\widehat{\varphi}$ since for each $\varepsilon > 0$ there exists some $0 < t < \min\{\varepsilon, t_0\}$ with $\|(x^0 + tr^0) - x^0\| < \varepsilon$ and $\widehat{\varphi}(x^0 + tr^0) < \widehat{\varphi}(x^0)$.

Now let the vector r^0 with $\|r^0\| = 1$ satisfy the second condition. Then it holds

$$\widehat{\varphi}(x^0) - \widehat{\varphi}(x^0; r^0) = \widehat{\varphi}(x^0) - \lim_{t\downarrow 0} \widehat{\varphi}(x^0 + tr^0) > 0.$$

Hence, there exists some $t_0 > 0$ such that $\widehat{\varphi}(x^0) > \widehat{\varphi}(x^0 + tr^0)\ \forall t \in (0, t_0)$. Thus, x^0 cannot be a local minimum of $\widehat{\varphi}$. □

Specifying the conditions of Theorem 5.15 by using Lemma 5.8 we obtain the following necessary optimality conditions:

1. If $x^0 \in \text{locmin } \varphi_p$, then

$$\varphi_p(x^0) = \varphi_p(x^0; r) \quad \text{and} \quad d\varphi_{p_{x^0}}(r) \ge 0 \quad \forall r.$$

2. If $x^0 \in \text{locmin } \varphi_o$, then

$$\varphi_o(x^0) < \varphi_o(x^0; r) \quad \text{or} \quad d\varphi_{o x^0}(r) \geq 0 \quad \forall r.$$

To prove the next theorem we will need the following lemma.

Lemma 5.9 *Consider problem* (5.24) *and a point* $(x^0, y^0) \in \mathbb{R}^n \times \mathbb{R}^n$ *with* $y^0 \in \Psi_D(x^0)$ *and* $\varphi_o(x^0) = F(x^0, y^0)$. *Then* $r \in C_{Y_{y^0}}(x^0)$ *implies*

$$\varphi_o(x^0) = \varphi_o(x^0; r).$$

Proof Since φ_o is radial-continuous and using Lemma 5.7, there exists $\tilde{y} \in \Psi_D(x^0)$ and a sequence $\{t_k\} \downarrow 0$ with $\tilde{y} \in \Psi_D(x^0 + t_k r)$, $\varphi_o(x^0 + t_k r) = F(x^0 + t_k r, \tilde{y})$ and

$$\varphi_o(x^0; r) = \lim_{k\to\infty} \varphi_o(x^0 + t_k r) = \lim_{k\to\infty} F(x^0 + t_k r, \tilde{y}) = F(x^0, \tilde{y}).$$

Clearly, it holds $r \in C_{Y_{\tilde{y}}}(x^0) \cap C_{Y_{y^0}}(x^0)$. Then, from the proof of Theorem 5.12 it follows that $r^\top(y^0 - \tilde{y}) = 0$. Furthermore, we know that $\tilde{y}, y^0 \in \Psi_D(x^0)$ and thus $h(y^0) - {x^0}^\top y^0 = h(\tilde{y}) - \tilde{y}^\top x^0$. Consequently, $h(y^0) - (x^0 + t_k r)^\top y^0 = h(\tilde{y}) - \tilde{y}^\top(x^0 + t_k r)\ \forall k$, i.e. $y^0 \in \Psi_D(x^0 + t_k r)\ \forall k$. Thus, we obtain $\varphi_o(x^0 + t_k r) \leq F(x^0 + t_k r, y^0)\ \forall k$, i.e.

$$\varphi_o(x^0; r) = \lim_{k\to\infty} \varphi_o(x^0 + t_k r) \leq \lim_{k\to\infty} F(x^0 + t_k r, y^0) = F(x^0, y^0) = \varphi_o(x^0).$$

Now, from Lemma 5.8, it follows the equality. □

Theorem 5.16 (Fanghänel [104]) *Consider problem* (5.24) *and let* x^0 *be a point which satisfies one of the following two conditions:*

1. $\widehat{\varphi}(x^0) < \widehat{\varphi}(x^0; r)\ \forall r \in \mathbb{R}^n$
2. $\widehat{\varphi}(x^0) \leq \widehat{\varphi}(x^0; r)\ \forall r \in \mathbb{R}^n$ *and* $d\widehat{\varphi}_{x^0}(r) > \gamma\ \forall r : \widehat{\varphi}(x^0) = \widehat{\varphi}(x^0; r),\ \|r\| = 1$ *with* $\gamma = 0$ *in the optimistic case and* $\gamma > 0$ *in the pessimistic case.*

Then, $\widehat{\varphi}$ *achieves a local minimum at* x^0.

Proof Suppose x^0 satisfies one of the two conditions of the theorem. Arguing by contradiction, we assume that there is a sequence $\{x^k\}_{k=1}^\infty$ with $\lim_{k\to\infty} x^k \to x^0$ and $\widehat{\varphi}(x^k) < \widehat{\varphi}(x^0)\ \forall k$. Since $\Psi_D(x^0) \supseteq \Psi_D(x)$ for all x in a neighborhood of x^0 and $\Psi_D(x^0)$ has finite cardinality, there exists some $y^0 \in \Psi_D(x^0)$ such that $Y_{y^0}(x^0)$ contains infinitely many of the points x^k. In the following, we consider the subsequence of points of the sequence $\{x^k\}_{k=1}^\infty$ belonging to the set $Y_{y^0}(x^0)$. Denote this subsequence by $\{x^k\}_{k=1}^\infty$ again. Because of the continuity of $F(\cdot, y^0)$, it follows that

$$F(x^0, y^0) = \lim_{k\to\infty} F(x^k, y^0) = \lim_{k\to\infty} \widehat{\varphi}(x^k) \leq \widehat{\varphi}(x^0).$$

Now we define $r^0 := \lim_{k\to\infty} \frac{x^k - x^0}{\|x^k - x^0\|}$. Hence, $r^0 \in C_{Y_{y^0}}(x^0)$. Due to $C_{Y_{y^0}}(x^0) \subseteq \text{cl int } C_{Y_{y^0}}(x^0)$, there exists some $\widehat{r} \in \text{int } C_{Y_{y^0}}(x^0)$ arbitrarily close to r^0 with $\|\widehat{r}\| = 1$. Now let $\{t_k\}_{k=1}^{\infty}$ be any sequence of positive real numbers with $t_k \downarrow 0$ for $k \to \infty$. Remember that

$$C_{Y_{y^0}}(x^0) = \text{cl}\{d : \exists\, x \in Y_{y^0}(x^0)\ \exists\, \alpha > 0 \quad \text{with} \quad d = \alpha(x - x^0)\}$$

by convexity and closedness of $Y_{y^0}(x^0)$. Hence, since $\widehat{r} \in \text{int } C_{Y_{y^0}}(x^0)$, it holds both $x^0 + t_k\widehat{r} \in Y_{y^0}(x^0)\ \forall k$ and $\lim_{k\to\infty} \|(x^0 + t_k\widehat{r}) - x^k\| = 0$. Continuity of g with respect to x leads to

$$\widehat{\varphi}(x^0; \widehat{r}) = \lim_{k\to\infty} \widehat{\varphi}(x^0 + t_k\widehat{r}) = \lim_{k\to\infty} F(x^0 + t_k\widehat{r}, y^0) = F(x^0, y^0) \le \widehat{\varphi}(x^0).$$

Hence, the first condition of the theorem cannot be valid.

Thus, x^0 satisfies the second condition of the theorem. Therefore, $\widehat{\varphi}(x^0) \le \widehat{\varphi}(x^0; r)\ \forall r \in \mathbb{R}^n$. Because of $Y_{y^0}(x^0) \subseteq \text{cl int } Y_{y^0}(x^0)$ for each k there exists a neighborhood $U^1(x^k)$ with $U^1(x^k) \cap \text{int } Y_{y^0}(x^0) \neq \emptyset$. On the other hand, there exists some neighborhood $U^2(x^k)$ with $F(x, y^0) < \widehat{\varphi}(x^0)\ \forall x \in U^2(x^k)$. Defining $U(x^k) := (U^1(x^k) \cap \text{int } Y_{y^0}(x^0)) \cap U^2(x^k)$, it holds both $U(x^k) \neq \emptyset\ \forall k$ and $\widehat{\varphi}(x) = F(x, y^0) < \widehat{\varphi}(x^0)\ \forall x \in U(x^k)$ due to $|\Psi_D(x)| = 1$ for $x \in \text{int } \mathscr{R}(x^0)$. Now, take a sequence $\{\widehat{x}^k\}_{k=1}^{\infty}$ with $\widehat{x}^k \in U(x^k)\ \forall k$ and $\widehat{x}^k \to x^0$ for $k \to \infty$. Let $\widehat{r}$ be an accumulation point of the sequence $\widehat{r}^k = \frac{\widehat{x}^k - x^0}{\|\widehat{x}^k - x^0\|}$. Since $\widehat{x}^k \in \text{int } Y_{y^0}(x^0)\ \forall k$ and $Y_{y^0}(x^0)$ is convex, there exists some $t_0 > 0$ such that $x^0 + t\widehat{r}^k \in \text{int } Y_{y^0}(x^0)\ \forall t \in (0, t_0)$. Hence, $\widehat{\varphi}(x^0) = \widehat{\varphi}(x^0; \widehat{r}^k) = F(x^0, y^0)\ \forall k$. Because the second condition is valid we have $d\widehat{\varphi}_{x^0}(\widehat{r}^k) > \gamma$ for all k. Consequently,

$$\begin{aligned}\gamma < d\widehat{\varphi}_{x^0}(\widehat{r}^k) &= \lim_{t\downarrow 0} \frac{\widehat{\varphi}(x^0 + t\widehat{r}^k) - \widehat{\varphi}(x^0; \widehat{r}^k)}{t} = \lim_{t\downarrow 0} \frac{F(x^0 + t\widehat{r}^k, y^0) - F(x^0, y^0)}{t} \\ &= \nabla_x F(x^0, y^0)\widehat{r}^k.\end{aligned}$$

Since F is continuously differentiable with respect to x, we have in the pessimistic case $\nabla_x F(x^0, y^0)\widehat{r} \ge \gamma > 0$.

Since $\varphi_o(x^0) = F(x^0, y^0)$ and $\widehat{r} \in C_{Y_{y^0}}(x^0)$ because of $\text{int } C_{Y_{y^0}}(x^0) \ni \widehat{r}^k \to \widehat{r}$ and $C_{Y_{y^0}}(x^0) \subseteq \text{cl int } C_{Y_{y^0}}(x^0)$ it follows from Lemma 5.9 that, in the optimistic case, we have $\varphi_o(x^0; \widehat{r}) = \varphi_o(x^0) = F(x^0, y^0)$.

Thus, it holds

$$\begin{aligned}0 < d\varphi_{o\,x^0}(\widehat{r}) &= \lim_{t\downarrow 0} \frac{\varphi_o(x^0 + t\widehat{r}) - \varphi_o(x^0; \widehat{r})}{t} \le \lim_{t\downarrow 0} \frac{F(x^0 + t\widehat{r}, y^0) - F(x^0, y^0)}{t} \\ &= \nabla_x F(x^0, y^0)\widehat{r}.\end{aligned}$$

Hence, we have for both the optimistic and the pessimistic case that

$$\nabla_x F(x^0, y^0)\widehat{r} > 0. \tag{5.30}$$

On the other hand, it holds

$$\widehat{\varphi}(x^0) > \widehat{\varphi}(\widehat{x}^k) = F(x^0, y^0) + \|\widehat{x}^k - x^0\| \nabla_x F(x^0, y^0)\widehat{r}^k + o(\|\widehat{x}^k - x^0\|)$$

which together with $F(x^0, y^0) = \widehat{\varphi}(x^0)$ and $\lim_{k\to\infty} \frac{o(\|\widehat{x}^k - x^0\|)}{\|\widehat{x}^k - x^0\|} = 0$ leads to

$$\nabla_x F(x^0, y^0)\widehat{r} \le 0.$$

This is a contradiction to inequality (5.30). □

Specifying the conditions of Theorem 5.16 by using Lemma 5.8, we obtain the following **sufficient optimality conditions**:

If

$$\varphi_p(x^0) = \varphi_p(x^0; r) \quad \text{and} \quad d\varphi_{p_{x^0}}(r) > \gamma > 0 \quad \forall r,$$

then $x^0 \in \text{locmin}\ \varphi_p$.

The condition

$$\forall r \quad \varphi_o(x^0) < \varphi_o(x^0; r) \quad \text{or} \quad d\varphi_{o_{x^0}}(r) > 0$$

implies $x^0 \in \text{locmin}\ \varphi_o$.

Example 5.17 Consider the bilevel optimization problem

$$\min\{F(x, y) : x \in \mathbb{R}^2, \ y \in \Psi_D(x)\}$$

where

$$\Psi_D(x) = \underset{y}{\text{Argmin}}\, \{\frac{1}{2}\|y\|^2 - y^\top x : y_1 \le 0, y_2 \ge 0, -y_1 + y_2 \le 1, \ y \in \mathbb{Z}^2\}$$

with $F(x, y) = y_2(x_2 - (x_1 + 0.5)^2 - 0.5) + (1 - y_2)(x_1 - x_2 + 1) + y_1(3x_1 + 1.5)$. We obtain

$$Y_D = \{y^1 = \begin{pmatrix}0\\1\end{pmatrix}, y^2 = \begin{pmatrix}0\\0\end{pmatrix}, y^3 = \begin{pmatrix}-1\\0\end{pmatrix}\} \text{ with}$$

$$\mathcal{R}(y^1) = \{x \in \mathbb{R}^2 : x_2 \ge 0.5, x_1 + x_2 \ge 0\},$$
$$\mathcal{R}(y^2) = \{x \in \mathbb{R}^2 : x_2 \le 0.5, x_1 \ge -0.5\} \text{ and}$$
$$\mathcal{R}(y^3) = \{x \in \mathbb{R}^2 : x_1 \le -0.5, x_1 + x_2 \le 0\}.$$

Then we have

$$\varphi_p(x) = \begin{cases} x_2 - (x_1+0.5)^2 - 0.5 & \text{if} \quad x_2 > 0.5,\ x_1 + x_2 > 0 \\ x_1 - x_2 + 1 & \text{if} \quad x_2 \leq 0.5, x_1 \geq -0.5 \\ -2x_1 - x_2 - 0.5 & \text{if} \quad x_1 + x_2 \leq 0,\ x_1 < -0.5. \end{cases}$$

Let $x^0 = (-1/2, 1/2)^\top$. Then it holds $\varphi_p(x^0) = \varphi_p(x^0; r) = 0\ \forall r \in \mathbb{R}^2$ and

$$0 < d\varphi_{p_{x^0}}(r) = \begin{cases} r_2 & \text{if} \quad r_2 > 0,\ r_1 + r_2 > 0 \\ r_1 - r_2 & \text{if} \quad r_2 \leq 0,\ r_1 \geq 0 \\ -2r_1 - r_2 & \text{if} \quad r_1 < 0,\ r_1 + r_2 \leq 0\,. \end{cases}$$

However x^0 is no local minimum of φ_p since $x(t) = (t-0.5, 0.5(1+t^2))^\top \to y^0$ for $t \downarrow 0$ but $\varphi_p(x(t)) = -\frac{1}{2}t^2 < \varphi_p(x^0)\ \forall t > 0$. Remark that in difference to Theorem 5.16 there does not exist any $\gamma > 0$ with $\gamma < d\varphi_{p_{x^0}}(r)\ \forall r$. □

5.4.6 Optimality Criteria Using Radial Subdifferential

Definition 5.7 Let $U \subseteq \mathbb{R}^n$, $x^0 \in U$ and $\widehat{\varphi} : U \to \mathbb{R}$ be radial-directionally differentiable at x^0. We say that $d \in \mathbb{R}^n$ is a *radial subgradient* of $\widehat{\varphi}$ at x^0 if

$$\widehat{\varphi}(x^0) + \langle r, d\rangle \leq \widehat{\varphi}(x^0; r) + d\widehat{\varphi}_{x^0}(r)$$

is satisfied for all r with $\widehat{\varphi}(x^0) \geq \widehat{\varphi}(x^0; r)$.

The set of all radial subgradients is called *radial subdifferential* and is denoted by $\partial_{rad}\widehat{\varphi}(x^0)$.

The following necessary criterion for the existence of a radial subgradient is valid:

Theorem 5.17 (Dempe and Unger [78]) *If there exists some $r \in \mathbb{R}^n$ with $\widehat{\varphi}(x^0; r) < \widehat{\varphi}(x^0)$, then it holds $\partial_{rad}\widehat{\varphi}(x^0) = \emptyset$.*

With this theorem we get the following equivalent definition of the radial subgradient:

$$\partial_{rad}\widehat{\varphi}(x^0) = \{d \in \mathbb{R}^n : \langle r, d\rangle \leq d\widehat{\varphi}_{x^0}(r)\ \forall r \text{ satisfying } \widehat{\varphi}(x^0) = \widehat{\varphi}(x^0; r)\}$$

if there is no direction such that the radial limit in this direction is less than the function value.

Using Lemma 5.8 we obtain that for the pessimistic solution function either $\partial_{rad}\varphi_p(x^0) = \emptyset$ if there exists some r with $\varphi_p(x^0) > \varphi_p(x^0; r)$ or $\partial_{rad}\varphi_p(x^0) = \{d \in \mathbb{R}^n : \langle d, r\rangle \leq d\varphi_{p_{x^0}}(r)\ \forall r\}$.

For the optimistic solution function the assumption of Theorem 5.17 is never valid.

Thus,

$$\partial_{rad}\varphi_o(x^0) = \{d \in \mathbb{R}^n : \langle r, d\rangle \le d\varphi_{x^0}(r)\ \forall r \text{ satisfying } \varphi(y^0) = \varphi(y^0; r)\}.$$

Now we derive optimality criteria in connection with the radial subdifferential.

Theorem 5.18 (Fanghänel [104]) *Let $\widehat{\varphi}$ denote the optimistic or pessimistic solution function for the bilevel optimization problem* (5.24). *If $x^0 \in locmin\,\widehat{\varphi}$ then $0 \in \partial_{rad}\widehat{\varphi}(x^0)$.*

Proof Suppose $x^0 \in locmin\,\widehat{\varphi}$. Then from Theorem 5.15 it follows $\widehat{\varphi}(x^0) \le \widehat{\varphi}(x^0; r)\ \forall r$ and $d\widehat{\varphi}_{x^0}(r) \ge 0$ for all r satisfying $\widehat{\varphi}(x^0) \ge \widehat{\varphi}(x^0; r)$. Due to the first inequality this is equivalent to $d\widehat{\varphi}_{x^0}(r) \ge 0\ \forall r : \widehat{\varphi}(x^0) = \widehat{\varphi}(x^0; r)$ which means $0 \in \partial_{rad}\widehat{\varphi}(x^0)$. □

Theorem 5.19 (Fanghänel [104]) *Let $\widehat{\varphi}$ denote the optimistic or pessimistic solution function for the bilevel optimization problem* (5.24). *If $0 \in int\, \partial_{rad}\widehat{\varphi}(x^0)$ then $\widehat{\varphi}$ achieves at x^0 a local minimum.*

Proof From Theorem 5.17 it follows $\widehat{\varphi}(x^0) \le \widehat{\varphi}(x^0; r)\ \forall r$. Since $0 \in int\, \partial_{rad}\widehat{\varphi}(x^0)$ there exists some $\varepsilon > 0$ with $\varepsilon r \in \partial_{rad}\widehat{\varphi}(x^0)\ \forall r : \|r\| = 1$. This means

$$\widehat{\varphi}(x^0) + \langle r, \varepsilon r\rangle = \widehat{\varphi}(x^0) + \varepsilon\|r\|^2 \le \widehat{\varphi}(x^0; r) + d\widehat{\varphi}_{x^0}(r)$$

for all r with $\widehat{\varphi}(x^0) \ge \widehat{\varphi}(x^0; r)$, $\|r\| = 1$. Thus, if for some $\|r\| = 1$ it holds $\widehat{\varphi}(x^0) = \widehat{\varphi}(x^0; r)$, it follows $d\widehat{\varphi}_{x^0}(r) \ge \varepsilon > 0$. Hence, all assumptions of Theorem 5.16 are satisfied and we conclude $x^0 \in locmin\,\widehat{\varphi}$. □

5.5 An Approach Using Monotonicity Conditions of the Optimal Value Function

5.5.1 Introduction

In this section an approach is presented which is inspired by an application in computing a best element for mean-variance models introduced by Markowitz [221]. For given assets with its expected return a_i and covariances q_{ij}, the problem is to find a portfolio of assets that has minimal variance by a given level of total return. The mathematical model of integer quadratic optimization is of the form:

$$\min_y\{y^T Qy : e^\top y = 1,\ Ay \ge d,\ y \in \mathbb{Z}^m_+\}, \tag{5.31}$$

where Q is the covariance matrix, d represents the lower bound of total return, and e is a m-dimensional all-one vector. We will use this problem as the lower level problem of a bilevel optimization problem (see Dempe et al. [70]).

Aim of the section is to describe a computational algorithm for solving mixed-integer bilevel problems of this kind. Since the assumptions for proving convergence of this algorithm are very restrictive, we suggest to weaken them which leads to the computation of a weak optimal solution, see Sect. 5.3.6. The results in this Section are taken from the report [70] by Dempe et al.

5.5.2 Problem Formulation

We consider the optimistic version of a mixed-integer nonlinear bilevel optimization problem:

$$\min_{x,y}\{F(x, y) : x \in X,\ y \in \Psi(x)\}, \tag{5.32}$$

where $X \subseteq \mathbb{R}^n$ is a bounded polyhedron with the finite set $x^1, \dots, x^q$ of vertices and $\Psi(x)$ is the solution-set mapping of the lower level parametric quadratic integer optimization problem (PIQP):

$$\Psi(x) := \operatorname*{Argmin}_y \{f(x, y) = y^\top Q y : Ay \geq x,\ y \in \mathbb{Z}^m_+\}. \tag{5.33}$$

Here $F, f : \mathbb{R}^n \times \mathbb{R}^m \to \mathbb{R}$, Q is a $\mathbb{R}^m \times \mathbb{R}^m$ symmetric positive semidefinite matrix, A is a $\mathbb{R}^n_+ \times \mathbb{R}^m_+$ matrix, and $y \in \mathbb{Z}^m_+$, $x \in \mathbb{R}^n$ are unknown lower and upper level variables.

Using the lower level optimal value function $\varphi(x)$, the bilevel problem can be replaced by a fully equivalent problem:

$$\min_{x,y}\{F(x, y) : y^\top Q y \leq \varphi(x),\ Ay \geq x,\ y \in \mathbb{Z}^m_+,\ x \in X\}, \tag{5.34}$$

where the optimal value function of the lower level problem is denoted by

$$\varphi(x) := \min_y\{y^\top Q y : Ay \geq x,\ y \in \mathbb{Z}^m_+\}. \tag{5.35}$$

5.5.3 Parametric Integer Optimization Problem

5.5.3.1 Parametric Integer Optimization Problem with a Single Parameter

We start with considering the (PIQP) with one parameter on the right-hand side, i.e. $n = 1$:

$$\varphi(x) = \min_y\{y^\top Q y :\ Ay \geq x,\ y \in \mathbb{Z}^m_+\}, \tag{5.36}$$

where $A^\top \in \mathbb{R}^m_+$ and $Q \in \mathbb{R}^{m\times m}$ is a quadratic positive semidefinite matrix on the set $x \in [d_l, d_u] \subseteq \mathbb{R}_+$, d_l and d_u are lower and upper bounds of x, respectively.

Theorem 5.20 *For the parametric integer quadratic problem* (5.36) *with a single parameter, the optimal value function* $\varphi(x)$ *is piecewise constant.*

Proof We consider the constraint set $M(x) = \{y \in \mathbb{Z}^m_+ : Ay \geq x\}$, where $A^\top \in \mathbb{R}^m_+$, $x \in [d_l, d_u] \subseteq \mathbb{R}_+$. For $A = (a_1, \ldots, a_n)$, we define the set of raster points

$$h(A) := \{h : h = \sum_{j=1}^{m} a_j y_j, \quad y_j \in \mathbb{Z}_+\}, \tag{5.37}$$

where a_j, y_j are components of A and y, respectively. We denote the elements of the raster point set by h_r, $r = 1, \ldots, q$. Note, that, since d_l, d_u are finite, we need to consider only a finite number of raster points. For simplicity of the presentation, in the following we will assume that the elements of the raster point set are sorted according to increasing efficiencies: $d_l \leq h_1 \leq h_2 \leq \cdots \leq h_q \leq d_u$. It is easy to see that for each $x \in (h_r, h_{r+1})$, $M(x)$ is constant, which implies that the optimal value function $\varphi(x)$ is constant on (h_r, h_{r+1}). □

Proposition 5.1 *The optimal value function* φ *is nondecreasing in* $x \in \mathbb{R}_+$.

Proof Because of $A^\top \in \mathbb{R}^m_+$, $y_j \in \mathbb{Z}_+$, if the parameter x increases, the feasible region will become smaller. This implies the statement of the proposition. □

Our observation is that φ is piecewise constant, but it is not convex. The formulation of a solution algorithm for problem (5.32) will follow the same lines as the one in Sect. 2.3.2. For that we need an approximation $\xi(x)$ of the optimal value function of the lower level problem $\xi(x) \geq \varphi(x)\ \forall x \in [d_l, d_u]$.

An upper bound for the optimal value function at each fixed d^i, $i = 1, \ldots, p$, is the optimal value function itself. Let

$$\xi_i(x) := \varphi(d^i) \quad \forall x \leq d^i. \tag{5.38}$$

Algorithm: `Step 0.` Select any $r \in \mathbb{R}$ with $\varphi(x) \leq r\,\forall x \in [d_l, d_u]$. Let $\xi(x)$ be a global upper approximation of the value function $\varphi(x)$. Set $\xi(x) := r\ \ \forall x \in [d_l, d_u]$.

`Step 1.` Select some points $d^i \in [d_l, d_u]$, $i = 1, \ldots, p$. Sort these points according to $d^p \leq d^{p-1} \leq \ldots \leq d^1$. Compute $\varphi(d^i)$ $\forall i = 1, \ldots, p$, and set $\xi_i(x) = \varphi(d^i)\forall x \leq d^i$, $\forall i = 1, \ldots, p$.

`Step 2.` For $i = 1$ to p, define the global upper bound recursively via

$$\xi(x) := \begin{cases} \min\{\xi_i(x), \xi(x)\} & \forall x \leq d^i \\ \xi(x) & \text{else.} \end{cases}$$

Figure 5.8 illustrates the optimal value function $\varphi(x) = v(x)$ and its approximation $\xi(x) = G(x)$.

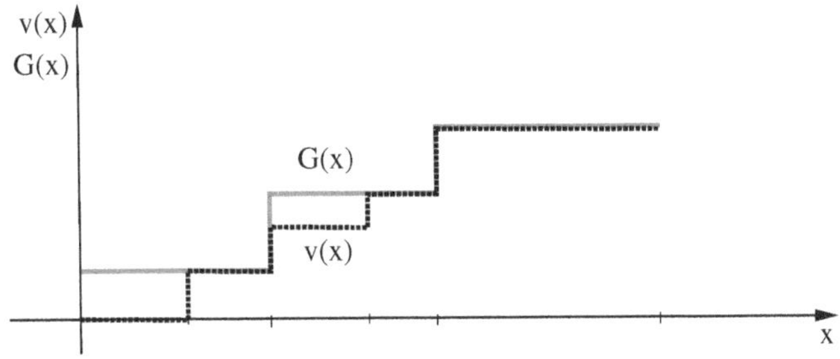

Fig. 5.8 Upper approximation $\xi(x) = G(x)$ of the optimal value function $\varphi(x) = v(x)$

5.5.3.2 Upper Bounding Approximation of a General PIQP

In this section, we will go back to the general case. We extend the result to higher dimensional spaces thus, we refer to (5.33) with $n > 1$.

Theorem 5.21 (Bank and Hansel [9]) *If the matrix A has rational elements only, then the optimal value function* $\varphi : [d_l, d_u] \to \mathbb{R}$ *defined by*

$$\varphi(x) = \min_y \{y^\top Qy : Ay \geq x, \ y \in \mathbb{Z}^m_+\} \tag{5.39}$$

is lower semicontinuous.

Here $x \in [d_l, d_u]$ means $d_l \leq x \leq d_u$ or $d^1_i \leq x_i \leq d^2_i \ \forall i = 1, \ldots, n$.

Recall that the optimal value function $\varphi(x)$ is discontinuous only at some points, which are linear combinations of non-negative integer parameters with the respective columns of the matrix A thus, the optimal value function φ of PIQP is piecewise constant.

Theorem 5.22 (Dempe et al. [70]) *If the matrix A has rational elements only, then the optimal value function is piecewise constant.*

Proof We consider the constraint set $M(x) := \{y \in \mathbb{Z}^m_+ : Ay \geq x\}$, where $A \in \mathbb{R}^n_+ \times \mathbb{R}^m_+$, $x \in [d_l, d_u] \subseteq \mathbb{R}^n_+$. For $A_i = (a_{i1}, \ldots, a_{im})$, $i \in 1, 2, \ldots, n$, we define the set of raster points

$$h(A) := \{h \in \mathbb{R}^n : h_i = \sum_{j=1}^m a_{ij} y_j, \quad y_j \in \mathbb{Z}_+, \ i \in 1, 2, \ldots, n\}, \tag{5.40}$$

where a_{ij}, y_j are components of A and y, respectively. We denote the elements of the raster point set by h_r, $r = 1, \ldots, q$.

Using these raster points, it is easy to find a finite number of (in general nonconvex, neither open nor closed) sets $\Xi_k, k = 1, \ldots, \kappa$ of parameters x such that $M(x)$ is constant over each of the sets $\Xi_k, k = 1, \ldots, \kappa$, see Fig. 5.9. Hence, the optimal value function $\varphi(x)$ is constant over Ξ_k. □

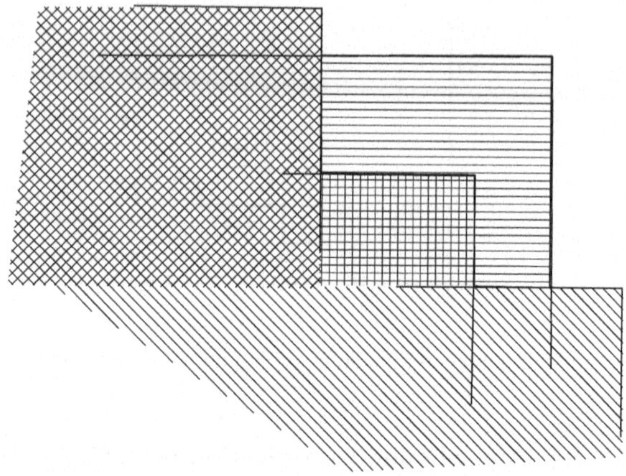

Fig. 5.9 Level sets of the function ξ

Now to extend the above algorithm to the general case, we define first the upper bound for each fixed d^i, $i = 1, \ldots, p$ (d^i are again raster points) as

$$\xi_i(x) := \varphi(d^i) \quad \forall x \leqq d^i. \tag{5.41}$$

The steps of constructing a global upper approximation of the optimal value function are now given in the following algorithm:

Algorithm: **Step 0.** Select any $r \in \mathbb{R}$ with $\varphi(x) \leq r \, \forall x \in [d_l, d_u]$. Let $\xi(x)$ be a global upper approximation of the value function $\varphi(x)$: $\xi(x) := r \;\; \forall x$.

Step 1. Select some points $d^i, i = 1, \ldots, p$. Compute $\varphi(d^i) \, \forall i = 1, \ldots, p$, and set $\xi_i(x) = \varphi(d^i) \quad \forall x \leqq d^i$.

Step 2. For $i = 1$ to p, define the global upper bound recursively by

$$\xi(x) := \begin{cases} \min\{\xi_i(x), \xi(x)\} \; \forall x \leqq d^i \\ \quad \xi(x) \qquad\qquad \text{else.} \end{cases}$$

Figure 5.9 can be used to visualize the regions where the function ξ is constant. These regions are used in the proof of Theorem 5.24.

The upper bounding function $\xi(x)$ derived by the algorithm above is not unique since the points d^i are not fixed. Adding new points, the quality of the approximation $\xi(x)$ is improved.

5.5.4 An Approximation Algorithm

In this section, we will describe an approximation algorithm for problem (5.34) by using the upper bound for the optimal value function of the lower level problem. Applying the ideas from Sect. 5.3 we need

Assumptions (A) (cf. Fanghänel [105]):

1. The set $\{(x, y) : y \in \Psi(x),\ x \in X\}$ is bounded and not empty.
2. For all $x \in X$, $\Psi(x) \neq \emptyset$.
3. $\text{card}(\bar{Y}) < \infty$, where $\bar{Y} := \{y \in \mathbb{Z}^m_+ : \exists x \in X \text{ with } y \in \Psi(x)\}$.
4. The matrix A has nonnegative, rational elements only.

If the matrix A has rational elements only, we can assume without loss of generality that all elements of the matrix A are either zero or natural numbers. Hence, the elements of the set $h(A)$ in Theorem 5.22 are vectors $h \in \mathbb{N}^n$.

Since

$$f(x, y) \leq \varphi(x) \leq \xi(x), \tag{5.42}$$

we can approximate problem (5.34) by:

$$\min_{x,y}\{F(x, y) : y^\top Q y \leq \xi(x),\ Ay \geq x,\ y \in \mathbb{Z}^m_+,\ x \in X\}. \tag{5.43}$$

We propose a global approximation algorithm based on the framework described in Dempe et al. [63] which solves the bilevel problem by iteratively updating an approximation of the lower level optimal value function.

Algorithm: `Computation of a global solution of problem (5.32)`

Step 0: Let $V^0 = \{x^1, \ldots, x^q\}$ be the set of vertices of the set X. Set $t := 0$.

Step 1: Compute the function $\xi(x)$ according to the above algorithm. If a global optimal solution of problem (5.43) exists, let $(\bar{x}^t, \bar{y}^t)$ be a global optimal solution of this problem. If no global optimal solution exists, GOTO Step 3.

Step 2: If the point $(\bar{x}^t, \bar{y}^t)$ is feasible for problem (5.34), STOP. The point is a global optimal solution with optimal function value $F(\bar{x}^t, \bar{y}^t)$. Otherwise GOTO Step 3.

Step 3: Let $(\bar{x}^t, \bar{y}^t)$ be a global optimal solution of problem

$$F(x, y) \to \min_{x,y} \tag{5.44}$$

$$(x, y) \in \text{cl}\,\{(u, v) : v^\top Q v \leq G(u),\ Av \geq u,\ v \in \mathbb{Z}^m_+,\ u \in X\}.$$

Compute a set W as large as possible such that $\xi(w) \geq \bar{y}^{t\top} Q \bar{y}^t$ for all $w \in W$ and select $\widehat{w} \in W$ with $\widehat{w} \geq \bar{x}^t$ and $\widehat{w}_i \geq \bar{x}^t_i + 1$ for at least one $i \in \{1, \ldots, n\}$. Set $V^{t+1} := V^\top \cup \{\bar{x}^t, \widehat{w}\}$. Update $\xi(x)$: First set

$$\xi(x) := \begin{cases} \min\{\xi(x), \varphi(\widehat{w})\} & \text{if} \quad x \leqq \widehat{w} \\ \xi(x) & \text{else.} \end{cases}$$

and then

$$\xi(x) := \begin{cases} \min\{\xi(x), \varphi(\bar{x}^t)\} & \text{if} \quad x \leqq \bar{x}^t \\ \xi(x) & \text{else.} \end{cases}$$

Set $t := t + 1$, GOTO Step 1.

Theorem 5.23 (Dempe and Khamisov [66]) *Let (x', y') be a global optimal solution of* (5.34) *and $(\bar{x}, \bar{y})$ be a global optimal solution of problem* (5.43). *Then $F(x', y') \geqslant F(\bar{x}, \bar{y})$.*

Proof We have $\{(x, y) : Ay \geq x,\ f(x, y) \leq \varphi(x),\ y \in \mathbb{Z}^m_+,\ x \in X\} \subseteq \{(x, y) : Ay \geq x,\ f(x, y) \leq \xi(x),\ y \in \mathbb{Z}^m_+,\ x \in X\}$, due to $\varphi(x) \leq \xi(x)$ for all x. This implies the statement of the theorem.

By construction, the feasible set of problem (5.43) needs not to be closed.

Theorem 5.24 *Let assumptions* (A) *be satisfied, assume that the above algorithm computes an infinite sequence $\{(x^t, y^t)\}_{t=1}^{\infty}$, and that the set $T := \{(x, y) : y \in \mathbb{Z}^m_+,\ x \in X,\ Ay \geq x\}$ is compact. Assume that all stability regions (cf. Definition* 5.1) *are closed. Let $(\widehat{x}, \widehat{y})$ be an accumulation point of the sequence $\{(x^t, y^t)\}_{t=1}^{\infty}$. Then the accumulation point $(\widehat{x}, \widehat{y})$ is a global optimal solution of problem* (5.34).

Proof (i) We assume that the set $T := \{(x, y) : y \in \mathbb{Z}^m_+,\ x \in X,\ Ay \geq x\}$ is compact and not empty and for all $x^t \in X$ the lower level problem can be solved. Let $(\bar{x}^t, \bar{y}^t)$ be the solution taken in Step 1 of the algorithm in iteration t. If the point $(\bar{x}^t, \bar{y}^t)$ is a global optimal solution of problem (5.43) and it is feasible for problem (5.34), it is a global optimal solution of problem (5.34) by Theorem 5.23.

(ii) Assume that the point $(\bar{x}^t, \bar{y}^t)$ is globally optimal for the problems (5.43) and not feasible for problem (5.34). This is only possible if

$$\varphi(\bar{x}^t) < \bar{y}^{t\top} Q \bar{y}^t \leq G(\bar{x}^t).$$

In this case,

$$\bar{x}^t \in W := \{x : \xi(x) = z\}$$

for some $z \in \mathbb{N}$ by assumptions (A). Hence, by the structure of the level sets of the function G (see Fig. 5.9), the set constructed in Step 3 of the algorithm exists.

(iii) If problem (5.43) has no solution, but $(\bar{x}^t, \bar{y}^t)$ is a global optimum of problem (5.44), a sequence $\{v^k\}_{k=1}^{\infty}$ converging to $\bar{x}^t$ exists such that

$$\lim_{k \to \infty} G(v^k) \geq \bar{y}^{t\top} Q \bar{y}^t > G(\bar{x}^t) \geq \varphi(\bar{x}^t).$$

By the structure of the function G,

$$\bar{x}^t \in \text{cl } W, \;\; W := \{x : \xi(x) = z\}$$

for some $z \in \mathbb{N}$ by assumptions (A). Hence, by the structure of the level sets of the function G (see Fig. 5.9) the set constructed in Step 3 of the algorithm again exists.
Then, the function G is updated as formulated in Step 3 of the algorithm and the set $\{x : \varphi(x) = \xi(x)\}$ is enlarged in each iteration. Hence, after a finite number of iterations both functions coincide. Then, a global optimum of problem (5.43) is also a global optimum of problem (5.34). Since the regions of stability are closed, problem (5.34) has a solution. It should be mentioned that the feasible set of problem (5.43) is closed if $\varphi(x) = \xi(x)$ for all x since the stability regions of the function $\mathscr{R}(y)$ are closed. This proves the Proposition. □

The assumption of the proposition that all regions of stability are closed is very restrictive. It can be weakened to the assumption that all stability regions $\mathscr{R}(x)$ with $\widehat{x} \in \text{cl } \mathscr{R}(y)$ satisfy $\widehat{x} \subset \mathscr{R}(y)$. But this again is a very restrictive assumption. If for some stability set $\mathscr{R}(y)$ with $\widehat{x} \in \text{cl } \mathscr{R}(y)$ we have $\widehat{x} \notin \mathscr{R}(y)$, the function $\varphi(x)$ has a jump at the point $x = \widehat{x}$.

5.5.4.1 Weak Solution

In this section, we will describe how to overcome the problem mentioned above by replacing $\mathscr{R}(\bar{x})$ by its closure $\text{cl}\mathscr{R}(\bar{x})$ to develop a weak solution for the bilevel optimization problem, see Sect. 5.3.6.

Definition 5.8 The set $\overline{\Psi}(x) := \{y \in \overline{Y} : x \in \text{cl}\mathscr{R}(y)\}$ is said to be *an extended solution set* of the lower level problem, and $\overline{\Psi}(\cdot) : \mathbb{R}^n \to 2^{\mathbb{R}^m}$ denote *the extended solution-set mapping* of the lower level problem.

It is shown in Fanghänel [105], that $\overline{\Psi}(x)$ is the smallest u.s.c. point-to-set mapping with $\Psi(x) \subseteq \overline{\Psi}(x)$ for each $x \in X$, see also Sect. 5.3. Obviously, if $\mathscr{R}(y)$ is a closed set, we have $\overline{\Psi}(x) = \Psi(x)$.

Definition 5.9 For all $\overline{x} \in X$, the function

$$\overline{\varphi}(\overline{x}) := \liminf_{x \to \overline{x}} \varphi(x) \tag{5.45}$$

is said to be the *extended optimal value function* of the lower level problem.

Replacing φ by $\overline{\varphi}$, we can solve the bilevel optimization problem by using the algorithm discussed above to obtain a global weak solution. We need only to replace the upper bound $\xi(\cdot)$ of $\varphi(\cdot)$ with $\overline{\xi}(\cdot)$ bounding $\overline{\varphi}(\cdot)$ from above. To compute this function replace $\varphi(x)$ in the algorithm on page xxx. Then, modify the algorithm on page xxx in the same way using $\overline{\xi}(\cdot)$ and $\overline{\varphi}(\cdot)$ in place of $\xi(\cdot)$ and $\varphi(\cdot)$. Since the level sets of the function $\overline{\xi}$ need not to be closed, problem (5.44) remains unchanged.

Then, using the same proof as in Theorem 5.24 with the obvious modifications we derive

Theorem 5.25 (Dempe et al. [70]) *Let assumptions (A) be satisfied, assume that the above algorithm computes an infinite sequence* $\{(x^t, y^t)\}_{t=1}^{\infty}$, *and that the set* $T := \{(x, y) : y \in \mathbb{Z}_+^m,\ x \in X,\ Ay \geq x\}$ *is compact. Let* $(\widehat{x}, \widehat{y})$ *be an accumulation point of the sequence* $\{(x^t, y^t)\}_{t=1}^{\infty}$. *Then the accumulation point* $(\widehat{x}, \widehat{y})$ *is a weak global optimal solution of problem* (5.34).

5.6 A Heuristic Algorithm to Solve a Mixed-Integer Bilevel Program of Type I

In this section, we consider a mixed-integer bilevel linear optimization (or the leader's) problem with one parameter in the right-hand side of the constraints in the lower level (or the follower's) problem. Motivated by the application to a natural gas cash-out problem (see Chap. 6), we consider a generalization of the particular case that consists in minimizing the cash-out penalty costs for a natural gas shipping company. The functions are linear at both levels, and the proposed algorithm is based upon an approximation of the optimal value function using a branch-and-bound method. Therefore, at every node of this branch-and-bound structure, we apply a new branch-and-bound technique to process the integrality condition. This section is an extension of the authors' previous paper Dempe et al. [63].

5.6.1 Introduction

The main goal of this section is to propose an efficient algorithm to solve the mixed-integer linear bilevel optimization problem of Type I. Knowing that this problem is hard to solve, we propose an algorithm generating approximations that converge to a global solution. The main novelty of the presented heuristic approach lies in the combination of a branch-and-bound (B&B) technique with a simplicial subdivision algorithm. The numerical experiments demonstrate the robust performance of the developed method for instances of small and medium size.

The section is organized as follows. The general formulation of the problem and the mathematical model is given in Sect. 5.6.2. The geometry of the problem is described in Sect. 5.6.3, whereas the approximation algorithm is presented in Sects. 5.6.4 and 5.6.5 illustrates the algorithm by a numerical example.

5.6.2 The Mathematical Model

The Mixed Integer Bilevel Linear Optimization Problem with a parameter in the right-hand side of the lower level is formulated as follows:

$$\min_{x,y} \left\{ a^\top y + b^\top x : \ Gx = d, \ y \in \Psi(x), \ x \in \mathbb{Z}_+^n \right\}, \tag{5.46}$$

which represents the upper level, where $a, \ y \in \mathbb{R}^m$, $b, \ x \in \mathbb{R}^n$, G is an $r \times n$ matrix, $d \in \mathbb{R}^r$. Note that we use here the optimistic version of the bilevel optimization problem (see Dempe [52]). For example, in the natural gas imbalance cash-out problem (cf. Chap. 6), the objective function of the upper level (the Shipper's problem) is the production cost plus the penalty that the gas shipper tries to minimize. In general, $\Psi(x)$ is defined as follows:

$$\Psi(x) = \operatorname*{Argmin}_{y} \left\{ c^\top y : \ Ay = x, \, y \geq 0 \right\}, \tag{5.47}$$

which describes the set of optimal solutions of the lower level problem (the set of rational reactions). Here $c, \ y \in \mathbb{R}^m$, A is an $n \times m$ matrix with $n \leq m$.

Let us determine the optimal value function of the lower level problem (in the natural gas imbalance cash-out model, the latter is called the Pipeline's problem, see Chap. 6) as follows:

$$\varphi(x) = \min_{y} \left\{ c^\top y : \ Ay = x, \, y \geq 0 \right\}. \tag{5.48}$$

We suppose that the feasible set of problem (5.47) is non-empty. Again, in the example of the natural gas imbalance cash-out problem, x is the parameter vector that can represent the values of different daily imbalances, the amount of gas, or the shipper's revenue. The lower level (the Pipeline's problem), depending on our objectives, may try to minimize the imbalance, the gas haul volumes, or the absolute value of the leader-follower cash transactions.

In this section, we consider a reformulation of (5.46)–(5.48) based upon an approach reported in the literature (see e.g. Ye and Zhu [325] or Dempe [52]) as a classical nondifferentiable optimization problem. If we take into account the lower level optimal value function (5.48), then problem (5.46)–(5.48) can be replaced equivalently by:

$$\min_{x,y} \left\{ a^\top y + b^\top x : \ Gx = d, c^\top y \leq \varphi(x), \ Ay = x, \ y \geq 0, x \in \mathbb{Z}_+^n \right\}. \tag{5.49}$$

Our technique is focused on the lower level objective value function (5.48). For this reason, we recall some important characteristics (see Grygarová [132] or Dempe and Schreier [77]) that will be helpful for solving problem (5.49).

5.6.3 The Problem's Geometry

Consider the parametric linear optimization problem (5.48)

$$\varphi(x) = \min_y \left\{ c^\top y | \ Ay = x, y \geq 0 \right\}.$$

In order to solve this problem, we can use the dual simplex algorithm, like in Dempe and Schreier [77]. Let us fix $x = x^*$ and let y^* be an optimal basic solution for $x = x^*$ with the corresponding basic matrix B. The latter is a quadratic submatrix of A having the same rank as A, and such that $y^* = (y_B^*, y_N^*)^\top$, with $y_B^* = B^{-1}x^*$ and $y_N^* = 0$. Then, we can say that $y^*(x^*) = (y_B^*(x^*), y_N^*(x^*))^\top = (B^{-1}x^*, 0)^\top$ is an optimal basic solution of problem (5.48) for the fixed parameter x^*. And if the following inequality holds:

$$B^{-1}x \geq 0,$$

then $y^*(x) = (y_B^*(x), y_N^*(x))^\top = (B^{-1}x, 0)^\top$ is also optimal for the parameter vector x.

It is possible to perturb x^* so that B remains a basic optimal matrix (Grygarová [132]). We denote by $\mathscr{R}(B)$ a set that we call the *region of stability* of B, which is defined as

$$\mathscr{R}(B) = \left\{ x \mid B^{-1}x \geq 0 \right\}.$$

For all $x \in \mathscr{R}(B)$, the point $y^*(x) = (y_B^*(x), y_N^*(y))^\top = (B^{-1}x, 0)^\top$ is an optimal basic solution of the problem (5.48). This region is nonempty because $y^* \in \mathscr{R}(B)$. Furthermore, it is closed but not necessarily bounded. If $\mathscr{R}(B)$ and $\mathscr{R}(B')$ are two different stability regions with $B \neq B'$, then only one of the following cases is possible.

1. $\mathscr{R}(B) \cap \mathscr{R}(B') = \{0\}$.
2. $\mathscr{R}(B) \cap \mathscr{R}(B')$ contains the common border of the regions $\mathscr{R}(B)$ and $\mathscr{R}(B')$.
3. $\mathscr{R}(B) = \mathscr{R}(B')$.

Moreover, $\mathscr{R}(B)$ is a convex polyhedral cone, on which the lower level optimal value function is a finite and linear function. To determine an explicit description of the function φ consider the dual problem to problem (5.48). If $\varphi(x)$ is finite, then

$$\varphi(x) = \max\{x^\top u : A^\top u \leq c\}.$$

Let $u^1, u^2, \ldots, u^s$ denote the vertices of the polyhedral set $\{u : A^\top u \leq c\}$. Then,

$$\varphi(x) = \max\{x^\top u^1, x^\top u^2, \ldots, x^\top u^s\},$$

whenever $\varphi(x)$ is finite. This shows that the function $\varphi(\cdot)$ is piecewise affine-linear and convex.

By duality, for some basic matrix B_i with $x \in \mathscr{R}(B_i)$, we have $B_i^\top u = c_{B_i}$ or $u = \left(B_i^\top\right)^{-1} c_{B_i}$, and thus,

$$x^\top u^i = x^\top (B_i^\top)^{-1} c_{B_i} = c_{B_i}^\top B_i^{-1} x.$$

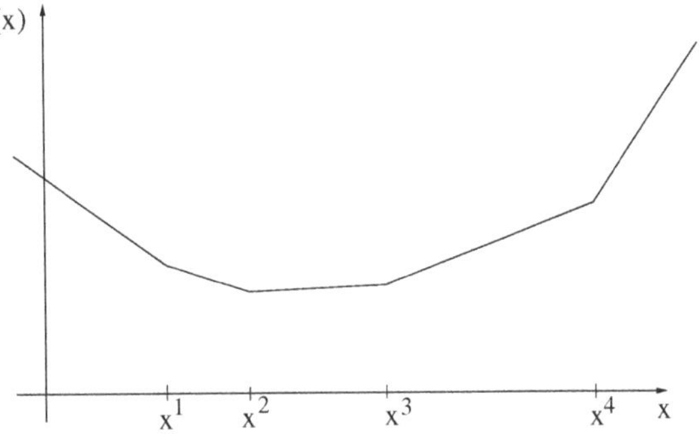

Fig. 5.10 Optimal value function $\varphi(x) = v(x)$ in linear optimization with *right-hand side* parameter

Setting $y^i(x) = ((B_i)^{-1}x, 0)^\top$ we derive

$$\varphi(x) = \max\left\{c^\top y^1(x), c^\top y^2(x), \ldots, c^\top y^q(x)\right\}.$$

As we can see in Fig. 5.10, the stability regions are represented by the segments on the x-axis. The function φ is nonsmooth, which makes this kind of problems hard to solve.

Recall the notion of a partially calm problem in Definition 3.7.

Theorem 5.26 *Let* (x^*, y^*) *solve problem* (5.46)–(5.48), *then* (5.49) *is partially calm at* (x^*, y^*).

Proof Fix an arbitrary value of $\delta > 0$ and assume that a vector

$$(x', y', u) \in (x^*, y^*, 0) + \delta B(0, 1),$$

where $B(0, 1)$ is the unit ball $\mathbb{R}^{n+m+m}$ centered at the origin, is feasible in problem (5.49), i.e.

$$c^\top y' - \varphi(x') + u = 0 \tag{5.50}$$
$$Ay' - x' = 0 \tag{5.51}$$

Let $y(x')$ be a solution to the linear lower level problem, i.e.

$$c^\top y(x') - \varphi(x') = 0 \tag{5.52}$$
$$Ay(x') - x' = 0. \tag{5.53}$$

Therefore $(x', y(x'))$ is feasible in the bilevel linear optimization problem. By the optimality of the solution (y^*, x^*) and by Cauchy-Schwarz-Buniakovski inequality, we get:

$$a^\top y' + b^\top x' - a^\top y^* - b^\top x^* \geq a^\top y' + b^\top x' - a^\top y(x') - b^\top x'$$

$$= a^\top y' - a^\top y(x') \geq -\|a\|_\infty \left\|y' - y(x')\right\|_\infty, \tag{5.54}$$

where $\|\cdot\|_\infty$ represents the infinity norm, i.e., $\|v\|_\infty = \max\{|v_i|,\ i = 1, \ldots, n\}$. By the definition of the lower level optimal value function, and because $(x', y', 0)$ is feasible in problem (5.49), u must be non-positive. If the optimal solution of the lower level problem $y = y(x')$ is not unique, we can select one solution closest to the point y', so that:

$$\left\|y' - y(x')\right\|_\infty = \min_{\varepsilon, y}\{\varepsilon : -\varepsilon e \leq y' - y \leq \varepsilon e,\ c^\top y - \varphi(x') = 0,\ Ay - x' = 0,\ y \geq 0\}.$$

Here, $e = (1, 1, \ldots, 1)^\top \in \mathbb{R}^n$, and inequalities are valid component-wise.

Now consider the dual to the latter linear optimization problem:

$$\begin{aligned} w = \max_{\xi \geq 0} &\Big\{(\xi_1 - \xi_2)^\top y' - \xi_3 \varphi(x') - \xi_4^\top x' : \\ &\xi_1 - \xi_2 - \xi_3 c - A^\top \xi_4 = 0,\ e^\top \xi_1 + e^\top \xi_2 = 1\Big\} \\ = \max&\{\xi_3(c^\top y' - \varphi(x')) + \xi_4^\top (Ay' - x') : \\ &e^\top \xi_1 + e^\top \xi_2 = 1, \xi_i \geq 0, i = 1, 2, 3, 4\Big\}. \end{aligned} \tag{5.55}$$

According to duality theory for linear optimization one has:

$$\begin{aligned} \left\|y' - y(x')\right\|_\infty &= \xi_3(x', y', 0)\left[c^\top y' - \varphi(x')\right] + \xi_4(x', y', 0)^\top (Ay' - x') \\ &= \xi_3(x', y', 0)\left[c^\top y' - \varphi(x')\right], \end{aligned}$$

where $\big(\xi_1(x', y', 0), \xi_2(x', y', 0), \xi_3(x', y', 0), \xi_4(x', y', 0)\big) \in \mathbb{R}^n_+ \times \mathbb{R}^n_+ \times \mathbb{R}_+ \times \mathbb{R}^m_+$ is a solution of the maximization problem (5.55). Hence, the following relationships hold:

$$\begin{aligned} \left\|y' - y\right\|_\infty &= \xi_3(x', y', 0)\left[c^\top y' - \varphi(x')\right] \\ &= \xi_3(x', y', 0)(-u) = \xi_3(x', y', 0)\,|u|. \end{aligned} \tag{5.56}$$

Since $\xi_3(x', y', 0)$ can be selected as a component of a vertex solution of the maximization problem (5.55), and because the feasible region is independent upon $(x', y', 0)$ and has a finite number of vertices, we come to:

$$\xi_3(x', y', 0) \leq L \tag{5.57}$$

where L is equal to

$$\max\{\xi_3 |\ (\xi_1, \xi_2, \xi_3, \xi_4) \text{ is a vertex of the set defined by the constraints in (5.55)}\}.$$

Finally, combining (5.54), (5.56) and (5.57), we conclude that problem (5.49) is partially calm with $\mu = \|a\|_\infty L$, since

$$a^\top y' + b^\top x' - a^\top y^* - b^\top x^* + \mu|u| \geq 0. \qquad \square$$

5.6.4 An Approximation Algorithm

We describe the algorithm based upon the above-mentioned theoretical insights. It is difficult to work with the objective value function (5.48) because we simply do not have it in an explicit form. This algorithm tries to approximate function (5.48) with a finite number of iterations. Additionally, the function φ in (5.48) is not differentiable: cf. Dempe and Zemkoho [81], Ye and Zhu [325] working with subdifferential calculus based upon the nonsmooth Mangasarian-Fromovitz constraint qualification.

The tools that we use in this section are mainly based on the fact that the function $\varphi(\cdot)$ in (5.48) is piecewise affine-linear and convex. Also, the basis for developing a good algorithm is provided by the following theorems, important for keeping on the convexity at every level of approximation.

Definition 5.10 The intersection of all the convex sets containing a given subset $W \subseteq \mathbb{R}^m$ is called the convex hull of W and is denoted by conv W.

Theorem 5.27 (Carathéodory's Theorem) *Let W be any set of points in $\mathbb{R}^m$, and let $C = \operatorname{conv} W$. Then $y \in C$ if and only if y can be expressed as a convex combination of $m + 1$ (not necessarily distinct) points in W. In fact, C is the union of all the generalized d-dimensional simplices whose vertices belong to W, where d equals the dimension of the set C: $d = \dim C$.*

Corollary 5.3 *Let $\{C_i \mid i \in I\}$ be an arbitrary collection of convex sets in $\mathbb{R}^m$, and let C be the convex hull of the union of this collection. Then every point of C can be expressed as a convex combination of $m + 1$ or fewer affinely independent points, each belonging to a different set C_i.*

The details and proofs of Theorems 5.27 and Corollary 5.3 can be found in Rockafellar [272].

Main parts of the proposed algorithm are the following.

In the initial step, we compute a first approximation $\Phi(\cdot)$ of the function $\varphi(\cdot)$ satisfying $\varphi(x) \leq \Phi(x)$ for all $x \in X$. Let

$$X = \{x : Gx = d,\ x \geq 0\}$$

be a convex polyhedron containing all the leader's strategies. Select $\widehat{m} + 1$ affine independent points x^i such that

$$X \subset \operatorname{conv}\{x^1, \ldots, x^{\widehat{m}+1}\} \subset \{x : |\varphi(x)| < \infty\}.$$

Here $\widehat{m} = m - rank(G)$, and $x^2 - x^1, x^3 - x^1, \cdots, x^{\widehat{m}+1} - x^1$ form a linearly independent system. We denote this set of points as

$$V = \{x^1, \ldots, x^{\widehat{m}+1}\}.$$

Then, we solve the lower level linear optimization problem (5.48) at each vertex, i.e., find

$$\varphi(x^1), \ldots, \varphi(x^{\widehat{m}+1})$$

and the corresponding solution vectors $(x^1, y^1), \ldots, (x^{\widehat{m}+1}, y^{\widehat{m}+1})$.

Now we build the first approximation of the optimal value function $\varphi(x)$:

$$\Phi(x) = \sum_{i=1}^{\widehat{m}+1} \lambda_i \varphi(x^i), \tag{5.58}$$

defined over

$$x = \sum_{i=1}^{\widehat{m}+1} \lambda_i x^i, \tag{5.59}$$

with $\lambda_i \geq 0, i = 1, \ldots, \widehat{m} + 1$, and

$$\sum_{i=1}^{\widehat{m}+1} \lambda_i = 1. \tag{5.60}$$

In (5.58), we have an expression with the variable λ that leads to variable x using (5.59) and (5.60). Essentially, this means that we replace the variable x by λ. Now since the function φ is convex, one has

$$c^\top y \leq \varphi(x) \leq \Phi(x),$$

and problem (5.49):

$$\min_{x,y}\{a^\top y + b^\top x : Gx = d, c^\top y \leq \varphi(x), Ay = x, y \geq 0, x \in \mathbb{Z}^m_+\}$$

can be approximated by the approximate integer problem (AIP) as follows:

$$\min_{x,y}\{a^\top y + b^\top x : Gx = d, c^\top y \leq \Phi(x), Ay = x, y \geq 0, x \in \mathbb{Z}^m_+\}. \tag{5.61}$$

In the following algorithm we will add new points $\widehat{x}$ to the set $\{x^1, \ldots, x^{\widehat{m}+1}\}$ such that the resulting set is affine dependent. This implies that the value for the function $\Phi(\cdot)$ in (5.58) in no longer uniquely determined. Hence, we use

$$\Phi(x) = \min_{\lambda}\{\sum_{i=1}^{\widehat{m}+1} \lambda_i \varphi(x^i) : x = \sum_{i=1}^{\widehat{m}+1} \lambda_i x^i,\ \lambda_i \geq 0,\ i = 1, \ldots, \widehat{m}+1\}. \quad (5.62)$$

Using the ideas in Sect. 5.6.3, $\Phi(\cdot)$ is a piecewise affine-linear and convex function. Moreover, the set conv $\{x^1, \ldots, x^{\widehat{m}+1}\}$ decomposes into a finite number p of simplices V_i such that:

1. $\Phi(\cdot)$ is affine-linear over V_i,
2. $\bigcup_{i=1}^{p} V_i = \text{conv}\,\{x^1, \ldots, x^{\widehat{m}+1}\}$,
3. $\text{int}\, V_i \cap \text{int}\, V_j = \emptyset,\ i \neq j$.

Then, problem (5.61) is replaced by

$$\min_{i=1,\ldots,p} \min_{x,y}\{a^\top y + b^\top x : Gx = d, c^\top y \leq \Phi(x), Ay = x, y \geq 0, x \in V_i \cap \mathbb{Z}_+^m\}, \quad (5.63)$$

the inner problems of which are again linear mixed-integer optimization problems. Let M be a list of all the active inner problems in problem (5.63). Using ideas in branch-and-bound algorithms, a problem is not active if it is solved.

Solving problem (5.63) (by solving each of the p inner problems using a certain method from discrete optimization as branch-and-bound or a cutting plane method) an optimal solution $(\overline{x}, \overline{y})$ is obtained. If this point is not feasible for the bilevel optimization problem (5.49), which means that $\varphi(\overline{x}) < \Phi(\overline{x})$, the point $\overline{x}$ is added to the set $\{x^1, \ldots, x^{\widehat{m}+1}\}$. Then, one of the sets V_i becomes affinely dependent. Excluding one element of the resulting set, affine independence can eventually be obtained (this is guaranteed if some correct element is dropped). When one uses this approach, at most $\widehat{m}+1$ new affine independent sets arise, each corresponding to a new linear approximation of the lower level objective function on the convex hull of these points. Call these sets again V_i.

Let $\widehat{V}$ be an arbitrary of these sets of affinely independent points and $\widehat{\Phi}(\cdot)$ the respective approximation of the function $\varphi(\cdot)$ over conv $\widehat{V}$, see (5.58).

Then, the resulting problem

$$\min_{x,y}\{a^\top y + b^\top x : Gx = d, c^\top y \leq \widehat{\Phi}(x), Ay = x, y \geq 0, x \in \mathbb{Z}_+^m \cap \text{conv}\,\widehat{V}\} \quad (5.64)$$

is added to the list of problems. In Step 4 of the subsequent algorithm all those problems are computed and added to the list of problems.

If one such simplex T is a subset of some region of stability, $T \subset \mathscr{R}(B_i)$, the feasible points (x, y) of problem (5.64) are also feasible for problem (5.49).

Now, we describe the proposed algorithm:

Algorithm: **Initialization.** Pick a tolerance value $\varepsilon > 0$ and compute the first approximation of the optimal value function of the lower level problem as given in (5.58).

Let $t = 1$, and $z_t = +\infty$, where z_t is the incumbent objective value. Put problem (5.61) into the problems list M.

Step 1. *Termination criterion.* Stop, if the problems list is empty. Else select one problem (corresponding to a set V_i) in the list M and delete this problem from that list. If all the points (x^i, y^i), $y^i \in \Psi(x^i)$, corresponding to vertices of V_i are close enough:

$$\max_{1 \le i \ne k \le \widehat{m}+1} \left\| (x^i, y^i) - (x^k, y^k) \right\| < \varepsilon$$

repeat Step 1.

Step 2. Solve the problem taken from the problems list using typical methods for integer optimization. Denote the set of optimal solutions as

$$S = \left\{ \left(\tilde{x}^1, \tilde{y}^1 \right), \ldots \right\}$$

and $\tilde{z}$ the objective function value. If the problem has no feasible solution, or if its objective function value is larger than z_t, then fathom this problem: set

$$z_{t+1} := z_t, \quad t := t + 1$$

and go to Step 1. Otherwise go to Step 3.

Step 3. If the components x of all the solutions belonging to S are elements of V, then store the solutions, set

$$z_{t+1} := \tilde{z}, \quad t := t + 1$$

and go to Step 1 (for such values of x, the point (x, y) is feasible for problem (5.49), their objective function value is better than the best one obtained so far). Otherwise, considering the solution $\left(\tilde{x}^j, \tilde{y}^j\right)$ from S such that the component $\tilde{x}^j$ is different from all the elements of V, we add $\tilde{x}^j$ to V_i, set $z_{t+1} := z_t$, $t := t + 1$ and go to Step 4.

Step 4. *Subdivision.* Make a subdivision of the set $V_i \cup \{x^j\}$ corresponding to this problem. Construct all new problems (5.64) as given above and add them to the list of problems. Go to Step 1.

Another idea of how to solve problem (5.49) is to work with the exact penalty function as described in Dempe and Kalashnikov [62], Dempe et al. [65], Dempe and Zemkoho [81]. Namely, we deduce a new reformulation of (5.49) using the facts that the objective value function (5.48) is piecewise affine-linear, convex and partially calm, as we showed in Sect. 5.6.3.

We suppose that there exists a $k_0 < \infty$ such that a point (x^0, y^0) is locally optimal for problem (5.49) if and only if it is locally optimal for the problem:

$$\min_{x,y}\{a^\top y + b^\top x + k[c^\top y - \varphi(x)] : Gx = d, Ay = x, y \geq 0, x \in \mathbb{Z}_+^m\}, \quad (5.65)$$

for all $k \geq k_0$.

The difficulty in dealing with (5.65) arises from the fact that the exact penalty function:

$$a^\top y + b^\top x + k[c^\top y - \varphi x)] \quad (5.66)$$

is not explicit due to the nature of the lower level optimal value function (5.48). Moreover, the penalty function (5.66) is also nonconvex. For this reason, we propose to use the algorithms presented in Gao [124, 125].

5.6.5 A Numerical Example

We consider the following bilevel parametric linear optimization problem, where the upper level is described as:

$$\min_{x,y}\{3y_1 + 2y_2 + 6y_3 + 2x_1 : 4x_1 + x_2 = 10, y \in \Psi(x), x_1, x_2 \in \mathbb{Z}_+\},$$

where

$$\Psi(x_1, x_2) = \operatorname*{Argmin}_{y_1, y_2, y_3}\{-5y_1 - 8y_2 - y_3 : 4y_1 + 2y_2 \leq x_1, 2y_1 + 4y_2 + y_3 \leq x_2,$$
$$y_1, y_2, y_3 \geq 0\}.$$

There the lower level optimal value function is given by:

$$\varphi(x_1, x_2) = \min_{y_1, y_2, y_3}\{-5y_1 - 8y_2 - y_3 : 4y_1 + 2y_2 \leq x_1, 2y_1 + 4y_2 + y_3 \leq x_2,$$
$$y_1, y_2, y_3 \geq 0\}.$$

The optimal solution of this problem is $(y_1^*, y_2^*, y_3^*; x_1^*, x_2^*) = (1/3, 1/3, 0; 2, 2)$. We start to solve the problem using the proposed algorithm.

Step 0. We choose the vertices $x^1 = (5/2, 0)$ and $x^2 = (0, 10)$ that belong to the convex hull of the leader's strategies at the upper level. Fix the tolerance value $\varepsilon = 0.1$. Now, we calculate $\varphi(x^1) = 0$ and $\varphi(x^2) = -10$, set $z_1 := +\infty$, then the first approximation is built as follows:

$$\Phi(x) = -x_2.$$

The approximate integer problem (AIP) that we add to the problems' list is given as follows:

$$\min_{x,y}\{3y_1 + 2y_2 + 6y_3 + 2x_1 : 4x_1 + x_2 = 10, 4y_1 + 2y_2 \leq x_1$$
$$2y_1 + 4y_2 + y_3 \leq x_2, -5y_1 - 8y_2 - y_3 \leq -x_2$$
$$y_1, y_2, y_3 \geq 0, x_1, x_2 \in \mathbb{Z}_+\}$$

Iteration 1
Step 1. We select (AIP) from the problems list.
Step 2. We solve problem (AIP) and obtain the (unique) solution $(\tilde{y}_1, \tilde{y}_2, \tilde{y}_3; \tilde{x}_1, \tilde{x}_2) = (0, 1/4, 0; 2, 2)$ with $\tilde{z} = 15/4$. Because $\tilde{z}$ is less than $+\infty$, we go to Step 3.
Step 3. As $\tilde{x} = (\tilde{x}_1, \tilde{x}_2) = (2, 2)$ is different from the elements of the set V, we add $\tilde{x} = (2, 2)$ to V, set $z_2 := +\infty$, $t := 2$ and go to Step 4.
Step 4. Make a subdivision at $\tilde{x} = (2, 2)$ thus obtaining two new problems: the first one corresponding to conv $\{x^2 = (0, 10), \tilde{x} = (2, 2)\}$, and the second one corresponding to conv $\{\tilde{x} = (2, 2), x^1 = (5/2, 0)\}$. Then we add these two new programs to the problems list, each one described as follows: the first one with the approximation

$$\Phi_1(x) = -17x_2/24 - 70/24,$$

and the second one with the approximation

$$\Phi_2(x) = -13x_2/6.$$

Finally, the new problems can be specified as follows:

$$(P^1) \quad \min_{x,y}\{3y_1 + 2y_2 + 6y_3 + 2x_1 \mid 4x_1 + x_2 = 10, 4y_1 + 2y_2 \leq x_1,$$

$$2y_1 + 4y_2 + y_3 \leq x_2, -5y_1 - 8y_2 - y_3 \leq \Phi_1(x), y_1, y_2, y_3 \geq 0, x_1, x_2 \in \mathbb{Z}_+\}.$$

(when removing x^1 from V), and

$$(P^2) \quad \min_{x,y}\{3y_1 + 2y_2 + 6y_3 + 2x_1 \mid 4x_1 + x_2 = 10, 4y_1 + 2x_2 \leq x_1,$$

$$2y_1 + 4y_2 + y_3 \leq x_2, -5y_1 - 8y_2 - y_3 \leq \Phi_2(x), y_1, y_2, y_3 \geq 0, x_1, x_2 \in \mathbb{Z}_+\}.$$

(when removing x^2 from V). Go to Step 1.
Iteration 2
Step 1. We select (P^1) from the problems list and go to Step 2.
Step 2. We solve (P^1) yielding the (unique) solution

$$(\tilde{y}_1, \tilde{y}_2, \tilde{y}_3; \tilde{x}_1, \tilde{x}_2) = (1/3, 1/3, 0; 2, 2)$$

with $\tilde{z} = 17/3$. And because $\tilde{z}$ is less than z_2, then we go to Step 3.
Step 3. As $\tilde{x} = (\tilde{x}_1, \tilde{x}_2) = (2, 2)$ coincides with one of the elements of V, we store the solution $(\tilde{y}_1, \tilde{y}_2, \tilde{y}_3; \tilde{x}_1, \tilde{x}_2) = (1/3, 1/3, 0; 2, 2)$, set $z_3 := 17/3$, $t := 3$, and go to Step 1.
Iteration 3
Step 1. We select (P^2) from the problems list and go to Step 2.
Step 2. We solve (P^2) obtaining the (unique) solution

$$(\tilde{y}_1, \tilde{y}_2, \tilde{y}_3; \tilde{x}_1, \tilde{x}_2) = (1/3, 1/3, 0; 2, 2)$$

with $\tilde{z} = 17/3$. And as $\tilde{z}$ is equal to z_3, then we go to Step 3.
Step 3. Because $\tilde{x} = (\tilde{x}_1, \tilde{x}_2) = (2, 2)$ coincides with one of the elements of V, we store the solution $(\tilde{y}_1, \tilde{y}_2, \tilde{y}_3; \tilde{x}_1, \tilde{x}_2) = (1/3, 1/3, 0; 2, 2)$, set $z_4 := 17/3$, $t := 4$, then go to Step 1.
Iteration 4
Step 1. The problems list is empty, so we finish the algorithm.

Therefore, the last stored solution $(\tilde{y}_1, \tilde{y}_2, \tilde{y}_3; \tilde{x}_1, \tilde{x}_2) = (1/3, 1/3, 0; 2, 2)$ with $z = 17/3$ is the solution obtained with our algorithm, and it coincides with the exact solution of the problem. □

Chapter 6
Applications to Natural Gas Cash-Out Problem

This chapter discusses several models in which bilevel programming has been applied to the natural gas industry. The analysis of done in the context of the US natural gas markets, in which regulations in the last 20 years have made it so that separations between supply chain agents is compulsory, in order to avoid monopolistic practices. As a result of this, natural gas shippers/traders and pipeline operators engage in business in a way that can be modeled as a bilevel problem.

This chapter is divided as follows: Sect. 6.1 describes the background for the problems, Sect. 6.2 details the formulation of the bilevel model that abstracts the problem, with an approximation to this given in Sect. 6.3. A preliminary direct solution method is sketched in Sect. 6.4, while Sect. 6.5 shows a penalty function approach to the solution of this model using a variational inequality. Section 6.6 presents a way to expand the problem and then solve it by first reformulating both levels with linear problems thus, obtaining a linear bilevel optimization problem. Then, the techniques in this chapter are used to provide numerical results in Sect. 6.7. Finally, in Sect. 6.8, one possible stochastic formulation is suggested, building upon the deterministic models presented before.

6.1 Background

During the early 1990s (Soto [293] and IHS Engineering [150]), several laws and regulations were passed in the United States and in the European Union, dictating, among other things, that some of the links of the Natural Gas Supply Chain were not to be controlled by a single party. In particular, pipeline ownership and natural gas shipping/trading activities, were not supposed to be performed by the same agent. This was decided under the rationale that a Transmit System Operator (TSO), acting also as a Natural Gas Shipping Company (NGSC), would have an unfair advantage over other NGSPs, using his pipeline. Under the new rules, the TSO could not own natural gas with the objective of selling it, whereas NGSCs were now required to

S. Dempe et al., *Bilevel Programming Problems*, Energy Systems,
DOI 10.1007/978-3-662-45827-3_6

employ the services of TSOs to move their natural gas through the Interstate Pipeline System (Energy Information Administration [98]).

Typically, a day of operations in this TSO-NGSC system has the NGSCs injecting natural gas at the production facilities (e.g. a wellhead, a refinery) and withdrawing a volume of natural gas, reported in advance to the TSO, at one of several **pool zones**, which are consumption points in which the fuel is sold to Local Distributing Companies (LDCs) for its delivery to final costumers. The TSO has the responsibility of guaranteeing that an amount of natural gas, equal in quantity and quality to the booked amount, is available at each pool zone for the NGSCs take out.

Operations in a pipeline network must be planned in advance in order to maintain proper operating conditions. This means the amount of gas that enters and leaves the system, as well as its characteristics, like pressure and exact chemical composition, must always remain within certain tolerance levels. For instance, withdrawing too much gas from a pipeline may cause a drop in pressure, which may impact operations; the TSO would seek to avoid this as much as possible. Also, since typically more than one NGSC use the same pipeline, the gas in there gets mixed and the POc must guarantee all receive natural gas that is just as good as that which they injected.

Unfortunately, NGSCs face uncertain demands from their customers in a daily basis. While they can more or less forecast the future consumption levels they will have to satisfy, the actual volumes of natural gas they take out from the pipeline network are often different from the amount they have initially injected miles away, at the wellheads. In order to satisfy this varying demanded volumes, NGSCs require an amount of flexibility when withdrawing natural gas. **Imbalances**, i.e. a difference in the amount of gas declared to be extracted from a pool zone and the actual extraction, are understood to occur. In fact, imbalance creation is a crucial part of the natural gas network industry.

However imbalances create problems for the TSOs as well as the possibility of unfair business practices like unjustified storages and speculation or hedging. Some form of regulation by the TSOs is needed to avoid such issues. While several types of regulatory options exist, this chapter focuses on penalization cash-out: a TSO charges the NGSC money for any gas extracted in addition to the forewarned amounts. The TSO will then use this money to pay for the costs of injecting gas to balance its network. Conversely, the TSO will agree to buy the gas not extracted, usually at a price lower than that the NGSC would get in the market.

Let us consider a situation in which one TSO and one NGSC engage in business as described above. The NGSC controls the booking of capacity in the pipeline, as well as the extraction at the pool zones. After several days of operation, the NGSC will have created (final day) imbalances which will likely not be zero (i.e. there is an imbalanced network). The TSO then takes over the control of the pipeline and the gas there, rearranging the volumes so that the absolute sum of the imbalances over all zones is as small as possible, charging the NGSC for this re-balancing procedure. The result of this are the **final imbalances**. The NGSC utilities come from the revenues of natural gas sold to its LDCs, considering that it has to pay the TSO for booked capacity and imbalances created. On the other hand, the TSO will only have an interests in reducing the imbalances, trying to make the amount of money cash-out is the smallest possible, regardless of whether he receives or delivers money.

Because of the nature of the process, the bilevel setting has the NGSC as the leader or upper level decision maker. The TSO, consequently, is deemed the follower or lower level decision maker. This does not mean that the TSO is of lesser value or has less authority than the NGSC. As we have seen, it is the TSO whom enforces the arbitrage over the NGSC. The designation of each level's agents is merely due to the timing of their decisions. Since the NGSC must create an imbalance configuration before the TSO attempts to rebalance it, it is the NGSC who acts as leader in the model.

One important (and complicating) factor is that, after re-arranging the NGSC imbalances, the TSO must make sure that all imbalances are either non-negative, or non-positive. A positive imbalance means that the NGSC is not taking all the natural gas it stated or booked for withdrawal at a given pool zone and day. Conversely, a negative imbalance happens when the NGSC has taken out more natural gas than originally declared. While the NGSC has certain liberties when some imbalances occur in the final configuration by the TSO, all imbalances must bear the same sign.

6.2 Formulation of the Natural Gas Cash-Out Model as a Mixed-Integer Bilevel Optimization Problem

In this section we will show an initial attempt to model the TSO-NGSC problem. This not only provides an introduction to the latter models presented in this chapter, but is also useful in showing the way we treat integrality in the lower level by moving it to the upper level, and the conditions under which this applies.

Table 6.1 shows the notation that will be used throughout this section. Notice that other sections in this chapter use equal or similar notation and symbols.

The original bilevel formulation in last section deals merely with modeling the cash-out, disregarding the NGSC's concern to maximize its profit. Because of its simplicity this model is presented first, before adding the more complicating factors we introduce in later sections. The preliminary upper level is a linear optimization problem if we consider y^g fixed; however, the lower level is considerably more complex. It contains an absolute value in the objective function, the binary variable θ, as well as logical operators such as max, min, and conditionals. Some of these are reduced to linear constraints when the upper level decision is fixed, which is important in some solution techniques. Others, however, require the model to be approximated in order to be simplified.

Table 6.1 Symbols for variables and parameters employed

Sets and indexes	
D	Maximum number of days in the contract period
Z	Number of pool zones
$\mathbf{D}$	Set of days; $\mathbf{D} = \{1, 2, \ldots, D\}$
$\mathbf{Z}$	Set of pool zones; $\mathbf{Z} = \{1, 2, \ldots, Z\}$
Upper level parameters	
$I^L, I^U \in \mathbb{R}_+^{DZ}$	Lower and upper bounds for daily imbalances at day t in zone i
$I_t^L, I_t^U \in \mathbb{R}_+^{D}$	Lower and upper bounds for the sum of the daily imbalances at day t
$S^L, S^U \in \mathbb{R}_+^{DZ}$	Bounds on imbalance swing from day $t-1$ to day t in pool zone i
$x_0^I \in \mathbb{R}_+^Z$	Imbalance at the beginning of day one in pool zone $i \in \mathbf{Z}$
Lower level parameters	
$L \in [0, 1]^{2Z}$	Fraction of gas consumed when moving one unit from pool zone i to zone j
$F \in \mathbb{R}_+^{2Z}$	Forward haul unit cost for moving one unit of gas from pool zone i to pool zone j
$B \in \mathbb{R}_+^{2Z}$	Backward haul credit for 'returning' a unit of gas from pool zone j to pool zone i
$R \in \mathbb{R}_+^Z$	Linear imbalance penalization coefficient in pool zone i
$\delta \in \mathbb{R}_+^Z$	Non-linear imbalance penalization coefficient in pool zone i
Upper level decision variables	
$x^I \in \mathbb{R}^{DZ}$	Imbalance at the end of day t in pool zone i
$x^s \in \mathbb{R}^{DZ}$	Imbalance swing from day $t-1$ to day t in pool zone i
Lower level decision variables	
$y^I \in \mathbb{R}^Z$	Final imbalance in zone i
$y^f \in \mathbb{R}^{Z(Z-1)/2}$	Volume of gas moved from zone i to zone j
$y^b \in \mathbb{R}^{Z(Z-1)/2}$	Volume of gas credited from zone j to zone i
$y^g \in \mathbb{R}$	Total cash-out for the NGSC
Auxiliary variables	
θ	binary variable equal to 1 if all final imbalances $y^I \geq 0$, and 0 if $y^I < 0$

6.2.1 The NGSC Model

NGSC cash-out: This is the total cost the NGSC incurs from creating imbalances. The TSO either charges or pays the NGSC for the balancing costs; this function is dependent of the lower level decision and is the reason the NGSC must consider the TSO's decisions (represented by variables $y = (y^I, y^f, y^b, y^g)$ and θ) before taking her own decision $x = (x^I, x^s)$

$$F_1(x; y) = y^g. \tag{6.1a}$$

Imbalance upper and lower bounds. While the NGSC enjoys certain liberties when dealing with the imbalances she creates, there are operational or contractual limits to this disruptions of the network balance.

$$I_{ti}^L \le x_{ti}^I \le I_{ti}^U, \quad t \in \mathbf{D}, \quad i \in \mathbf{Z}. \tag{6.1b}$$

Total imbalance upper and lower bounds. Not only the individual imbalances in each pool zone are bounded, but also their sum, as a way to equalize the total imbalance in the network caused by the NGSC.

$$I_t^L \le \sum_{i \in \mathbf{Z}} x_{ti}^I \le I_t^U, \quad t \in \mathbf{D}. \tag{6.1c}$$

Imbalance swing definition. Naturally, the imbalance swing variables s are defined as the difference between one day's imbalance and the former's at any given pool zone:

$$x_{ti}^I = x_{t-1,i}^I + x_{ti}^s, \quad t \in \mathbf{D}, \quad i \in \mathbf{Z}. \tag{6.1d}$$

Imbalance swing upper and lower bounds. Just as the imbalances themselves are bounded, so is the ability of the NGSC to change an imbalance from one day to another. This means that the NGSC cannot switch from taking a large extra volume from the pipeline one day, to leaving another very large amount the next day.

$$S_{ti}^L \le x_{ti}^I - x_{t-1,i}^I \le S_{ti}^U, \quad t \in \mathbf{D}, \quad i \in \mathbf{Z}. \tag{6.1e}$$

It can bee seen that this upper level is rather simple; since all constraints are linear, after some basic manipulations, one can define a matrix A^U and a vector, C^U such that (6.1b)–(6.1e) can be expressed as $G(x) \le 0$, with $G(x) = A^U x - C^U$. In this case, the upper level decision set is defined as

$$\mathbb{X} := \{x : G(x) \le 0\}.$$

However, as Theorem 2.2 has made clear, even with linear problems in each level, a bilevel problem is considerably complex and hard to solve.

6.2.2 The TSO Model

The TSOs objective. The revenue the TSO obtains is exactly the negative of the NGSC objective, that is, $-y^g$. However, it is not the TSO's objective to outrightly try to take in as much money in the cash-out as possible. The TSO is interested in balancing its network and it will do so trying to make the cash-out as little as possible for either party. This is of course achieved by minimizing the absolute value of y^g:

$$f_1(y, \theta) = |y^g|; \tag{6.2a}$$

Final imbalance definition. The relationship between the final imbalances and the final day imbalances is straightforward: a final imbalance is the result of any (final day) imbalance already in the pool zone, plus the amount of gas moved toward there from positive imbalance zones, minus the amount of gas moved from that pool zone to another pool zone. Notice the factor $(1 - L_{ij})$ in the received natural gas volumes due to transportation costs that happens when the TSO moves gas downstream to that zone (and only in that case).

$$y_j^I = x_{Dj}^I + \sum_{i:i<j} \left[(1 - L_{ij}) y_{ij}^f - y_{ij}^b \right] + \sum_{k:k>j} \left(y_{jk}^b - y_{jk}^f \right), \ j \in \mathbf{Z}. \tag{6.2b}$$

Gas conservation. This follows directly from summing up the latter equations. This constraint indicates that no natural gas loss should occur.

$$\sum_{i \in \mathbf{Z}} y_j^I + \sum_{(i,j):i<j} L_{ij} y_{ij}^f = \sum_{i \in \mathbf{Z}} x_{Di}^I. \tag{6.2c}$$

No cyclic movement of gas. The amount of gas the TSO can move from pool zone i, either upstream or downstream, must be less or equal than the positive imbalance initially in that pool zone.

$$\sum_{j:j>i} y_{ij}^f + \sum_{k:k<i} y_{ki}^b \leq \max\{0, x_{D,i}^I\}, \quad i \in \mathbf{Z}. \tag{6.2d}$$

Forward haul bounds. For any given pair of pool zones $(i, j) : i < j$, natural gas can only be moved downstream from i to j if the former pool zone bears a positive imbalance, and the latter has a negative imbalance, and never more than the positive imbalance in the originating pool zone. If the upper level variables have already been fixed, these constraints become linear.

$$y_{ij}^f \leq \begin{cases} x_{D,i}^I & \text{if } x_{D,i}^I > 0 \text{ and } x_{D,j}^I < 0; \\ 0 & \text{otherwise.} \end{cases}, \quad i \in \mathbf{Z}. \tag{6.2e}$$

Backward credit bounds. Since gas cannot be moved upstream, only a credit is given to the NGSC, with no physical movement occurring though otherwise identical to a lossless forward gas haul. For any given pair of pool zones $(i, j) : i < j$, this credit can only be given if the upstream pool zone j bears a positive imbalance, and the downstream one has a negative one, also never more than the positive imbalance in the originating pool zone. These constraints similarly become linear once the upper level decision is made.

$$y_{ij}^b \leq \begin{cases} x_{D_j}^I & \text{if } x_{D,j}^I > 0 \text{ and } x_{D,i}^I < 0; \\ 0 & \text{otherwise.} \end{cases}, \quad i \in \mathbf{Z}. \tag{6.2f}$$

Imbalance sign matching The final imbalances created by the TSO must bear the same sign as the final day imbalances defined by the NGSC. Just as with the past two sets of inequalities, these constraints become linear once the NGSC has fixed her decision.

$$\min\{0, x_{Di}^I\} \leq y_i^I \leq \max\{0, x_{Di}^I\}, \quad i \in \mathbf{Z}. \tag{6.2g}$$

Signs across all pool zones. This constraint represents a business rule in which no imbalance may bear a different sign than the others. The TSO must make sure that all imbalances are either non-negative, or non-positive. If $\mathbf{M}_1$ is a very large scalar (i.e. significantly larger than the largest possible absolute imbalance), then:

$$-\mathbf{M}_1(1-\theta) \leq y_i^I \leq \mathbf{M}_1\theta, \quad i \in \mathbf{Z}. \tag{6.2h}$$

NGSC's cash-out costs. The last constraint shows how the NGSC costs from the penalization are calculated. Basically, it pays the TSO for any positive imbalance created and for movements of natural gas downstream, and receives another amount from negative imbalances and from natural gas credited upstream. The addition of the term with δ_i is necessary to avoid the creation of pseudo-storage (Dempe et al. [65]), its quadratic nature guarantees differentiability. Unfortunately, this adds another nonlinearity in the model, one which cannot be as easily fixed as the past ones (when the model is extended, as described later in Sect. 6.6.3, the nonlinear term becomes non-essential and may be dropped.) If $(y_i^I)_+ = \max\{0, y_i^I\}$, then

$$y^g = -\sum_{i \in \mathbf{Z}} \left[R_i y_i^I - \delta_i (y_i^I)_+^2 \right] + \sum_{(i,j):i<j} \left[F_{ij} \left(1 - L_{ij}\right) y_{ij}^f - B_{ij} y_{ij}^b \right]. \tag{6.2i}$$

Types of variables:

$$y_i^I, y^g \text{ free}, \quad i \in \mathbb{Z}; \tag{6.2j}$$

$$y_{ij}^f, y_{ij}^b \geq 0, \quad i, j \in \mathbf{Z}; \tag{6.2k}$$

$$\theta \in \{0, 1\}. \tag{6.2l}$$

Again, with an adequate selection of a matrices A^L, B^L and a vector C^L, we can summarize the constrains (6.2b)–(6.2h), (6.2j), (6.2k) as

$$g(y, \theta, x) = A^L x + B^L \begin{bmatrix} y \\ \theta \end{bmatrix} - C^L \leq 0 \tag{6.3}$$

Unfortunately, the nonlinearity caused by constraints (6.2i) and (6.2l) prevents us from formulating a purely matrix-form definition of the feasible set as in the upper level. Therefore, set $\mathbb{Y}(x)$ should be defined simply as the set of all points (y, θ) such that the lower level constraints hold for a fixed upper level decision, i.e.

$$\mathbb{Y}(x) := \{(y, \theta) : g(y, \theta; x) \leq 0, \ (6.2\mathrm{i}), \ (6.2\mathrm{l})\}. \tag{6.4}$$

6.2.3 The Bilevel Model

Let $\varphi : \mathbb{R}^{DZ} \to \mathbb{R}$ be the optimal value function of the lower level problem:

$$\varphi(x^I) := \min_{y,\theta}\{f_1(y, \theta) : (y, \theta) \in \mathbb{Y}(x)\} \tag{6.5}$$

The solution set $\Psi(x)$ for the lower level is therefore defined as

$$\Psi(x) := \{(y, \theta) \in Y(x)) : f_1(y, \theta) \leq \varphi(x^I)\}; \tag{6.6}$$

and the graph for this function is consequently

$$\mathbf{gph}\, \Psi := \{(x, y, \theta) : (y, \theta) \in \Psi(x)\}. \tag{6.7}$$

Using the above definitions, we can formulate the mixed integer bilevel optimization problem for the NGSC-TSO model as

$$\mathbf{MIBP1} : \min_{x,y,\theta}\{F_1(x, y, \theta) : (x, y, \theta) \in \mathbf{gph}\, \Psi(x), x \in \mathbb{X}\}. \tag{6.8}$$

6.3 Approximation to a Continuous Bilevel Problem

As discussed before, this model presents several complications. Some of them arise from modeling necessities, like the nonlinear $\delta_i(y_i^I)_+^2$; others come from the abstraction itself, like the absolute value in f_1.

The nonlinearities present can be addressed somewhat straightforwardly: either by reformulating the problem, or by using specific solution methods that bypass them. This will be addressed more thoroughly later in this chapter. Nevertheless, one particularly important issue is the binary variable in the lower level. The TSO optimizes a function that is, implicitly, a function of a binary variable. This turns the lower level objective function into a piecewise continuous linear function which is remarkably different from that in the upper level (Dempe et al. [65]). In order to make this situation more tractable, the binary variable will be moved from the lower to the upper level before proposing any solution technique.

With the binary variable out of the lower level, we redefine the TSO feasible region as $\mathbb{Y}^{\beta}(x) := \{(y) : (6.2\mathrm{b})\text{–}(6.2\mathrm{k}) \text{ hold for } \theta = \beta\}$.

The bilevel model discussed in the rest of this chapter will therefore be slightly different from the original version. The new formulation requires calculating the optimal solution for not one, but for two (albeit continuous) bilevel optimization problems. Then, we simply need to chose the value of θ for which the NGSC's decision is optimal:

$$\mathbf{BP2}(\beta) : \min_{x,y,\theta}\{F_1(x, y, \theta) : (x, y, \theta) \in \mathbf{gph}\,\Psi,\ x \in \mathbb{X}, \theta = \beta\}. \tag{6.9}$$

This new problem **is not** equivalent to **MIBP1**. The optimal solution to the original problem is inside an interval formed by the two solutions of the modified problem. Specifically speaking, if the parameters $L_{i,j}$ are all zero, then the value of θ the TSO chooses is implicitly determined by the NGSC's final day imbalance configuration (i.e., 1 if the sum of all the final day imbalances is positive, 0 otherwise.) However, the lower level may chose between $\theta = 1$ and $\theta = 0$ if at least some $L_i j$ are positive. As it is shown by Dempe et al. [65], this also implies that the final imbalances x^I are all close to zero. Unless the transportation costs y^f, y^b are remarkably disproportionate, y^g will be close to zero too, hence the error in the approximation is small, and may be considered good enough for practical purposes. This can be interpreted as saying that the penalization the TSO would impose to the NGSC is so small that it can be disregarded.

6.4 A Direct Solution Approach

There are many ways to solve the (continuous) bilevel approximation **BP2**. Some of these approaches are more refined, others take into account the structure of the problem itself, etc. This is because bilevel programs, even those of the simplest kinds, do not lend themselves well to general solution techniques.

The first solution method we propose is deemed a "direct approach", because of its arguably simple procedure. After initialization, we can summarize the approach in two iterating steps:

- Find an upper level vector $x \in \mathbb{X}$ (e.g. using any non-gradient nonlinear optimization method), and
- Evaluate the upper level objective function by minimizing the lower level, parameterized by x.

Notice how, using this approach, all the logical constraints in the TSO problem, namely (6.2d)–(6.2h), become linear, since x^I are all fixed at the times the lower level is solved.

The direct method, simple as it is, faces fundamental difficulties. Due to the non-convexity of the upper level objective function, and the disconnectedness of the feasible region (both traits inherent to general bilevel optimization problems) there are risks that the proposed NGSC solution is not feasible (it voids the TSO's feasible region), or that the problem reaches a local optimum.

However, this direct method serves us well as a point of comparison to more elaborated algorithms, like the one presented in Sect. 6.5. Before we move into that, we can start applying some reformulations to the model, some of which will be re-used later in the chapter.

6.4.1 Linear TSO Objective Function

The objective of the the TSO is to minimize the cash flow, whether the cash goes from itself to the NGSC, or the other way around. This is naturally modeled with an absolute value. To avoid this nonlinearity, we can easily add one more variable and a pair of constraints. Let y^d be an unconstrained, continuous variable, and consider the inequalities

$$-y^d \leq y^g \leq y^d. \tag{6.10}$$

In the optimum, y^d, being otherwise unrestricted, will force one of the constraints to be active, hence y^d will be positive, and equal to the absolute value of y^g. It is then possible to replace the objective function in (6.2a) with

$$f_2(y) = f_2(y, x, \theta) = y^d. \tag{6.11}$$

6.5 A Penalty Function Approach to Solve the Natural Gas Cash-Out Problem

This section presents a solution approach to problem **BLP2**, based mainly in modeling the TSO problem as a variational inequality and then introducing it as a penalty function to the NGSC problem.

Let us represent the NGSC cost, y^g, as function

$$\widehat{F}(y) = y^g = -\sum_{i \in \mathbf{Z}} \left[R_i y_i^I - \delta_i (y_i^I)_+^2 \right] + \sum_{(i,j):i<j} \left[F_{ij} \left(1 - L_{ij}\right) y_{ij}^f - B_{ij} y_{ij}^b \right].$$

Then, for any vectors $\mathbf{x} \in \mathbb{X}$, $\mathbf{y}^\beta \in \mathbb{Y}^\beta(\mathbf{x})$, the TSO reaction to the NGSC decision is a solution to the equilibrium problem represented by the variational inequality

$$\left\langle \widehat{F}(\mathbf{y}^\beta) \nabla \widehat{F}(\mathbf{y}^\beta), \mathbf{y} - \mathbf{y}^\beta \right\rangle \geq 0, \quad \text{for all} \quad \mathbf{y} \in \mathbb{Y}^\beta(x), \tag{6.12}$$

where $\nabla \widehat{F}(\mathbf{y}^\beta)$ is the usual gradient of the function $\widehat{F}$.

Using this variational inequality, we can reduce problem **BLP2** to the *generalized bilevel program*:

$$\mathbf{GBP3}(\beta) : \min_{\mathbf{x},\mathbf{y},\theta}\{\widehat{F}(y) : \mathbf{x} \in \mathbb{X}, \mathbf{y} \in \mathbb{Y}^{\beta}(\mathbf{x}), \theta = \beta, \text{ and } (6.12)\}. \tag{6.13}$$

While expression (6.12) in **GBP3**(β) makes it difficult to implement and solve, there exists the possibility to 'move' the variational inequality constraint to the objective function as a penalty term. Thus, the TSO problem becomes:

$$\Gamma_{\alpha}^{\beta}(\mathbf{x}, \mathbf{y}^{\beta}) = \max_{\mathbf{y}\in\mathbb{Y}^{\beta}(\mathbf{x})} \phi(\mathbf{x}, \mathbf{y}^{\beta}, \mathbf{y}), \tag{6.14}$$

where

$$\phi(\mathbf{x}, \mathbf{y}^{\beta}, \mathbf{y}) = \left\langle \widehat{F}(\mathbf{y}^{\beta})\nabla \widehat{F}(\mathbf{y}^{\beta}), \mathbf{y}^{\beta} - \mathbf{y}\right\rangle - \frac{1}{2}\alpha\|\mathbf{y} - \mathbf{y}^{\beta}\|^{2},$$

and α is a non-negative number (cf. Marcotte [217], Marcotte and Dussault [219]).

The gap function Γ_{α}^{β} is non-negative over its domain, and can only be 0 when the vector $\mathbf{y}^{\beta}$ is a solution to the TSO problem (otherwise the variational inequality is not satisfied and the penalty term is not zero, which makes the gap function negative). Because of this, we can add

$$\Gamma_{\alpha}^{\beta}(\mathbf{x}, \mathbf{y}^{\beta}) \leq 0$$

as a constraint to problem **GBP3**(β) and use Γ_{α}^{β} as a penalty term for the upper level. This renders problem **GBP3** into the standard nonlinear mathematical program

$$\mathbf{NLP4}(\beta) : \min\{Q_{\alpha}^{\beta}(\mathbf{x}, \mathbf{y}^{\beta}, \mu) : \mathbf{x} \in \mathbb{X}, \mathbf{y}^{\beta} \in \mathbb{Y}^{\beta}(\mathbf{x}), \theta = \beta\}, \tag{6.15}$$

with $Q_{\alpha}^{\beta}(\mathbf{x}, \mathbf{y}^{\beta}, \mu) = \widehat{F}(\mathbf{y}^{\beta}) + \mu\Gamma_{\alpha}^{\beta}(\mathbf{x}, \mathbf{y}^{\beta})$.

The relations between problems **NLP4**(β) to **GLP3**(β) depend on the value of the penalty weight μ, which needs to be large enough to lead to $\Gamma(\mathbf{x}, \mathbf{y}^{\beta})$ being zero.

The set of feasible values for problem **NLP4**(β), $\mathbb{C}_{\beta} = \{(\mathbf{x}, \mathbf{y}) : \{\mathbf{x} \in \mathbb{X}, \mathbf{y} \in \mathbb{Y}^{\beta}(\mathbf{x})\}$, is clearly compact since

(a) all variables are bounded either implicitly or explicitly by I^{L}, I^{U}, and
(b) all bounds are closed expressions.

Therefore, we can define a sequence of iterations of problem **NLP4**(β) over increasingly larger, unbounded weights μ that leads us to the global optimal solution of problem **GBP3**(β).

Let $(\mathbf{x}(\mu), \mathbf{y}(\mu))$ be a global optimal solution to **NLP4**(β) (which always exists thanks to the compactness of $\mathbb{C}_{\beta}$ and the continuity of Q_{α}) corresponding to weight μ. Lemma 6.1 can be used to construct an inexact penalization algorithm to solve **GBP3**(β) using **NLP4**(β).

Lemma 6.1 *Let* $(\mathbf{x}^{*}, \mathbf{y}^{\beta *})$ *be a global optimum solution to problem* **GBP3**(β)*, and denote* $\widehat{F}^{\beta *} = \widehat{F}(\mathbf{x}^{*}, \mathbf{y}^{\beta *})$.

Let $\{(\mathbf{x}^k, \mathbf{y}^{\beta,k})\}_{k=1}^{\infty}$ *be a sequence of iterated optimal solutions to problem* **NLP4**(β) *corresponding to successively increasing weights* μ^k, *where* $\mathbf{x}^k$, *and* $\mathbf{y}^{\beta,k}$ *are optimal vectors for each level and weight* μ^k. *Then the following is valid:*

$$Q_{\alpha}^{\beta}(\mathbf{x}^k, \mathbf{y}^{\beta,k}, \mu^k) \le Q_{\alpha}^{\beta}(\mathbf{x}^{k+1}, \mathbf{y}^{\beta,k+1}, \mu^{k+1}); \tag{6.16a}$$
$$\Gamma_{\alpha}^{\beta}(\mathbf{x}^k, \mathbf{y}^{\beta,k}) \le \Gamma_{\alpha}^{\beta}(\mathbf{x}^{k+1}, \mathbf{y}^{\beta,k+1}); \tag{6.16b}$$
$$\widehat{F}(\mathbf{x}^k, \mathbf{y}^{\beta,k}) \le \widehat{F}(\mathbf{x}^{k+1}, \mathbf{y}^{\beta,k+1}); \tag{6.16c}$$
$$Q_{\alpha}^{\beta}(\mathbf{x}^k, \mathbf{y}^{\beta^k}, \mu^k) \le \widehat{F}^{\beta *}. \tag{6.16d}$$

Proof From the optimality of the iterated solutions of problem **NLP4**(β), and since $\mu^{k+1} \ge \mu^k$, we have

$$\begin{aligned} Q_{\alpha}^{\beta}(\mathbf{x}^{k+1}, \mathbf{y}^{\beta,k+1}, \mu^{k+1}) &= \widehat{F}(\mathbf{x}^{k+1}, \mathbf{y}^{\beta,k+1}) + \mu^{k+1}\Gamma_{\alpha}^{\beta}(\mathbf{x}^{k+1}, \mathbf{y}^{\beta,k+1}) \\ &\ge \widehat{F}(\mathbf{x}^{k+1}, \mathbf{y}^{\beta,k+1}) + \mu^{k}\Gamma_{\alpha}^{\beta}(\mathbf{x}^{k+1}, \mathbf{y}^{\beta,k+1}) \\ &\ge \widehat{F}(\mathbf{x}^{k}, \mathbf{y}^{\beta,k}) + \mu^{k}\Gamma_{\alpha}^{\beta}(\mathbf{x}^{k}, \mathbf{y}^{\beta,k}) = Q_{\alpha}^{\beta}(\mathbf{x}^k, \mathbf{y}^{\beta,k}, \mu^k), \end{aligned}$$

which implies (6.16a). Combining the above inequalities and the fact (again, due to the optimality of the solutions in the corresponding problems) that

$$Q_{\alpha}^{\beta}(\mathbf{x}^{k+1}, \mathbf{y}^{\beta,k+1}, \mu^{k+1}) \le Q_{\alpha}^{\beta}(\mathbf{x}^k, \mathbf{y}^{\beta,k}, \mu^{k+1}),$$

we have (after simplifying and dividing by $(\mu^{k+1} - \mu^k)$) line (6.16b):

$$\Gamma_{\alpha}^{\beta}(\mathbf{x}^k, \mathbf{y}^{\beta,k}) \le \Gamma_{\alpha}^{\beta}(\mathbf{x}^{k+1}, \mathbf{y}^{\beta,k+1}).$$

The latter inequality, together with the optimality of the solution $(\mathbf{x}^k, \mathbf{y}^k, \mu_2^k)$ for problem **NLP4**(β), yields (6.16c):

$$\widehat{F}(\mathbf{x}^{k+1}, \mathbf{y}^{\beta,k+1}) \ge \widehat{F}(\mathbf{x}^k, \mathbf{y}^{\beta,k+1}).$$

Once again, the optimality of $(\mathbf{x}^k, \mathbf{y}^k, \mu^k)$, combined with the fact that $\Gamma_{\alpha}^{\beta}(\mathbf{x}^k, \mathbf{y}^{\beta,k}) = 0$ at $(\mathbf{x}^*, \mathbf{y}^{\beta *})$, gives (6.16d):

$$\widehat{F}^{\beta *} = Q_{\alpha}^{\beta}(\mathbf{x}^*, \mathbf{y}^{\beta *}, \mu^k) \ge Q_{\alpha}^{\beta}(\mathbf{x}^k, \mathbf{y}^{\beta,k}, \mu^k)$$

thus completing the proof of the lemma. □

Using Lemma 6.1 and the compactness of the set $\mathbb{C}$, we can easily prove the following theorem:

Theorem 6.1 *Let* $\{(\mathbf{x}^k, \mathbf{y}^{\beta,k})\}_{k=1}^{\infty}$ *be a sequence corresponding to iterated optimal solutions to problem* **NLP4**(β), *using unbounded, increasing large weights* μ^k. *Then every limit point of the sequence* $\{(\mathbf{x}^k, \mathbf{y}^{\beta,k})\}_{k=1}^{\infty}$ *is a solution to problem* **GBP3**(β).

Proof The compactness of the set $\mathbb{C}$ guarantees the existence of a subsequence $\{(\mathbf{x}^k, \mathbf{y}^{\beta,k})\}_{k=1}^{\infty}$ that converges to a limit point $(\bar{\mathbf{x}}^\beta, \bar{\mathbf{y}}^\beta)$. Without loss of generality, assume that the sequence itself is converging. The continuity of function $\widehat{F}$ and Lemma 6.1 yield

$$\lim_{k\to\infty} \widehat{F}(\mathbf{x}^k, \mathbf{y}^{\beta,k}) = \widehat{F}(\bar{\mathbf{x}}^\beta, \bar{\mathbf{y}}^\beta),$$

and

$$Q_\alpha^{\beta*} := \lim_{k\to\infty} \left[Q_\alpha^\beta(\mathbf{x}^k, \mathbf{y}^k, \mu^k) \right] \leq \widehat{F}^{\beta*}.$$

Subtracting the limits above, and recalling the optimality of $\widehat{F}^{\beta*}$, we have

$$\lim_{k\to\infty} \mu^k \Gamma_\alpha^\beta(\mathbf{x}^k, \mathbf{y}^{\beta,k}) = Q_\alpha^{\beta*} - \widehat{F}(\bar{\mathbf{x}}^\beta, \bar{\mathbf{y}}^\beta) \leq \widehat{F}^{\beta*} - \widehat{F}(\bar{\mathbf{x}}^\beta, \bar{\mathbf{y}}^\beta) \leq 0.$$

Remember that function Γ_α^β has non-negative values, which implies

$$\lim_{k\to\infty} \Gamma_\alpha^\beta(\mathbf{x}^k, \mathbf{y}_k^\beta) = 0.$$

Since Γ_α^β is continuous, this limit indicates that the limit point $(\bar{\mathbf{x}}^\beta, \bar{\mathbf{y}}^\beta)$ is a feasible solution for problem **GBP3**(β). This, together with (6.16d), leads to

$$\widehat{F}(\bar{\mathbf{x}}^\beta, \bar{\mathbf{y}}^\beta) \leq \widehat{F}^{\beta*},$$

which proves the optimality of the limit point $(\bar{\mathbf{x}}^\beta, \bar{\mathbf{y}}^\beta)$ for problem $P1(\beta)$. □

Solving **GLP3**(β) (and by extension, **BLP2**(β)), for both values of β, and taking the best one for the NGSC delivers a good approximation to **MIBP1**. However, the actual process of finding a solution vector is left open. Unlike the direct method in Sect. 6.4, we must provide a solution vector for both levels at the same time in each step, which causes the logical constraints in the TSO problem to remain nonlinear. Later, in Sect. 6.7, we will provide a number of experimental results comparing these and other approaches.

6.6 An Expanded Problem and Its Linearization

Up to this point, the programs we have dealt with have been solely focused in the cash-out penalization the TSO charges the NGSC. However, the latter is clearly interested in maximizing her profits, not just on the penalization costs. The TSO, on the contrary, and because of the nature of its business, needs not to worry much about its operation costs beyond the periodic re-balancing, since the network costs are low and highly constant (Juris [162]).

These facts compel us to expand the model in order to include additional variables, doing so in the upper level only. Interestingly, they have the added benefit of discarding the model necessity for a nonlinear cash-out function in the lower level.

The addition of new items to the upper level will understandably complicate the model. In order to remedy this, we will introduce later changes that render both levels linear (once the other level is fixed), which delivers a bilevel linear optimization problem, i.e. a more general and simpler type of multi-level optimization problem.

6.6.1 Upper Level Expansion

In this version of the NGSC-TSO bilevel model, the NGSC has a number of new variables to consider. Basically, she has to fulfil a contract, delivering a minimum amount of gas to her client, a LDC.

The upper level includes several new variables and constants, summarized in Table 6.2.

NGSC cash-out: The objective function for the NGSC consists now on the negative revenues of the company. The cash-out costs are joined by the costs of using the pipeline ($C^b_{ti} x^P_{ti}$,) as well as those for not meeting contractual obligations ($C_{ti} \max\{0, E_{ti} - x^E_{ti}\}$). The NGSC also earns profits. These come from fulfilling contractual obligations ($P_{ti} \min\{x^E_{ti}, E_{ti}\}$,) as well as from extra gas sales she accommodates in each pool zone, beyond the contractual amounts ($P^e_{ti} \min\{0, x^E_{ti} - E_{ti}\}$). If we redefine x as (x^I, x^E, x^P), then:

Table 6.2 Symbols for variables and parameters in the expanded model

Upper level parameters	
$E \in \mathbb{R}^{ZD}$	Expected demand at day t in pool zone i
$E^M \in \mathbb{R}^{ZD}$	Maximum amount of gas that is possible to extract at day t in pool zone i
$P \in \mathbb{R}^{ZD}$	Unit price for the first E_{ti} units of gas extracted/sold (contracted gas) at day t in and pool zone i
$P^e \in \mathbb{R}^{ZD}$	Unit price for whatever units of gas extracted/sold beyond E_{ti} at day t in pool zone i
$C \in \mathbb{R}^{ZD}$	Cost for undelivered contracted gas unit on day t in pool zone i
$C^b \in \mathbb{R}^{ZD}$	Cost for reserved gas capacity on day t in pool zone i
Upper level decision variables	
$x^E \in \mathbb{R}^{ZD}$	Amount of gas extracted and sold by the NGSC at day t in pool zone i
$x^P \in \mathbb{R}^{ZD}$	Amount of gas planned to be extracted (or equivalently, the booked capacity) by the and NGSC at day t in pool zone i

$$F_2(\mathbf{x}; y^g) = y^g - \sum_{t\in\mathbf{D}, i\in\mathbf{Z}} \left[P_{ti} \min\{x_{ti}^E, E_{ti}\} + P_{ti}^e \min\{0, x_{ti}^E - E_{ti}\} \right]$$
$$+ \sum_{t\in\mathbf{D}, i\in\mathbf{Z}} \left[C_{ti} \max\{0, E_{ti} - x_{ti}^E\} + C_{ti}^b x_{ti}^P \right] \tag{6.17a}$$

Total Gas Volumes: The former formulation dealt with how large *imbalances in each zone are*; yet it said nothing about the actual extraction, nor about the booked volumes. Once we introduce variables for these, the relationship is quite straightforward: the imbalance equals the booked capacity minus the extraction in a given pool zone.

$$x_{ti}^I = x_{ti}^P - x_{ti}^E \tag{6.17b}$$

Extraction Limits: The NGSC cannot extract more gas from a given pool zone than what is physically possible, and this extraction must be a non-negative number.

$$0 \le x_{ti}^E \le E_{ti}^M \tag{6.17c}$$

Booking Limits: The NGSC cannot book more extraction privileges from a given zone than what is physically possible, and this booked volume must be a non-negative number.

$$0 \le x_{ti}^P \le E_{ti}^M \tag{6.17d}$$

All other upper level constants remain the same, and so the matrix A_L^U and the vector C_2^U can be redefined to accommodate these new constrains, with a correspondent function $G_2(x) = A_2^U x - C_2^U$. Therefore, the upper level feasible set for the expanded NGSC model can be defined as:

$$\mathbb{X}_2 = \{x : G_2(x) \le 0\}$$

6.6.2 Lower Level Expansion

Since there are no changes in the abstraction of the TSO model, the lower level remains highly the same as in Sect. 6.2. Nevertheless, there are two changes we do introduce: first, the objective function will be that one described in Sect. 6.4.1, that is, the one using the linear term y^d. This is because, considering the work we will be doing later linearizing the entire model, this expression is convenient for us now.

Secondarily, and arguably more interesting, we will drop the term $\delta(y_i^I)_+^2$ from the lower level penalization definition. As we stated before, this term serves a mere modelling purpose. Without it, the NGSC may chose to create unjustifiably large imbalances, consequently the solution of the problem would not be necessarily the actual best decision for the NGSC, but rather the one that created the largest positive imbalances.

However, once we have added new terms to the lower level, there is no longer a need for defining the penalization like this. If the NGSC were to create extraordinarily large imbalances, then either she would not be meeting her contracts, or she'd be booking very large volumes. Both of these options convey a cost for the NGSC, so that she has now to decide whether this cost is a good trade-off on each of the pool zones.

The penalization, without the quadratic auxiliary term, is

$$y^g = -\sum_{i \in \mathbf{Z}} R_i y_i^I + \sum_{(i,j):i<j} \left[F_{ij} \left(1 - L_{ij}\right) y_{ij}^f - B_{ij} y_{ij}^b \right]. \qquad (6.18)$$

Taking into consideration these changes, we can now write the (all linear) lower level constraints (6.2b)–(6.2h), (6.2j), (6.10), (6.18), as the matrices A_2^L, B_2^L, C_2^L, and have

$$g_2(y\theta) = A_2^L x + B_2^L \begin{bmatrix} y \\ \theta \end{bmatrix} - C_2^L.$$

The lower level feasible set then becomes:

$$\mathbb{Y}_2^\beta(\mathbf{x}) = \{y : g_2(x) \leq 0, \theta = \beta\}. \qquad (6.19)$$

One could argue that, since the NGSC is paying the TSO $C_{ti}^b x_{ti}^P$ for using the network, such term could be included in the lower level objective function (as opposed to listing it simply as y^d). Logically speaking, though, there is no relationship between what the TSO controls (namely, the imbalance re-arrangement) and the amount it is charged to the NGSC by the usage of the pipeline. Consequently, while a final report of the TSO finances could include this term, it might in fact not be advantageous trying to implement it at this point. The variables x_{ti}^P are not controlled by the TSO, as the cost C_{ti}^b is not subject to renegotiations at the time of the balancing procedure.

Now we can formulate a new bilevel problem. If we consider, as in past sections, that the binary variable θ is moved to the upper level, then can define the following series of expressions.

$$\varphi_2^\beta(\mathbf{x}) := \min_{\mathbf{y}}\{f_2(\mathbf{y}) : \mathbf{y} \in \mathbb{Y}_2^\beta(\mathbf{x})\};$$

$$\Psi_2^\beta(\mathbf{x}) := \{\mathbf{y} \in Y^\beta(\mathbf{x}) : f_2(\mathbf{y}) \leq \varphi_2^\beta(\mathbf{x})\};$$

$$\mathbf{gph}\, \Psi_2^\beta := \{(\mathbf{x}, \mathbf{y}) : \mathbf{y} \in \Psi_2^\beta(\mathbf{x})\}.$$

So the expanded bilevel optimization problem for the NGSC-TSO model is:

$$\mathbf{BP5}(\beta) : \min\{F_2(\mathbf{x}) : (\mathbf{x}, \mathbf{y}) \in \mathbf{gph}\, \Psi_2^\beta, \mathbf{x} \in \mathbb{X}_2\}.$$

Additionally, the individual problems for this formulation can be defined as

$$\min\{F_2(\mathbf{x}) : \mathbf{x} \in \mathbb{X}_2\} \tag{6.20}$$

and

$$\min\{f_2(\mathbf{y}) : \mathbf{y} \in \mathbb{Y}_2^{\beta}(\mathbf{x})\}. \tag{6.21}$$

6.6.3 Linearization of the Expanded NGSC Model

The model for problem **BP5**(β), while having more explicative power, is considerably more complex than any of the other problems presented, due to the existence of the max and min operators in the objective of the NGSC.

This, however, gives us the chance to implement what we call "linearization" techniques. In this and the next sections, we will make even further reformulations to each level, obtaining in the end a bilevel linear optimization problem, the simplest kind of bilevel programs (Bard [10], Dempe [52], Wen and Hsu [313]). The linear model for the NGSC, as well as its equivalence to (6.20), is provided in Lemma 6.2 below. It does include an important hypothesis: basically, that the extra price the NGSC can obtain for beyond-contract sales is not "very large", at least not as large as the sum of the contract prices and the booking costs. This is indeed a rather weak requirement; if not fulfilled, the NGSC could simply overbook as much as she can, since the extra prices would more than cover any booking costs.

The new model also contains two new sets of artificial variables, x^{a_1} and x^{a_2}. These artificial variables are used to work around the min and max operators in the objective function.

Lemma 6.2 *Let $P_{ti} - P_{ti}^{e} + C_{ti} > 0$ $\forall t, i$. Consider the function*

$$\begin{aligned} F_3(x; y) = y^g - &\sum_{t\in\mathbf{D}, i\in\mathbf{Z}} \left[P_{ti}^{e} x_{ti}^{a_1} + \left(P_{ti} - P_{ti}^{e}\right) x_{ti}^{a_2}\right] \\ &+ \sum_{t\in\mathbf{D}, i\in\mathbf{Z}} \left[C_{ti}\left(E_{ti} - x_{ti}^{a_2}\right) + C_{ti}^{b}\left(x_{ti}^{P} + x_{ti}^{E}\right)\right] \end{aligned} \tag{6.22}$$

and the set of constraints

$$I_{ti}^{L} \le x_{ti}^{I} \le I_{ti}^{U}, \quad t \in \mathbf{D}, \quad i \in \mathbf{Z}; \tag{6.23a}$$

$$S_{ti}^{L} \le x_{ti}^{I} - x_{t-1,i}^{I} \le S_{ti}^{U}, \quad t \in \mathbf{D}, \quad i \in \mathbf{Z}; \tag{6.23b}$$

$$I_{t}^{L} \le \sum_{i\in\mathbf{Z}} x_{ti}^{I} \le I_{t}^{U}, \quad t \in \mathbf{D}; \tag{6.23c}$$

$$0 \le x_{ti}^{E} \le E_{ti}^{M}, \quad t \in \mathbf{D}, \quad i \in \mathbf{Z}; \tag{6.23d}$$

$$0 \le x_{ti}^P + x_{ti}^E \le E_{ti}^M, \quad t \in \mathbf{D}, \quad i \in \mathbf{Z}; \tag{6.23e}$$

$$x_{ti}^{a_1} \le x_{ti}^E, \quad t \in \mathbf{D}, \quad i \in \mathbf{Z}; \tag{6.23f}$$

$$0 \le x_{ti}^{a_2} \le E_{ti}, \quad t \in \mathbf{D}, \quad i \in \mathbf{Z}; \tag{6.23g}$$

$$0 \le x_{ti}^{a_2} \le x_{ti}^{a_1}, \quad t \in \mathbf{D}, \quad i \in \mathbf{Z}; \tag{6.23h}$$

(a) *Let* $(x_1^*, y^{g*}) = (x^{I*}, x^{s*}, x^{E*}, x^{P*}, y^{g*})$ *be a feasible solution to problem* (6.20)*, where* y^{g*} *solves* (6.21)*. Then, there exist values* x^{a_1*}, x^{a_2*} *for the new variables such that* $(x_2^*, y^{g*}) = (x^{I*}, x^{E*}, x^{a_1*}, x^{a_2*}, y^{g*})$ *is also feasible for problem*

$$\min\{F_3(x; y^g) : (6.23\text{a})-(6.23\text{h})\}, \tag{6.24}$$

and their objective function values are equal.

(b) *Conversely, let* $(x_2^{**}, y^{g**}) = (x^{I**}, x^{E**}, x^{a_1**}, x^{a_2**}, y^{g**})$ *be a feasible solution to problem* (6.24)*, where* y^{g**} *solves problem* (6.21)*. Then there exist values of* x^{s**} *and* x^{P**} *such that* $(x_1^{**}, y^{g**}) = (x^{I**}, x^{s**}, x^{E**}, x^{P**}, y^{g**})$ *is a feasible solution to problem* (6.20)*, and their objective function values are equal.*

(c) *If* (x_1^*, y^{g*}) *is an optimal solution to problem* (6.20)*, then* (x_2^*, y^{g*})*, is an optimal solution to* (6.24)*. Conversely, if* (x_2^{**}, y^{g**}) *is an optimal solution to* (6.24)*, then* (x_1^{**}, y^{g**}) *is an optimal solution to* (6.20)*.*

Proof (a) Let (x_1^*, y^{g*}) be a feasible vector for problem (6.20). Lines (6.23a), (6.23c), and (6.23d) are trivially satisfied by x^I, x^E. Combining (6.1e) and (6.1d), we have that (6.23b) also holds. Similarly, (6.17b) together with (6.17d) make x^{I*}, x^{E*} satisfy (6.23e).

Define the auxiliary variables x^{a_1*}, x^{a_2*} as:

$$x_{ti}^{a_2*} = \min\{x_{ti}^{E*}, E_{ti}^*\}, \; t \in \mathbf{D}, \; i \in \mathbf{Z}; \tag{6.25}$$

$$\begin{aligned} x_{ti}^{a_1*} &= \max\{0, x_{ti}^{E*} - E_{ti}\} + x_{ti}^{a_2*} \\ &= \max\{0, x_{ti}^{E*} - E_{ti}\} + \min\{x_{ti}^{E*}, E_{ti}\}, \quad t \in \mathbf{D}, \quad i \in \mathbf{Z}. \end{aligned} \tag{6.26}$$

Then, if $x_{ti}^{E*} \ge E_{ti}$, $x_{ti}^{a_1*} = (x_{ti}^{E*} - E_{ti}) + E_{ti} = x_{ti}^{E*}$. Otherwise, $x_{ti}^{E*} < E_{ti}$ and again $x_{ti}^{a_1*} = 0 + x_{ti}^{E*} = x_{ti}^{E*}$. Therefore, we have that $x^{a_1*} = x^{E*}$. Thus, constraint (6.23f) is satisfied by x^{a_1*}.

The fact that $E_{ti} \ge 0 \, \forall t, i$, means that the definition of x^{a_1*}, x^{a_2*} makes these satisfy (6.23g) and (6.23h). Hence (x_2^*, y^{g*}) is a feasible solution to problem (6.24). Let us now demonstrate the equality of the correspondent objective function values. By definition, we can use x^{a_1*}, x^{a_2*} in the first double sum in (6.17a) as follows:

$$\sum_{t \in \mathbf{D}, i \in \mathbf{Z}} \left[P_{ti} x_{ti}^{a_2*} + P_{ti}^e (x_{ti}^{a_1*} - x_{ti}^{a_2*})\right] = \sum_{t \in \mathbf{D}, i \in \mathbf{Z}} \left[P_{ti}^e x_{ti}^{a_1*} + (P_{ti} - P_{ti}^e) x_{ti}^{a_2*}\right].$$

Furthermore,

$$\max\{0, E_{ti} - x_{ti}^{E*}\} = -\min\{0, x_{ti}^{E*} - E_{ti}\} = E_{ti} - \min\{E_{ti}, x_{ti}^{E*}\}, \quad (6.27)$$

together with (6.17b), imply that the second sum in (6.17a) becomes:

$$\sum_{t\in\mathbf{D}, i\in\mathbf{Z}} \left[C_{ti}(E_{ti} - x_{ti}^{a_2*}) + C_{ti}^{b}(x_{ti}^{I*} + x_{ti}^{E*})\right].$$

This means that both functions (6.17a) and (6.22) have the same value at their corresponding feasible solutions (x_1^*, y^{g*}) and (x_2^*, y^{g*}).

(b) Consider now a feasible solution (x_2^{**}, y^{g**}) to problem (6.24).
Lines (6.1b),(6.1c), and (6.17c) are trivially satisfied, for they are basically the same as (6.23a), (6.23c), and (6.23d).
Define $x_{ti}^{S**} = x_{ti}^{I**} - x_{t-1,i}^{I**}$, and $x^{P**} = x^{I**} + x^{E**}$. Making use of (6.23b) and (6.23e), we can see that constraints (6.1e), (6.1d), (6.17b), and (6.17d) hold for x_1^{**}. Therefore, vector (x_1^{**}, y^{g**}) is feasible for (6.20).
Let us now prove the equality of both solutions' objective function values.
The coefficient of variable x^{a_1**} in (6.22) is $(-P^e)$, which is always non-positive. This means that, when the problem is minimized, each of the components of the variable will grow as much as possible in order to minimize the objective function value. Since this growth is solely constrained by the value of x^{E**}, the optimality of the problem implies that we will have

$$x^{a_1**} = x^{E**} = \max\{0, x^{E**} - E\} + \min\{E, x^{E**}\}. \quad (6.28)$$

In the same manner, the coefficients of variables x^{a_2**} in (6.22) are $-(P^e - P^e + C)$ which are, by hypothesis, negative. Hence, when we minimise the problem, $x_{ti}^{a_2**}$ will grow as much as possible. The maximum growth for this variable is bounded by constraints (6.23g) and (6.23h), and by (6.28) as follows:

$$x^{a_2**} = \min\{E, x^{a_1**}\} = \min\{E, x^{E**}\}. \quad (6.29)$$

Combining (6.28) and (6.29) with the objective function (6.22) we have:

$$\begin{aligned}
&\sum_{t\in\mathbf{D}, i\in\mathbf{Z}} \left[P_{ti}^{e}\left(\max\{0, x_{ti}^{E**} - E_{ti}\} + \min\{x_{ti}^{E**}, E_{ti}\}\right)\right] \\
&\quad + \sum_{t\in\mathbf{D}, i\in\mathbf{Z}} \left[\left(P_{ti} - P_{ti}^{e}\right)\min\{E_{ti}, x_{ti}^{E**}\}\right] \\
&\quad = \sum_{t\in\mathbf{D}, i\in\mathbf{Z}} \left(P_{ti}^{e}\max\{0, x_{ti}^{E**} - E_{ti}\} + P_{ti}\min\{E_{ti}, x_{ti}^{E**}\}\right).
\end{aligned}$$

This corresponds to the first double sum in the objective function (6.17a) of the nonlinear problem.

The second double sum of (6.22), put together with (6.27), (6.29), and the definition of x^{P**}, yields:

$$\sum_{t \in \mathbf{D}, i \in \mathbf{Z}} \left[C_{ti} \left(E_{ti} - x_{ti}^{a_2**} \right) + C_{ti}^{b} \left(x_{ti}^{I**} + x_{ti}^{E**} \right) \right]$$
$$= \sum_{t \in \mathbf{D}, i \in \mathbf{Z}} \left(C_{ti} \max\{0, E_{ti} - x_{ti}^{E**}\} + C_{ti}^{b} x_{ti}^{P**} \right).$$

When combined, the two latter equalities make it clear that both objective functions have equal values at the corresponding feasible solutions (x_1^{**}, y^{g**}) and (x_2^{**}, y^{g**}).

(c) Finally, we prove the optimality of the constructed solutions in the last two items. Let (x_1^{*}, y^{g*}) be an optimal solution for problem (6.20) with an optimal value of $F_2(x_1^{*}, y^{g*}) = F_2^{*}$. If the feasible solution (x_2^{*}, y^{g*}), as given in item (a), is not optimal for problem (6.24), then there exists a feasible vector (x_2^{**}, y^{g**}) such that

$$F_3(x_2^{**}, y^{g**}) = F_3^{**} < F_2^{*}.$$

Because of (b), (x_1^{**}, y^{g**}), with $x^{P**} = x^{I**} + x^{E**}$, problem (6.20) has an objective function value of $F_2^{**} = F_3^{**} < F_2^{*}$, which contradicts the optimality of F_2. Therefore, for any optimal solution to the nonlinear problem, we can construct an optimal solution to the linear problem such that their objective functions' optimal values are equal.

The same argument can be used for the converse statement. □

Lemma 6.3 *For a given upper level vector x, a fixed θ, and the variable $y = (y^I, y^f, y^b, y^g, y^d, y^{a_1}, y^{a_2})$, consider the function:*

$$f_3(y; x^I, \theta) = y^d + \mathbf{M_2} \sum_{i \in \mathbf{Z}} (y_i^{a_1} + y_i^{a_2}) \tag{6.30}$$

and the set of constraints:

$$\left(1 - L_{ij}\right) y_{ij}^{f} \le y_i^{a_1}, \quad i, j \in \mathbf{Z}, \quad i < j; \tag{6.31a}$$

$$y_{ij}^{f} \le y_j^{a_2}, \quad i, j \in \mathbf{Z}, \quad i < j; \tag{6.31b}$$

$$y_{ij}^{b} \le y_j^{a_1}, \quad i, j \in \mathbf{Z}, \quad i < j; \tag{6.31c}$$

$$y_{ij}^{b} \le y_i^{a_2}, \quad i, j \in \mathbf{Z}, \quad i < j; \tag{6.31d}$$

$$-y_i^{a_2} \le y_i^I \le y_i^{a_1}, \quad i \in \mathbf{Z}; \tag{6.31e}$$

$$y_i^{a_1} \ge x_{D,i}^I, \quad i \in \mathbf{Z}; \tag{6.31f}$$

$$y_i^{a_1} \ge 0, \quad i \in \mathbf{Z}; \tag{6.31g}$$

$$y_j^{a_2} \ge -x_{D,i}^I, \quad j \in \mathbf{Z}; \tag{6.31h}$$

$$y_j^{a_2} \ge 0, \quad j \in \mathbf{Z}, \tag{6.31i}$$

$$y_j^I = x_{D,j}^I + \sum_{i:i<j} \left[(1 - L_{ij}) y_{ij}^f - y_{ij}^b \right] + \sum_{k:k>j} \left(y_{jk}^f - y_{jk}^b \right), \quad j \in \mathbf{Z}. \tag{6.31j}$$

$$y^g = -\sum_{i \in \mathbf{Z}} R_i y_i^I + \sum_{(i,j):i<j} \left[F_{ij} \left(1 - L_{ij} \right) y_{ij}^f - B_{ij} y_{ij}^b \right]. \tag{6.31k}$$

$$-\mathbf{M}_1(1-\theta) \le y_i^I \le \mathbf{M}_1 \theta, \quad i \in \mathbf{Z} \tag{6.31l}$$

$$-y^d < y^g < y^d \tag{6.31m}$$

$$y_i^I, y^g, y^d \textit{ free}, \quad i \in \mathbf{Z}; \tag{6.31n}$$

$$y_{ij}^f, y_{ij}^b \ge 0, \quad i, j \in \mathbf{Z}; \tag{6.31o}$$

where M_2 is a sufficiently large scalar number.

(a) *Let $(y_1^*; x^I, \theta) = \left(y^{I*}, y^{f*}, y^{b*}, y^{g*}, y^{d*}; x^I, \theta\right)$ be a feasible solution the problem* (6.21) *for the fixed values x^I, θ. Then there exists (y^{a_1*}, y^{a_2*}) such that*

$$(y_2^*; x^I, \theta) = \left(y^{I*}, y^{f*}, y^{b*}, y^{g*}, y^{d*}, y^{a_1*}, y^{a_2*}; x^I, \theta\right)$$

is a feasible solution to

$$\min_y \{ f_3(y; x^I, \theta) : \text{(6.31a)–(6.31o) hold for fixed } x^I, \theta \} \tag{6.32}$$

(b) *Let $(y_2^{**}; x^I, \theta) = (y^{I**}, y^{f**}, y^{b**}, y^{g**}, y^{d**}, y^{a_1**}, y^{a_2**}; x^I, \theta)$ be an optimal solution to* (6.32). *Then $(y_1^{**}; x^I, \theta) = (y^{I**}, y^{f**}, y^{b**}, y^{g**}, y^{d**}; x^I, \theta)$ is a feasible solution to problem* (6.21).

(c) *If $(y_1^*; x^I, \theta)$ is optimal for* (6.21), *then $(y_2^*; x^I, \theta)$ is optimal for* (6.32) *and vice-versa.*

Proof (a) Let $\left(y_1^*\right)$ solve the nonlinear problem (6.21) for fixed x^I, θ. If we define $y_i^{a_1*} = \max\{x_{D,i}^I, 0\}$, $y_i^{a_2*} = \max\{-x_{D,i}^I, 0\}$, $i \in \mathbf{Z}$, then it is clear that y^{a_1*}

and y^{a_2*} satisfy (6.31f)–(6.31i). Variables y^{I*}, y^{f*}, y^{b*}, y^{g*}, y^{d*} will trivially satisfy the constraints (6.31j)–(6.31o) for θ.
If (6.2e) and (6.2f) hold true for y^{f*}, y^{b*}, then the latter variables will also fulfil (6.31a), (6.31c) and, as y^{I*}, y^{f*}, y^{b*} satisfy (6.2b) and (6.2g), then (6.31b), (6.31d) are also valid for these y^{f*}, y^{b*}.
With y^{a_1*}, y^{a_2*} defined as above, constraint (6.2g) can be rewritten as (6.31e), therefore, as y^{I*} satisfies the former, so it will the latter.
The point $(y_2^*; x^I, \theta)$ is then feasible for problem (6.32). The objective value of the linear problem coincides with

$$\begin{aligned} &f_3(y_2^*; x^I, \theta) \\ &\quad = f_2(y_1^*; x^I, \theta) + \mathbf{M_2} \sum_{i \in \mathbf{Z}} \left(\max\{x^I_{D,i}, 0\} + \max\{-x^I_{D,i}, 0\} \right). \end{aligned} \tag{6.33}$$

(b) Consider now an optimal solution $(y_2^{**}; x^I, \theta)$ to problem (6.32) for fixed values x^I, θ. If $\mathbf{M_2}$ is large enough, a minimization process will force the variables y^{a_1**}, y^{a_2**} to take their minimum values in order to minimize their contribution to the objective function. Thus, we will have

$$y_i^{a_1**} = \max\{x^I_{D,i}, 0\}, \quad y_i^{a_2**} = \max\{-x^I_{D,i}, 0\}, \quad i, j \in \mathbf{Z}. \tag{6.34}$$

The variables $y_i^{a_1}$ represent the *amount of gas that can be drawn from zone* i, whereas variables $y_j^{a_2}$ represent the *amount of gas that can be deposited into zone* j. If $y_i^{a_1**}$ [resp., $y_j^{a_2**}$] is 0, then y_{ij}^{f**} [resp., y_{ij}^{b**}] will be equal to 0 because of (6.31a)–(6.31d). Hence y^{f*}, y^{b*} satisfy (6.2e) and (6.2f).
With y^{a_1**}, y^{a_2**} defined in (6.34), constraint (6.31e) can be rewritten as (6.2g). Therefore, if the former is true for x^{**}, the latter will also hold.
Let us now prove that y^{f**}, y^{b**} satisfy (6.2d). If $x^I_{D,i} \geq 0$, for any $i \in \mathbf{Z}$, then expression (6.31e) becomes

$$- y_i^{a_2**} = 0 \leq y_i^I \leq x^I_{D,i} = y_i^{a_1**}. \tag{6.35}$$

Constraint (6.2b) can be transformed as follows:

$$\sum_{j:j>i} y_{ij}^f + \sum_{k:k<i} y_{ki}^b = x^I_{D,i} + \sum_{j:j<i} (1 - L_{ji}) y_{ji} + \sum_{k:k>i} y_{ik}^b - y_i^I; \quad i \in \mathbf{Z}. \tag{6.36}$$

By (6.31b) and (6.31d), the sums in the right hand side of this equation become 0, which yields:

$$\sum_{j:j>i} y_{ij}^f + \sum_{k:k<i} y_{ki}^b = x^I_{D,i} - y_i^I \leq x^I_{D,i} = y_i^{a_1**}. \tag{6.37}$$

Now, on the contrary, suppose that $I_{D,i} < 0$ for an arbitrary $i \in \mathbf{Z}$. In this case, the left-hand side sums in (6.36), when combined with (6.31a) and (6.31c), become zero:

$$\sum_{j:j>i} y_{ij}^{f} + \sum_{k:k<i} w_{ki} = 0 = y_i^{a_1 **}. \tag{6.38}$$

Lines (6.37), (6.38) show that constraint (6.2d) is valid and hence the values $(y_1^*; x^I, \theta)$ are feasible for problem (6.32). The objective value of the nonlinear problem is

$$f_2\left(y_1^*; x^I, \theta\right) = f_3\left(y_2^{**}; x^I, \theta\right) - \mathbf{M_2} \sum_{i \in \mathbf{Z}} \left(\max\{x_{D,i}^I, 0\} + \max\{-x_{D,i}^I, 0\}\right). \tag{6.39}$$

(c) We have shown that for any feasible solution for either problem, one can find a corresponding feasible solution for the other problem with an explicit relationship between both problems' objective function values. It should also be clear that, if a vector solves one problem, so does its counterpart to the other problem. Indeed, let

$$\kappa = \mathbf{M_2} \sum_{i \in \mathbf{Z}} (\max\{x_{D,i}^I, 0\} + \max\{-x_{D,i}^I, 0\}),$$

then if the nonlinear problem has an optimal solution with an objective function value σ^* strictly less than $(\tau^* - \kappa)$, where τ^* is the optimal objective function value of the linearised problem, then by item (a), the linearised problem has a feasible solution with the objective function value $\sigma^* + \kappa < \tau^*$, which contradicts to the optimality of τ^*. The same argument can be easily applied to the converse statement. Hence, an optimal solution for any problem may be constructed from a likewise optimal solution to the other problem. This completes the proof. □

Notice that, whereas Lemma 6.2 provides an exact reformulation of the NGSC's original nonlinear model, the same is not necessarily true for the TSO model (6.32). The equivalence of models (6.21) and (6.32) depends on the value of $\mathbf{M_2}$ being large enough to 'force' variables y^{a_1} and y^{a_2} to take their minimum possible values at the optimal solution. Luckily, just as we did with parameter $\mathbf{M_1}$, the boundedness of the original TSO variables allows us to calculate a suitable lower bound for $\mathbf{M_2}$ using Lemma 6.4.

Lemma 6.4 *Consider problem* (6.32). *If*

$$\mathbf{M_2} \gg \max_{n \in \mathbf{Z}} \left(R_n + \max\left\{ \sum_{i:i<n} B_{in}, \sum_{i:i<n} F_{in} \right\} \right)$$

then any optimal solution to problem (6.32) *is also an optimal solution to problem* (6.21).

Proof First, notice how the entire second term in the lower level objective function, along with the usage of the artificial variables y^{a_1} and y^{a_2}, is an alternative to avoid setting directly

$$y_i^{a_1} = \max(x_{D,i}^I, 0), \tag{6.40a}$$

$$y_i^{a_2} = \max(-x_{D,i}^I, 0), \tag{6.40b}$$

since the operators would break linearity whenever the upper level decisions x_{ti}^I have not been already fixed, as it is the case with the penalization algorithm in Sect. 6.5.

In order to make the latter equations hold true, it is imperative that the artificial variables attain their lower bound value, as demanded by constraints (6.31f), (6.31i). To achieve this, the value of $\mathbf{M_2}$ should be large enough so as to guarantee that the optimization of f_2 minimizes the second term of the function rather than the first term. In other words, the maximum decrease of f_2 after increasing one unit of y^{a_1} or y^{a_2} should be less than $\mathbf{M_2}$:

$$-\frac{dy^d}{dy_n^{a_1}} < \frac{d\left[\mathbf{M_2}\sum_{i\in\mathbf{Z}}(y_i^{a_1}+y_i^{a_2})\right]}{dy_n^{a_1}} \le \mathbf{M_2},$$

and

$$-\frac{dy^d}{dy_n^{a_2}} < \frac{d\left[\mathbf{M_2}\sum_{i\in\mathbf{Z}}(y_i^{a_1}+y_i^{a_2})\right]}{dy_n^{a_2}} \le \mathbf{M_2},$$

for $n \in \mathbf{Z}$.

From (6.31b), (6.31e), (6.31n), and (6.31k), we get

$$-y^g \le \sum_{i\in\mathbf{Z}} R_i y_i^{a_1} + \sum_{(i,j):i<j} B_{ij} y_j^{a_1};$$

which, together with (6.31m), delivers:

$$-\frac{dfy^d}{dy_n^{a_1}} \le -\frac{dy^g}{dy_n^{a_1}} = R_n + \sum_{i:i<n} B_{in}; \quad n \in \mathbf{Z}.$$

Furthermore, from (6.31e), (6.31b), (6.2k), (6.31i), and (6.31k) we can state

$$y^g \le \sum_{i\in\mathbf{Z}} R_i y_i^{a_2} + \sum_{(i,j):i<j} F_{ij} y_j^{a_2}. \tag{6.41}$$

analysis, are not a direct concern to an optimization model, so they won't be discussed very thoroughly.

However, the TSO's operations are not heavily affected by randomness. The costs for operating the network are rather constant (Juris [162]), and occasional disruptions, as discussed before, are resolved through imbalance-based cash-outs, whose costs are based in mid-term contracts. Hence, the TSO in our models needs not to consider randomness in its decision.

The stochastic bilevel nonlinear problem discussed in detail below borrows heavily from problem **BP5**. The most noticeable modifications are in the upper level's objective function and on the way periods t relate to each other. As with the first expanded model, we will later formulate linear versions for both levels, culminating with an stochastic bilevel linear optimization problem.

Next, we present the upper level NGSCs stochastic model derived from (6.20). It belongs to the class of stochastic models with *recourse*, that is, it contains both variables that are set before any random outcome is known (non-recursive variables) and variables that may be divided upon after some or all of the randomness is realized (Kall and Wallace [183]).

The random parameters in this formulation are volumes demanded and prices the NGSC faces. The data for this may either be already available from the same sources from which the deterministic figures were retrieved, or it can be estimated using for example forecasts (Brockwell and Davis [29]), regression analysis [99], or simulation. It goes without saying that the quality of the stochastic estimations will impact directly in the effectiveness of the model as a decision-making tool.

In order to model randomness, we make use of an *scenario tree* (Kall and Wallace [183]). Each branch of the tree represent a set of possible outcomes to the random variables involved. We call a "node" the set variables and parameters that belong to time periods with identical estimates on all stochastic variables. The first node, or root node of the tree, represents the period in which more certainty exists. In our case, the root node contains the non-recursive variables. From the root node, one or mode child nodes branch out, which in turn will sprawl more branched nodes. The farther from the root node a node is, the less certain the estimations of the stochastic parameters are in that node. This branch process continues until the nodes have covered all time periods in the model; the final nodes in each branch, the ones without any branching child nodes, are customarily called leaf nodes or simply leaves.

In the left figure in Fig. 6.1, we can see how a simple tree with two stages of recursion looks like. The root node contains the parameters and variables that are deterministic, i.e., they cannot be changed later, such as booked volumes. The first stage of recursion represents, say, the first seven days of NGSC-TSO system operation. There is arguably more certainty here than in future estimations, so just two nodes are enough. We can think of these nodes as "high prices", "mean prices" and "low prices", with projected demands reflecting a reaction to these prices.

For the second stage of recursion, there is more uncertainty, so more nodes are branched out from the nodes in the first stages. This means that, for the second week, we have accumulated nine different possible outcomes for the prices and demands in the two week-long process. Each of these is a scenario, and each scenario's recourse

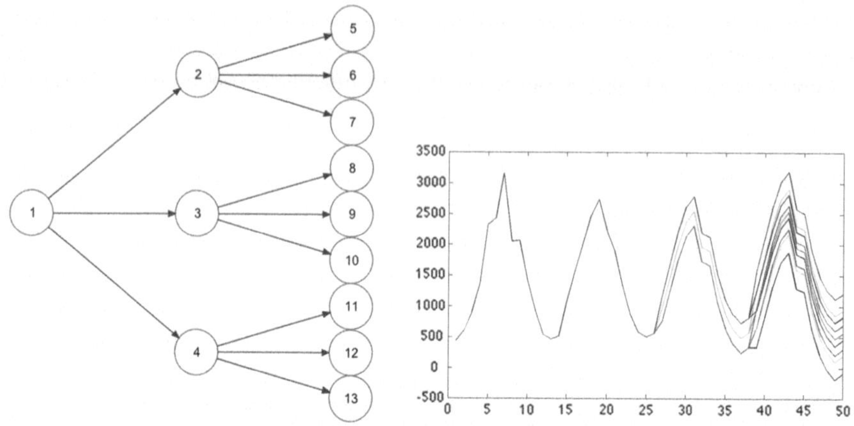

Fig. 6.1 A scenario tree and accompanying time series. As the number of nodes grow, so does the uncertainty, implying that more time series are considered in the long run

variables (in our case, the imbalances created by the NGSC) are likely to take different values, as shown at the right of Fig. 6.1. This implies that the last-day imbalances will also differ among scenarios, and consequently each will create a different situation for the TSO to react to.

Since each time period may be represented in several different nodes, likely a large number as we approach the final stages of the process, it follows that this stochastic model contains a considerable larger number of variables than its deterministic counterpart. If each node has, in general, branched on more than one child, we will have an exponentially growing scenario tree.

The notation for the variables used from here on in the upper level changes slightly, in order to accommodate the newly added node scheme. This is specially true with the indexing format. Taking into consideration Tables 6.1 and 6.2, the meaning of the variables is intuitive enough so as to not needing a new notation table.

Objective Function. The objective function in the stochastic model does away with revenues from out-of-contract sales. This is done to emphasize the impact of variability on the demands faced by the NGSC; being able to sell extra gas decreases the likelihood that the NGSC will plan accordingly to the demands forecast. The only other major change is the introduction of the probability of a particular node k happening, symbolized by $\mathbf{p}_k$. Summarily, this new objective function is the expected value of all nodes' contributions, plus the expected value of all TSO responses to each leaf node. Make $x = (x^I, x^E, x^P)$; then

$$F_4(x) = \sum_{k \in \mathbf{K}^S} \mathbf{p}_k y^g(x_k^I) + \sum_{k \in \mathbf{K}} \mathbf{p}_k \sum_{t \in \Theta^k, i \in \mathbf{J}} \Big[C_{kti} \max\{0, E_{kti} - x_{kti}^E\} - P_{kti} \min\{e_{kti}, E_{kti}\} + C_{kti}^b x_{kti}^P \Big]. \tag{6.46}$$

There are many things to note here. Each node $k \in \mathbf{K}$ represents a set of time periods which have a reasonably congruent estimation of the random variables; it might be one week, or five days, or one month. The number of periods in a node needs not to be equal among nodes in different stages, hence t represent now the tth day in a given node. Nodes belonging to the same stage must, however, have the same number of days.

Counter-intuitively, $\sum_k \mathbf{p}_k > 1$. The root node always happens, so $p_1 = 1$. Indeed, one and only one node in the sth stage of recursion may occur, hence $\sum_{k \in K^s} \mathbf{p}_k = 1$. If we have exactly S stages, then K^S represents the stage with the leaf nodes, which explains the expected value for the TSO response, where $y^g(x_k^I) \in \Psi_3^{\beta}(x_k^I)$ at the kth node, which must be a leaf node.

The terms in the long sum are mostly analogous to the deterministic expanded model.

Imbalance, Imbalance Totals, and Imbalance Swing Limits. These constraints haven't changed beyond the addition of the cluster indexing:

$$I_{kti}^L \le x_{kti}^I \le I_{kti}^L, \quad k \in \mathbf{K}, \quad t \in \mathbf{D}^k, \quad i \in \mathbf{J}; \tag{6.47a}$$

$$I_{kt}^L \le \sum_{i \in \mathbf{J}} x_{kti}^I \le I_{kt}^L, \quad k \in \mathbf{K}, \quad t \in \mathbf{D}^k; \tag{6.47b}$$

$$S_{kti}^L \le x_{kti}^s \le S_{kti}^L, \quad k \in \mathbf{K}, \quad t \in \mathbf{D}^k, i \in \mathbf{J}; \tag{6.47c}$$

Imbalance Swing Relationship. These constraints change a bit. Remember that the nodes in a single stage represent different realizations of several stochastic variables, and as such, I_k gives the chance to act in accord to that realizations. However, all imbalances in the first day of each node are constrained, via the imbalance swing limit, by the last day of the former period:

$$x_{kti}^I = \begin{cases} x_{0i}^I + s_{kti} & \text{if } \ k = 1, \quad t = 1; \quad i \in \mathbf{J} \\ x_{a(k), N^{a(k)}, i}^I + x_{kti}^s & \text{if } \ k > 1, \quad t = 1, \quad i \in \mathbf{J} \\ x_{k,t-1,i}^I + s_{kti} & \text{otherwise.} \end{cases} \tag{6.47d}$$

Here, $a(k)$ is a node-to-node mapping that refers to the node from which k branched (its parent node).

Imbalance Definition. Imbalances are the difference between the booked volumes and the extracted volumes.

$$x_{kti}^I = x_{kti}^P - x_{kti}^E, \quad k \in \mathbf{K}, \quad t \in \mathbf{D}^k, \quad i \in \mathbf{J}. \tag{6.47e}$$

Non-recursion on Booked Volumes. The booked volumes p are common to all nodes with an equal day t in a single stage. The NGSC decides upon them before obtaining information about any stochastic outcome. Hence, all nodes k in a given stage $\mathbf{K}^s$ must respond to the same booked volumes:

$$x^P_{kti} = x^P_{k'ti}, \quad k, k' \in \mathbf{K}^s, \quad t \in \mathbf{D}^k, \quad i \in \mathbf{J}, \quad s \in \mathbf{S}. \tag{6.47f}$$

Sign of the Booked and Extracted Volumes. All volumes booked and sold by the NGSC must be non-negative.

$$x^E_{kti}, x^P_{kti} \geq 0, \quad k \in \mathbf{K}, \quad t \in \mathbf{D}^k, \quad i \in \mathbf{J}. \tag{6.47g}$$

The lower level model remains unchanged, in the understanding that the variables we formerly referred to as x^I_{Di} are now $x^I_{kD^k i}$, where $k \in \mathbf{K}^S$, due to the changes in which the NGSC decides upon the nodes in the scenario tree.

One very important difference is, of course, that we now have $\mathbf{K}^S$ lower levels problems to solve; the expected responses of the TSO to each final stage node. For a binary scenario tree with four stages of recursion, this means eight different lower level problems have to be solved for each upper level vector.

Using the new set of constrains and the new equations, we define new upper level matrices A^U_4, C^U_4, function $G_4(x; \theta, y)$, and feasible set $\mathbb{X}_4$:

$$A^U_4 x \leq C^U_4 \Leftrightarrow \text{(6.47a)–(6.47g) hold}; \tag{6.48}$$

$$G_4(x) := A^U_4 x - C^U_4; \tag{6.49}$$

$$\mathbb{X}_4 := \{x : G_4(x) \leq 0\}. \tag{6.50}$$

And its corresponding bilevel model, using the same lower level as in the original deterministic expanded model, but adapting lower level feasible space to account for the responses to each final stage node:

$$\varphi^\beta_4(x^I_k) := \min_y \{f_3(y_k; x^I_k) : y_k \in \mathbb{Y}^\beta_3(\mathbf{x}_k)\};$$

$$\Psi^\beta_4(x) := \left\{\{y_k\}_{k \in \mathbf{K}^S}, y_k \in \mathbb{Y}^\beta_3(\mathbf{x}_k), f_3(y_k; x^I_k, \theta) \leq \varphi^\beta_4(x^I_k)\right\};$$

$$\mathbf{gph}\, \Psi^\beta_4 := \{(x, \overline{y}) : \overline{y} \in \Psi^\beta_4(x_k)\}.$$

This makes the stochastic bilevel optimization problem:

$$\mathbf{BP7}(\beta) : \min\{F_4(\mathbf{x}) : (x, \overline{y}) \in \mathbf{gph}\, \Psi^\beta_4, x \in \mathbb{X}_4\}.$$

The last model in this chapter is the linear equivalent version of the stochastic problem **BP7**(β). The formulation is analogous to the one done with the deterministic model, in which we use auxiliary variables to circumvent the max and min operators in the upper level function, as explained in Lemma 6.5.

Lemma 6.5 *Consider the objective function*

$$F_5(x) = F_5(x^I, x^E, x^{a_1}) = \sum_{k \in \mathbf{K}^S} \mathbf{p}_k y^g(y^I_k)$$

$$+\sum_{k\in\mathbf{K}}\mathbf{p}_k\sum_{t\in\Theta^k,i\in\mathbf{J}}\Big[C_{kti}\left(E_{kti}-x^{a_1}_{kti}\right)$$
$$-P_{kti}x^{a_1}_{kti}+C^b_{kti}\left(x^I_{kti}+x^E_{kti}\right)\Big], \tag{6.51}$$

and the set of constraints

$$I^L_{kti}\le x^I_{kti}\le I^L_{kti},\quad k\in\mathbf{K},\quad t\in\mathbf{D}^k,\quad i\in\mathbf{J}; \tag{6.52a}$$

$$I^L_{kt}\le\sum_{i\in\mathbf{J}}x^I_{kti}\le I^L_{kt},\quad k\in\mathbf{K},\quad t\in\mathbf{D}^k; \tag{6.52b}$$

$$\begin{cases} S^L_{11i}\le x^I_{11i}-x^I_{0i}\le S^L_{11i} & \textit{if}\ \ i\in\mathbf{J},\\ S^L_{kti}\le x^I_{k1i}-x^I_{a(k),N^{a(k)},i}\le S^L_{kti} & \textit{if}\ \ i\in\mathbf{J},\\ S^L_{kti}\le x^I_{k,t-1,i}+x^s_{kti}\le S^L_{kti} & \textit{if}\ \ i\in\mathbf{J}; \end{cases} \tag{6.52c}$$

$$x^{a_1}_{kti}\le E_{kti},\quad k\in\mathbf{K},t\in\mathbf{D}^k,i\in\mathbf{J}; \tag{6.52d}$$

$$x^{a_1}_{kti}\le x^E_{kti},\quad k\in\mathbf{K},t\in\mathbf{D}^k,i\in\mathbf{J}; \tag{6.52e}$$

$$x^I_{kti}+x^E_{kti}=x^I_{k'ti}+x^E_{k'ti},\quad k,k'\in\mathbf{K}^s,t\in\mathbf{D}^k,t'\in\mathbf{D}^{k'},i\in\mathbf{J},s\in\mathbf{S}; \tag{6.52f}$$

$$x^E_{kti},x^{a_1}_{kti}\ge 0,\quad k\in\mathbf{K},t\in\mathbf{D}^k,i\in\mathbf{J}. \tag{6.52g}$$

(a) *Let* $(x_1^*;y^{g*})=(x^{I*},x^{s*},x^{E*},x^{P*};y^{g*})=(\mathbf{x}^*;y^{g*})$ *be a feasible solution to problem* (6.48), *where* y^{g*} *is such that* $\mathbf{y}\in\Psi_4^{\beta}(\mathbf{x}^*)$ *for a given* θ.
There exists x^{a_1*} *such that* $(x_2^*;y^{g*})=(x^{I*},x^{E*},x^{a_1*};y^{g*})$ *is also feasible for problem*
$$\min\{F_5(x_2^*;y^g):(6.52\text{a})-(6.52\text{g})\ \text{hold}\}, \tag{6.53}$$
and their objective values are equal.

(b) *Let* $(x_2^{**};y^{g*})=(x^{I**},x^{E**},x^{a_1**};y^{g**})$ *be a feasible solution to problem* (6.53), *where* y^{g**} *is such that* $y\in\Psi_4^{\beta}(\mathbf{x})$ *for a fixed* θ. *Then there exist values of* x^{s**} *and* x^{P**} *such that* $(x_1^{**};y^{g*})=(x^{I**},x^{s**},x^{E**},x^{P**};y^{g**})$ *is a feasible solution to problem* (6.48), *and their objective values are equal.*

(c) *If* $(x_1^*;y^{g*})$ *is an optimal solution to problem* (6.48), *then the corresponding* $(x_2^*;y^{g*})$, *is an optimal solution to* (6.53).
Conversely, if $(x_2^{**};y^{g**})$ *is an optimal solution to* (6.48), *then* $(x_1^{**};y^{g**})$ *is an optimal solution to* (6.53).

The proof of Lemma 6.5 is mostly analogous to that of the upper level linearization of problem (6.24) given in Lemma 6.2 (once the leaf-node considerations are taken into account) so it is not presented.

Using the new upper level model (6.53), we can define once more a feasible set and lower level equations to formulate the stochastic bilevel linear problem **SBLP8**(β):

$$A_5^U x \leq C_5^U \Leftrightarrow \text{(6.52a)–(6.52g) hold;} \tag{6.54}$$

$$G_5(x) := A_5^U x - C_5^U; \tag{6.55}$$

$$\mathbb{X}_5 := \{x : G_5(x) \leq 0\}; \tag{6.56}$$

$$\mathbf{SBLP8}(\beta) : \min\{F_5(\mathbf{x}) : (\mathbf{x}, \bar{\mathbf{y}}) \in \mathbf{gph}\, \Psi_4^{\beta}(x), \mathbf{x} \in \mathbb{X}_5\}. \tag{6.57}$$

6.9 Natural Gas Market Classification Using Pooled Regression

Econometric studies about natural gas emerged as an important research object since natural gas may now be sold and traded in a number of stock markets, each one responding to potentially different behavioral drives. In this section, we present a method to differentiate sets of time series based on a regression model relating price, consumption, supply, and other factors. Our objective is to develop a method to classify different areas, regions or states into groups or classes that share similar regression parameters. Once obtained, these groups may be used to make assumptions about corresponding natural gas prices in further studies. This section is based mainly on the recently published paper [178] by Kalashnikov et al.

As was mentioned in the introduction of this chapter, in the early 1990s, several regulations were passed in the US and the European Union [98, 100, 293] changing the way natural gas was marketed and traded. Particularly, this liberalization [150] effectively ended a period in which natural gas was a state-driven industry. The liberalization has also created the emergent natural gas markets, as well as a strong demand for models to better tackle the new problems and profit from this new setting ([101] and Midthun [230]).

Owing not only to this liberalization, but also to the new local conditions that are more open to competition, new small players entered the natural gas industry, especially at the local scale. Indeed, the US has over 80 interstate, long-distance pipelines (see Doane and Spulber [93]), serving different regions with various climatic, demographic, economic and political circumstances. Natural gas usage in Alabama, for example, intuitively is not the same as in Oregon, thus the market dynamics of the fuel are also different, and this, we presume, should be reflected in some way in the econometric data of the states.

Not only macro-economic trends, however, are affected by this setting. When doing cross-regions studies of various aspects of the supply chain, such as the forecasting of demand (Gutiérre [135] and Lyness [209]), the balancing of the pipelines after imbalances have been created by the natural gas shippers (Dempe et al. [65],

Kalashnikov and Ríos-Mercado [181], Keyaerts et al. [186]), or the dynamics of interstate-intrastate systems (see Huntington [149]), one has to take into account the existence of different markets.

The existence of a common relationship between price and consumption of natural gas across several zones allows for strong claims of uniformity, which are useful when, for example, we are building scenarios for a stochastic problem. Indeed, if we manage to group the regions in clusters with similar price and consumption functions, we can reduce the number of variables needed in a scenario tree formulation (Midthun [230] and Tomasgard et al. [300]).

As such, we specify a regression function that relates many of the most relevant econometric figures for each of the 48 contiguous states of the American Union, modeling price as a function of explicative variables such as natural gas consumption, supply and storage levels, as well as population (number of costumers), oil prices, temperatures, and production. The regression coefficients are then used to divide the set of states into several subsets, or groups, obtaining a partition in which all the states in a group share the same regression parameters, and thus can be classified as an (implicit) market. The partition is made considering both statistical and non-statistical characteristics of the obtained regression coefficients. The resulting partitions are next compared with others in their similitude and statistical significance, which would validate the goodness of the combination of the dendrogram and GRASP grouping methods.

This section is organized as follows: the motivation and literature review on natural gas econometric regression is given in Sect. 6.9.1. Section 6.9.2 describes the way the regression function is derived, while Sect. 6.9.3 details the method for using the said function to perform the classification. Section 6.9.4 presents and discusses the results of the study.

6.9.1 Natural Gas Price-Consumption Model

This work was motivated by our previous research in the natural gas supply chain, specifically developing an optimization model that addresses issues in interstate pipelines (see all the previous sections of this chapter). The data used in this model, however, came from different regions, and therefore the time series involved did not necessarily behave in the same way.

As an example, suppose we are trying to model a certain problem that involves forecasting the residential consumption and price of natural gas in the states of Washington and Oregon, i.e., four time series. If the robustness of the model is also a concern, then we should additionally consider different forecasting scenarios. Even with only two possible forecasting scenarios for each series (high/low consumption or prices) this translates into 2^4 possible behaviors of the econometric parameters. If consumption is expressed as a function of price, however, then the scenario tree has only 2^2 branches. Furthermore, if the regression function for both states is the same, then the number of scenarios can be reduced to just two. As the number of states being modeled scales up, i.e., there are more than two parameters of interest, common assumptions like those mentioned above help reduce greatly the amount of scenarios in a stochastic model, optimization or otherwise.

As we studied particular sets of data, it was noted that historical data of consumption and price showed conspicuous properties that could be used for the sake of our aims. Even though these data collections were taken from different states, all pairs of time series showed elastic consumption/demand (Nelder [249] and [102]), exponentially growing price averages (MacAvoy [211] and Nelder [249]) and both series in every pair seemed to be highly correlated to each other.

Indeed, the possibility of characterizing one set of series as a (regression) function of the other was interesting, as it would reduce the amount of data we needed to consider when modeling optimization problems. It is, of course, a common practice in economic and managerial sciences to do that since, for example, demand data is simpler to work with than price data (Talluri and Van Ryzin [298]). The latter is mainly because the demand is usually easier to predict, and its behavior is less chaotic than that of prices. Such historical relationship between price and consumption is a common topic of study in time series economic analysis (Keat and Young [185]), which is mostly performed with the inclusion of other explicative variables, such as the price of substitutes (electricity, coal), weather conditions, etc.

This is the case of several models where the calculation of elasticities is the primary goal of the study by Gowdy [131]. Log-linear models (Beierlein [16], Lin et al. [203], Krichene [193] and Yoo et al. [326]) are generally favored because of the ease they provide when computing elasticity figures. However, linear models also have applications in the natural gas industry, like the Short-Term Integrated Forecasting system (STIFS) used by the United States Energy Information Agency in order to estimate natural gas demand as a function of several types of important variables related to the energy industry [99].

6.9.1.1 Former and Current Approaches

As explained in Kalashnikov's previous work [174], a carefully designed regression function can help to achieve the strong assumptions mentioned above. Nevertheless, the study of such relationships and the possibility of forming state clusters based merely upon time series data analysis turned out to be interesting by itself, and we developed two different approaches to partition the collection of states. As we observed, neighboring states showed a large amount of diversity, yet different methods of grouping seemed to place certain states consistently together.

Two major areas of opportunity discovered were the design of the regression function, and the trade-off that each partition algorithm made use of.

Kalashnikov's previous paper [174] aimed at a very definite objective regarding the qualities of the regression model: it had to correlate consumption and price of residential natural gas series, using the former as the explicative variable because of the ease in its forecasting. The expression thus obtained served its purpose well, as demonstrated in its application to the optimization models by Kalashnikov et al. in

[169]; nevertheless, a more inclusive approach would involve series that comprise more information. Following the examples found in the literature and our own experience, we revealed that including more explicative series provided very good results in terms of regression fit. This has led to the model presented in the next subsection.

Coming back to the partitioning method, the two approaches presented before were:

- the Dendrogram Grouping Method, which "cuts" a binary tree (whose nodes represent regression parameters) based on how close to each other the parameters are with respect to a given metric function and a weight scheme for the entries. This method proved replicative and fast, but it does not provide statistical significance to the grouped states' parameters (i.e., one state might find that temperature is a significant regressor, whereas some other state in the same group may not).
- Another one is a greedy heuristic that starts with a number of states called "group leaders", and iteratively selects for each remaining state the group that suits the state best, based on its regression coefficient $\mathbb{R}^2$. Because of the large amount of regressions performed, this method was reported to be slower and subject to accidental fluctuations, but the final results always guaranteed that all states in one group shared the same significance in their parameters.

In the following subsections, we explain how we have improved over our latest approach, adding explicative power and robustness to the partitioning method and, ultimately, creating a better technique to identify similar regions based on their econometric data.

6.9.2 Regression Analysis

6.9.2.1 Individual Multiple Linear Regression (IMLR)

Let n be the total number of states, m the number of observations per time series (months, in this case), $I = \{1, 2, \ldots, n\}$ be the set of the 48 contiguous states of the American Union, $t \in \Theta = \{1, 2, \ldots, m\}$ the (discrete) time parameter, $\{P'_{i,t}\}$ the differenced residential natural gas price in state $i \in I$ at time $t \in \Theta$, $\{T'_{i,t}\}$ the differenced temperature, in Kelvin, shifted so that the minimum figure is e, $\{O'_t\}$ the differenced average spot price of oil in the US at time $t \in \Theta$, $\{N'_{i,t}\}$ the differenced number of residential consumers of natural gas in state $i \in I$ at time $t \in \Theta$, and $\{C'_{i,t}\}$ the differenced consumption of natural gas in state i at time t.

Notice that all these series are *differenced*, or more precisely, lag-(-1)-differenced from the original values. This is because the said original values all tested positive for unit roots in the advanced Dickey-Fuller test. In contrast to the original series, the differenced series prove to be stationary, hence we make use of the latter.

This is the linear model we devised to relate the above-described series:

$$\widehat{C'}_{i,t} = \alpha_{0,i} + \alpha_{1,i} P'_{i,t} + \alpha_{2,i} C'_{i,t-12} + \alpha_{3,i} T'_{i,t} + \alpha_{4,i} O'_t + \alpha_{5,i} N'_{i,t}; \quad t \in \Theta; \quad i \in I. \tag{6.58}$$

We choose a Robust Regression Analysis using Huber weights to fit the series over traditional least-squares method due to non-normality of the residuals experienced with the latter. Furthermore, due to the steps described in the next sections, heteroskedasticity would likely appear in the residuals once the pooling regression is carried on.

While most of the series were reasonably fit by (6.58), a couple of them showed very erratic behavior in either their natural gas price or consumption series. This is expected insofar economic forecasting is commonly subject to the large instability at time t. As the driving force behind short term fluctuations in natural gas pricing is consumer demand rather than production supply, price was shown to be a significant factor when describing market consumption.

The selection of the descriptive variables was made considering other consumption models in the literature, the available data, and the significance found in the preliminary regression analysis. In particular, electricity prices, the natural gas supply and production, as well as a time index, were tested but found not to be significant in most of the states. This was especially interesting in the case of electricity prices, which certain sources cite as usual descriptors for the natural gas demand, but which were found to be 0.05 significant in only 12 of the 48 cases, thus dropped from the model.

The consumption and price of natural gas are endogenous variables as both are correlated to system shocks, such as unstable governments or weather-related events. As an alternative to the use of least squares regression to fit the model given in (6.58), a two-stage least squares approach could be employed with such instrumental variables as the number of gas producing wells, reserve estimates, and underground storage, to name only a few. However, this approach is not considered here, because the response (reaction) time of consumers' consumption habits to the shocks is much longer than that to the spot prices set by the market every day.

6.9.2.2 Pooled Multiple Linear Regression (PMLR)

Now we address the issue of how one can use the same regression formula for more than one state, which would create several classes of states where demand responds to changes in the descriptors in a similar mode.

Assume that we have split n collections of state time series into several classes, with the members of each class sharing a common set of regression parameters. Then the pooled data from the groups would be regressed at the same time, creating *pooled regressions*.

Let $I = \{I_1, I_2, \ldots, I_K\}$ be a partition of I, and consider the model:

$$\begin{aligned} \widehat{C'}_{i,t} &= \beta_{0,i} + \beta_{1,k} P'_{i,t} + \beta_{2,k} C'_{i,t-12} + \beta_{3,k} T'_{i,t} \\ &\quad + \beta_{4,k} O'_t + \beta_{5,k} N'_{i,t}; \quad t \in T, \quad \forall i \in I_k, \quad k = 1, 2, \ldots, K. \end{aligned} \tag{6.59}$$

Note that this model—called the Pooled Multiple Linear Regression (PMLR) model—has K sets of parameters for each regressor variable, except for the intercepts

a_0^i, which we allow to be different for each state. In comparison, model (6.58) has n sets of parameters.

How should one define the partition **I** of the set of states? A good partition is expected to deliver groups of more or less congruent sizes, while maintaining a high individual $\mathbb{R}^2$ value for each state. A good partition method should also be replicative (i.e., the same partition is obtained for the same group of states), fast enough, and support the statistical significance.

6.9.3 Dendrogram-GRASP Grouping Method (DGGM)

In this subsection, a combination of both grouping methods mentioned in Kalashnikov et al. [176] into a GRASP heuristic is proposed. The resulting technique inherits the replicative property of the dendrogram method, while retaining the statistical significance of the heuristic algorithm.

6.9.3.1 Dendrograms

Dendrograms are binary trees in which two observation vectors a and b form the (sub-)branches of a higher branch c, so that

(i) these two observation vectors are "closer" to each other than to any other observation d, and

(ii) c is not an observation per se, but a new, artificial vector formed by some linear combination of a and b.

The term "closer" is interpreted with respect to some metric (e.g., the Euclidean metric), while the artificial observations are produced by the weighted combination method. Once the dendrogram is formed, it is cut down from the root thus generating (sub-)dendrograms with the branches resulting from the cut. The height of the cut is determined according to one of several criteria (the number of sub-dendrograms produced, the maximum allowed membership for the sub-dendrogram, etc). The leaves pertaining to a given sub-dendrogram will pool their regression data together and form one group for the PMLR.

Previous experiments by Kalashnikov et al. [176] have shown that what is called the "average Euclidean" metric [151] delivers satisfactorily high $\mathbb{R}^2$ levels with a better homogeneity in the resulting groups than other linkage function options.

6.9.3.2 GRASP Heuristics

GRASP stands for Greedy Randomized Adaptive Search Process; it is a meta-heuristic, that is, a general method designed to provide good—but not necessarily optimal—results in problems otherwise too complicated to find an optimal solution, especially combinatorial problems, see Festa and Resende [109].

Summarily, our GRASP approach will start with a seed formed by several one-state-groups; then, for each state, it will identify those groups that deliver higher $\mathbb{R}^2$ values once the data for the current state is pooled with that of the group. This is called the Restricted List of Candidates, or RLC. A group I_k from the RLC is chosen at random, and the current state is added to I_k, pooling its data with those already in the group. A number of swaps and movements are performed once the states are all in place, in order to try to improve the values of the resulting statistics $\mathbb{R}^2$.

It is important to note that setting the values for the GRASP routine is rather subjective, since there is no definite objective to be achieved. Indeed, one cannot determine what number of groups is optimal, or which way is the best to define the greedy function. For example, one could prefer to increase the grouped $\mathbb{R}^2$ value in each group rather than the average of the individual $\mathbb{R}^2$s in that group, or might do vice-versa. This is exemplified by the function

$$F_w(I_k) = \omega R^2_{I_k} + \frac{1-\omega}{|I_k|} \sum_{i \in I_k} R^2_i,$$

where

$$R^2_{I_k} = 1 - \frac{\sum_{t \in T, i \in I_k} (y_{it} - \hat{y}_{it})^2}{\sum_{t \in T, i \in I_k} (y_{it} - \bar{y}_{I_k})}, \; R^2_i = 1 - \frac{\sum_{t \in \mathbf{T}} (y_{it} - \hat{y}_{it})^2}{\sum_{t \in \mathbf{T}} (y_{it} - \bar{y}_i)}.$$

Here, $y_{it} = \ln C'_{it}$, and $\bar{y}_\iota$ is understood as the average of all of the observations belonging to ι if the latter is a state (e.g., $\iota = i$), or as the average of the observations of the states in ι, if the latter is a set of states (e.g., $\iota = I_k$).

For the local search, we handle the improvement function $G_\tau(I_k, I_\ell, I_i)$, which is used when deciding if it is convenient to move state i from group k to group ℓ. It is parameterized by the improvement weight τ.

$$G_\tau(I_k, I_l, I_i) = (1-\tau)\frac{R^2_{I_k} + R^2_{I_\ell}}{2} + \tau R^2_i.$$

6.9.3.3 Dendrogram-GRASP Algorithm

The following algorithm is used to classify the set of 48 contiguous states of the United States into groups that share a common regression function:

1. Initialize the values for each of the time series in each of the 48 states. Set a seed size s_{Seed}, a maximum number of groups s_{Max}, a RLC size s_{RLC}, an individual/grouped R^2 weight $\omega \in [0, 1]$, an individual/grouped threshold $\varphi \in (0, 1)$, an improvement weight $\tau \in (0, 1)$, a relative improvement threshold $\psi \in [0, 1]$, and a maximum number of local search steps, s_{ls}.

Seed Selection

2. Perform an IMLR on each of the 48 sets of time series, obtaining α_{ji}, $j = 0, \ldots, 6$, $i \in I$.
3. Form a dendrogram of 48 leaves with the vectors α, using the average Euclidean mean as the linkage function, and cut it so that there are exactly s_{Seed} sub-trees.
4. Select the state with the highest R_i^2 from each of the obtained groups and call it the kth group's leader. Define the one-state groups obtained as the partition $\mathbf{I}_k$. All the non-selected (spare) states form the set *Active*.

Greedy Process

5. For each state x in the set *Active*:
   a. Pool the data of x with the data of each of the formed groups and perform a pooled regression. Select a number of s_{RLC} groups with the highest value of the greedy function F_w and form the RLC.
   b. Choose randomly one of the groups from the RLC, for example, I_a.
      - If none of the candidate groups in the RLC delivers $F_w(I_k) > \varphi$, and we haven't yet reached the maximum number of groups s_{Max}, create a new group $I_x = \{x\}$ containing only x, remove x from the active set, and update all the parameters.
      - Otherwise, assign x to I_a, remove x from the active set, and update all the parameters.
6. All of the states are now partitioned into the groups, and we can begin the local search.

Local Search

7. For $l = i$ to $l = s_{ls}$, do
   a. Randomly select one of the formed groups, I_a, and one state in that group, x. Select another group, I_b. Compute $g_1 = G_\tau(I_a, I_b, x)$
   b. Remove x's data from I_a and pool the same data of x with I_b. Compute $g_2 = G_\tau(I_a, I_b, x)$.
   c. If $g_1 \geq (1 + \psi)g_2$, remove x from I_b and return it to I_a. Otherwise, continue.
8. Report the obtained groups as the desired partition.
9. End.

6.9.3.4 Partition Similarity

To determine the similitude of two partitions, we will use an expression that, roughly speaking, counts the number of coincidences found in two partitions and divides it by the number of total possible coincidences, given the sizes of the groups in each partition. While there are many disputable ways to measure the similitude between partitions with a different number of elements, this method was chosen because of its normality. Indeed, it will always return 1 when both partitions are identical, and will always return 0 when there are no coincidences between two partitions, that is, when no two states share a group in both partitions, and no state is single-grouped in both partitions.

Let $I = \{I_1, I_2, \ldots, I_K\}, J = \{J_1, J_2, \ldots, J_L\}$ be two arbitrary partitions of the set of states, with $I_i = \{I_1^i, I_2^i, \ldots, I_{k_i}^i\}, i = 1, \ldots, K$, and $J_j = \{J_1^j, J_2^j, \ldots, J_{l_j}^j\}$, $j = 1, \ldots, L$.

The function a_{IJ} defined by

$$a_{IJ}(I_i) = \begin{cases} 1 & \text{if } \ I_i = \{m\} = J \quad \text{for any } J \in J, \\ 0 & \text{otherwise,} \end{cases} \tag{6.60}$$

for $I_i \in I$, assumes the value 1 if group I_i contains a single state in partition I *and* this state also forms a group-singleton in partition J.

For every pair of states, we will asses if they share a group in a given partition using the following function b_J:

$$b_J(m, n) = \begin{cases} 1 & \text{if } \ m, n \in J_j, \quad \text{for any } j; \\ 0 & \text{otherwise,} \end{cases} \tag{6.61}$$

for $m, n \in I$.

In case the function a_{IJ} has the value of 1, we say that we have a (one-state) coincidence, which means that the state has been found incompatible with other states twice, no matter which method formed partitions I, J.

Similarly, if the function b_J returns 1 for two states *in a group from the partition* I, we say that we have a (two-state) coincidence, that is in both partitions, the two states are members of the same group.

To measure the number of coincidences between two partitions, we use the function:

$$C_q(I_i, IJ) = a_{IJ}(I_i) + (1 - a_{IJ}(I_i)) \left(\sum_{m \in I_i} \sum_{n \in I_i, n \neq m} \frac{q b_J(m, n) + (1 - q)}{2} \right), \tag{6.62}$$

for $I_i \in I, q = \{0, 1\}$.

If the parameter q equals 1, then the function C_q counts the number of either type of coincidences that couples of states reveal in the group I_i in comparison to the groups they belong to in the partition J. Conversely, if $q = 0$, then we simply count the total number of possible coincidences for the states in the group $I_i \in I$. Note that the function C_q is not necessarily symmetric with respect to the pairs of partitions: $C_q(I_i, IJ)$ need not have the same value as $C_q(I_i, JI)$.

The similitude function *Sim* used is defined as follows:

$$Sim(I, J) = \frac{\sum_{I_i \in I} C_1(I_i, IJ) + \sum_{J_j \in J} C_1(J_j, JI)}{\sum_{I_i \in I} C_0(I_i, IJ) + \sum_{J_j \in J} C_0(J_j, JI)}. \tag{6.63}$$

Notice that if there is at least one group in either partition containing more than one element, then C_0 for that group is at least 1, whereas if there does not exists such group in either partition, then $a_{IJ}(a) = 1$ and consequently $C_0(a, IJ) = 1$ for any $a \in I \cap J$. Therefore, the denominator is never 0, which makes this function well-defined.

Lemma 6.6 *Let I and J be two partitions of the set $I = \{1, 2, \ldots, n\}$, and let function Sim be defined by* (6.63). *The following statements are true:*

1. $Sim(I, J) = Sim(J, I)$.
2. $Sim(I, J) = 1$ *if and only if* $I = J$.
3. $Sim(I, J) = 0$ *if and only if there are neither one-state nor two-state coincidences between I and J.*
4. $0 \leq Sim(I, J) \leq 1$.

Proof 1. This is easy to see from the structure of the function.

2. Let $I = J$. If $I_i = \{m\} = J_k$ for some i and k, then $C_1(I_i, IJ) = C_0(I_i, IJ)$. Otherwise, if the order of I_i is greater than one, then the second term in (6.62) (the definition of C_q) assumes the same value no matter whether $q = 1$ or $q = 0$. Therefore, the numerator and denominator in *Sim* are equal.
   Conversely, if there exists one I_i such that $I_i \neq J \; \forall J \in J$, then $C_1(I_i, IJ)$ is strictly less than $C_0(I_i, IJ)$. Since $C_1(J_i, IJ) \leq C_0(J_i, IJ)$, it follows that the numerator in (6.63) (defining *Sim*) is strictly smaller than the denominator, and therefore $Sim(I, J) < 1$.
3. If there is at least one one-state coincidence, or a two-state coincidence, then the numerator in *Sim* is larger than 0, and therefore $Sim(I, J) > 0$.
   Conversely, since C_q is nonnegative for every value of q, $Sim(I, J) = 0$ means that both terms in the numerator are zero, which is only possible if $a_{IJ}(I_i) = a_{IJ}(J_j) = 0$ for every member of I and J, *and* $b_J(m, n) = 0$ for every $m, n \in I$, which means that there is no coincidence of any type between these two partitions.
4. The first inequality follows from the fact that both the numerator and denominator in (6.63) are positive. The second inequality comes from the same argument as in item 2, i.e., the numerator is either equal or strictly less than the denominator. □

6.9.4 Experimental Results

This subsection presents the results of the numerical experimentation performed on a number of times series pertaining to each of the 48 data sets. The values for the historical natural gas prices, consumption, number of consumers, as well as the oil spot prices were taken from the US Energy Information Agency, whereas the temperature figures for each state were obtained from the US Department of Commerce, National Oceanographic and Atmospheric Agency [279].

6.9.4.1 IMLR Results

The first step was to perform the IMLR for the 48 sets of time series; this provided the regression parameters for the dendrogram formation. The five time series corresponding to every state had 240 monthly observations each.

Individual regression models showed regression $\mathbb{R}^2$ coefficients with the average of 0.77, and the minimum of 0.61. The normality and heteroskedasticity were not tested due to the use of Robust Regression with Huber weights. Randomness of the residuals was tested, and high p-values were found for many states.

6.9.4.2 Dendrogram-GRASP Grouping Results

There are two main aspects we wanted to consider when evaluating the effectiveness of the Dendrogram-GRASP approach: how replicative it is, and how good a partition is produced. The first issue is evaluated by examining how good and how similar the partitions are that come from the same seed (as opposed to those that come from randomly generated seeds). The goodness of one partition is measured with the average group [state] coefficient of determination, $\mathbb{R}^2_{I_k}$ [$\mathbb{R}^2_i$], calculated across all the groups [states] of the partitions.

There are, however, a number of different design parameters that should be included in the experimentation. Each experimental observation consists of the generation of 10 partitions, using the following parameters:

- A seed choice: the dendrogram seed (DDR), a random seed common to all 20 partitions (FIX), and a random seed for each partition (RND).
- The individual versus grouped $\mathbb{R}^2$ weight, ω, which determines what is more important when adding a new state to an existing group in the GRASP routine. Values considered in the experimentation are $\omega = 0$ (only the single states' $\mathbb{R}^2$s are considered), 0.5, and 1 (only the groups' $\mathbb{R}^2$s are important).
- The new group threshold, φ: the closer to one it is, the more likely new single-state groups will be created in the GRASP routine. The tested values are $\varphi \in \{0.90, 0.95\}$.
- The length of the restricted candidate list, s_{RCL}. The values considered are $s_{RLC} \in \{1, 5\}$.
- The number of local search moves, $s_{ls} \in \{0, 100\}$.
- The local search individual/grouped $\mathbb{R}^2$ weight, τ. Considered values are $\tau \in \{0, 0.66, 1\}$.

The starting number of groups was fixed at 10, and the maximum number of groups allowed was set at 15. Each combination of levels was replicated 20 times. This resulted in 5760 experimental observations.

In each observation, we calculated the average similitude between the various partitions involved, as well as their similitude with a randomly created partition. The compared similitudes were:

Chapter 7
Applications to Other Energy Systems

7.1 Consistent Conjectural Variations Equilibrium in a Mixed Oligopoly in Electricity Markets

Results described in this section are based mainly upon paper of Kalashnikov et al. [165], which also included applications to an oligopolistic market of electricity. Even if the main models and tools developed in the paper are not directly related to Bilevel Programming, they can be used to construct more complicated schemes involving the Stackelberg equilibrium and other bilevel-type concepts.

In more detail, this section deals with a model of mixed oligopoly with *conjectured variations equilibrium* (CVE). The agents' conjectures concern the price variations depending upon their production output's increase or decrease. We establish existence and uniqueness results for the conjectured variations equilibrium (called an *exterior equilibrium*) for any set of feasible conjectures. In order to introduce the notion of an *interior equilibrium*, we develop a *consistency criterion* for the conjectures (referred to as *influence coefficients*) and prove the existence theorem for the interior equilibrium (understood as a CVE with consistent conjectures). To prepare the base for the extension of our results to the case of non-differentiable demand functions, we also investigate the behavior of the consistent conjectures depending upon a parameter that represents the demand function's derivative with respect to the market price.

7.1.1 Introduction

In recent years, investigation of behavioral patterns of agents of mixed markets, in which state-owned (public, domestic, etc.) welfare-maximizing firms compete against profit-maximizing private (foreign) firms, has become more and more popular. For pioneering works on mixed oligopolies (see Merril and Schneider [225], Ruffin [276], Harris and Wiens [139], and Bös [24, 25]). Excellent surveys can be found in Vickers and Yarrow [308], De Frajas and Delbono [118], Nett [253].

S. Dempe et al., *Bilevel Programming Problems*, Energy Systems,
DOI 10.1007/978-3-662-45827-3_7

The interest in mixed oligopolies is high because of their importance to the economies of Europe (Germany, England and others), Canada and Japan (see Matsushima and Matsumura [224], for an analysis of "herd behavior" by private firms in many branches of the economy in Japan). There are examples of mixed oligopolies in the United States such as the packaging and overnight-delivery industries. Mixed oligopolies are also common in the East European and former Soviet Union transitional economies, in which competition among public and private firms existed or still exists in many industries such as banking, house loan, life insurance, airline, telecommunication, natural gas, electric power, automobile, steel, education, hospital, health care, broadcasting, railways and overnight-delivery. Moreover, according to Bös [25], Fershtman [107], Matshumura and Kanda [222], in many cases the government has held, or even still holds, a non-negligible proportion of shares in privatized firms, and there are firms with a mixture of private and public ownership. Since privatized firms with mixed ownership must respect the interests of private shareholders, they cannot be pure domestic social surplus maximizers. At the same time they must respect the interests of the government, so they cannot be pure profit-maximizers. By controlling the shares that it holds, the government may be able to indirectly control the activities of the privatized firm.

In the majority of the above-mentioned papers, the mixed oligopoly is studied in the framework of classical Cournot, Hotelling or Stackelberg models (cf. Matsushima and Matsumura [224], Matsumura [223], Cornes and Sepahvand [44]). It is well known (cf. for instance, Figuières et al. [110]) that the Nash equilibrium (including Cournot equilibrium as a particular case) is the outcome consistent with rational agents who take rival decisions as given when they optimize. Alternately, in the Stackelberg equilibrium there are two agents who take their decisions sequentially; the first agent to move is referred to as the leader, whereas the second mover is called the follower. The Stackelberg equilibrium is an outcome consistent with the follower's rational behavior given that he has observed the leader's move, and the leader's rational behavior who can infer what will be the follower's rational reaction to his current decision.

Conjectural variations equilibria (CVE) were introduced by Bowleyl [26] and Frisch [119, 120], as another possible solution concept in static games. According to this concept, agents behave as follows: each agent chooses his/her most favorable action taking into account that every rival's strategy is a conjectured function of her own strategy.

In the works by Bulavsky and Kalashnikov [37, 38, 152], a new scale of conjectural variations equilibria (CVE) was introduced and investigated, in which the conjectural variations (represented via the influence coefficients of each agent) affected the structure of the Nash equilibrium. In other words, we considered not only a classical Cournot competition but also a Cournot-type model with influence coefficient values different from 1 (as the influence coefficient 1 corresponds to the classical Cournot model). Various equilibrium existence and uniqueness results were obtained in the above-cited works.

For instance, in Isac et al. [152], the classical oligopoly model was extended to the conjectural oligopoly as follows. Instead of the classical Cournot assumptions, all producers $i = 1, 2, \ldots, n$, used the conjectural variations described below:

$$G_i(\eta) = G + (\eta - q_i)\, w_i\,(G, q_i)\,.$$

Here, G is the current total quantity of the product cleared in the market, q_i and η are, respectively, the present and the expected supplies by the ith agent, whereas $G_i(\eta)$ is the total cleared market volume *conjectured* by the ith agent as a response to changing his/her own supply from q_i to η. The conjecture function w_i was referred to as the ith agent's *influence quotient* (*coefficient*). Notice that the classical Cournot model assumes $w_i \equiv 1$ for all i. Under general enough assumptions concerning properties of the influence coefficients $w_i = w_i\,(G, q_i)$, cost functions $f_i = f_i(q_i)$, and the inverse demand (price) function $p = p(G)$, new existence and uniqueness results for the conjectural variations equilibrium (CVE) were obtained. This approach was further developed in Kalashnikov et al. [168, 175] with application to the mixed oligopoly model. Here again, all agents (both public and private companies) make their decisions based upon the model's data (inverse demand and cost functions) and their influence coefficients (conjectures) $w_i = w_i\,(G, q_i)$.

As is mentioned in Figuières et al. [110], Giocoli [130], the concept of conjectural variations has been the subject of numerous theoretical controversies (see e.g. Lindhi [204]). Nevertheless, economists have made extensive use of one form or the other of the CVE to predict the outcome of non-cooperative behavior in several fields of economics. The literature on conjectural variations has focused mainly on two-player games (cf. Figuières et al. [110]). The central concept of the theory is the notion of *conjecture*. The *variational conjecture* r_j usually describes player j's reaction, as anticipated by player i, to an *infinitesimal variation* of player i's strategy. This mechanism leads to the notion of a *conjectured reaction function* of the opponent. Given these conjectured reactions on part of the rivals, each agent optimizes his/her perceived payoff. This leads to the concept of a *conjectural best response function*. An equilibrium is obtained when no player has an interest in deviating from his/her strategy, i.e., his/her conjectural best response to the strategies of the other player.

The *consistency* (or, sometimes, "rationality") of the equilibrium is defined as the *coincidence* between the conjectural best response of each agent and the conjectured reaction function of the same. A conceptual difficulty arises when one considers consistency in the case of many agents (see, Figuières et al. [110]). The strongest notion of consistency requires that the conjectural best response of player i coincides with what the other players have conjectured about his/her reaction, that is, with one of their conjectured reaction functions. However, when n agents are present, there are n best response functions and $n(n-1)$ conjectures. Therefore, if $n > 2$, equilibrium is consistent only if all players have the same conjectures about player i's reaction. This is the approach followed explicitly by Başar and Olsder [7]; this assumption can be also found in Fershtman and Kamien [108] dealing with conjectures in differential games. In the literature on conjectural variations in static n-player games, the problem is usually implicitly addressed by assuming a complete identity of all the agents (cf. Laitner [198], Bresnahan [28] and references therein, Novshek [256]). Using a bit different approach, Perry [263] for oligopoly, Cornes and Sandler [43] and

Sugden [295] for public goods, consider a class of games where for each agent, the contributions of all other players to her payoff are aggregated. It is as if each agent plays against a unique (virtual) player representing the remaining agents.

To cope with this conceptual difficulty arising in many players models, Bulavsky [36] proposed a completely new approach. Instead of assuming the identity of the agents in the conjectural variation model of a homogeneous good market, it is supposed that each player makes conjectures not about the (optimal) response functions of the other players but only about the variations of the market price depending upon his infinitesimal output variations. Knowing the rivals' conjectures (called influence coefficients), each agent can realize certain verification procedure and check out if his influence coefficient is consistent with the others. Exactly the same verification formulas were obtained independently in Liu et al. [206] establishing the existence and uniqueness of consistent conjectural variation equilibrium in electricity market. However, they applied a much more difficult optimal control technique, searching only steady states as a final result (a similar technique was used in Driskill and McCafferty [95]. Moreover, they restricted the inverse demand function to a linear one, and the agents' cost functions to quadratic ones in their model, whereas the approach in Bulavsky [36] allows nonlinear and even non-differentiable demand functions and arbitrary (twice continuously differentiable) convex cost functions of the agents.

In this section, we extend the results obtained in Bulavsky [36] to a mixed oligopoly model. In the same manner as in Bulavsky and Kalashnikov [37, 38], we consider a conjectural variations oligopoly model, in which the degree of influence on the whole situation by each agent is modeled by special parameters (influence coefficients). However, in contrast to the models defined in Bulavsky and Kalashnikov [37, 38] and Kalashnikov et al. [168, 175], here, we follow the ideology of Bulavsky [35, 36] selecting the market clearing price p, rather than the producers' output, as an observable variable.

The section is organized as follows. In Sect. 7.1.2, we describe the mathematical model from Bulavsky [36] extended to the mixed oligopoly case and then, in Sect. 7.1.3, we define the concept of exterior equilibrium, i.e., a conjectural variations equilibrium (CVE) with the influence coefficients fixed in an exogenous form. The existence and uniqueness theorem for this kind of CVE ends the subsection. Section 7.1.4 deals with the more advanced concept of interior equilibrium, which is defined as the exterior equilibrium with consistent conjectures (influence coefficients). The consistency criterion, the consistency verification procedure, and the existence theorem for the interior equilibrium are formulated in the same Sect. 7.1.4. To provide the tools for the future research concerning the interrelationships between the demand structure (with not necessarily smooth demand function) and the CVEs with consistent conjectures (influence coefficients), the behavior of the latter as functions of certain parameter (governed by the derivative by p of the demand function $G = G(p)$) is studied in Theorem 7.3 completing Sect. 7.1.4. Finally, Sect. 7.1.5 contains the results of numerical experiments with a test model of an electricity market from Liu et al. [206], with and without a public company among the agents.

7.1.2 Model Specification

Consider a market of a homogeneous good (natural gas, oil, electricity, timber, etc.) with no less than 3 producers/suppliers with cost functions $f_i = f_i(q_i)$, $i = 0, 1, \ldots, n$, where $n \geq 2$, and q_i is the output/supply brought by producer i, $i = 0, 1, \ldots, n$. Consumers' demand is described by a demand function $G = G(p)$, whose argument p is the market clearing price. An active demand value D is nonnegative and does not depend upon the price. We will reflect the equilibrium between the demand and supply for a given (clearing) price p by the following balance equality

$$\sum_{i=0}^{n} q_i = G(p) + D. \tag{7.1}$$

We assume the following properties of the model's data.

A1. The demand function $G = G(p) \geq 0$ defined for the (clearing) price values $p \in (0, +\infty)$ is non-increasing and continuously differentiable. □

A2. For each producer/supplier $i = 0, 1, \ldots, n$, its cost function $f_i = f_i(q_i)$ is quadratic, i.e.,

$$f_i(q_i) = \frac{1}{2}a_i q_i^2 + b_i q_i, \tag{7.2}$$

with $a_i > 0$, $b_i > 0$, $i = 0, 1, \ldots, n$. Moreover, we assume that

$$b_0 \leq \max_{1 \leq i \leq n} b_i. \tag{7.3}$$

Each private (or, foreign) producer i, $i = 1, \ldots, n$, chooses his/her output volume $q_i \geq 0$ so as to maximize his/her net profit function $\pi(p, q_i) := p \cdot q_i - f_i(q_i)$. On the other hand, the public (or, domestic) company number $i = 0$ selects its production value $q_0 \geq 0$ so as to maximize domestic social surplus defined as the difference between the consumer surplus, the private (foreign) companies' total revenue, and the public (domestic) firm's production costs:

$$S(p; q_0, q_1, \ldots, q_n) = \int_0^{\sum_{i=0}^{n} q_i} p(x)dx - p \cdot \left(\sum_{i=1}^{n} q_i\right) - b_0 q_0 - \frac{1}{2}a_0 q_0^2. \tag{7.4}$$

Now we postulate that the agents (both public and private) assume that their variation of production volumes may affect the price value p. The latter assumption could be implemented by accepting a conjectured dependence of fluctuations of the price p upon the variations of the (individual) output values q_i. Having that done, the first order maximum condition to describe the equilibrium would have the form: For the public company (with $i = 0$)

$$\frac{\partial S}{\partial q_0} = p - \left(\sum_{i=1}^{n}\right)\frac{\partial p}{\partial q_0} - f_0'(q_0)\begin{cases} = 0, & \text{if } q_0 > 0; \\ \leq 0, & \text{if } q_0 = 0; \end{cases} \tag{7.5}$$

and

$$\frac{\partial \pi_i}{\partial q_i} = p + q_i\frac{\partial p}{\partial q_i} - f_i'(q_i)\begin{cases} = 0, & \text{if } q_i > 0; \\ \leq 0, & \text{if } q_i = 0, \end{cases} \quad \text{for } i = 1, \ldots, n. \tag{7.6}$$

Therefore, we see that to describe the behavior of agent i and treat the maximum (equilibrium) conditions, it is enough to trace the derivative $\partial p/\partial q_i = -v_i$ rather than the full dependence of p upon q_i. (We introduce the minus here in order to deal with nononegative values of v_i, $i = 0, 1, \ldots, n$.) Of course, the conjectured dependence of p on q_i must provide (at least local) concavity of the ith agent's conjectured profit as a function of its output. Otherwise, one cannot guarantee the profit to be maximized (but not minimized). As we suppose that the cost functions $f_i = f_i(q_i)$ are quadratic and convex, then, for $i = 1, \ldots, n$, the concavity of the product $p \cdot q_i$ with respect to the variation η_i of the current production volume will do. For instance, it is sufficient to assume the coefficient v_i (from now on referred to as the ith agent's *influence coefficient*) to be nonnegative and constant. Then the conjectured local dependence of the agent's net profit upon the production output η_i has the form $[p - v_i(\eta_i - q_i)]\,\eta_i - f_i(\eta_i)$, while the maximum condition at $\eta_i = q_i$ is provided by the relationships

$$\begin{cases} p = v_i q_i + b_i + a_i q_i, & \text{if } q_i > 0; \\ p \leq b_i, & \text{if } q_i = 0. \end{cases} \tag{7.7}$$

Similarly, the public company conjectures the local dependence of domestic social surplus on its production output η_0 in the form

$$\int_0^{\eta_0 + \sum_{i=1}^{n} q_i} p(x)dx - [p - v_0(\eta_0 - q_0)] \cdot \left(\sum_{i=1}^{n} q_i\right) - b_0 - a_0 q_0, \tag{7.8}$$

which allows one to write down the (domestic social surplus) maximum condition at $\eta_0 = q_0$ as follows:

$$\begin{cases} p = -v_0 \sum_{i=1}^{n} q_i + b_0 + a_0 q_0, & \text{if } q_0 > 0; \\ p \leq -v_0 \sum_{i=1}^{n} q_i + b_0, & \text{if } q_0 = 0. \end{cases} \tag{7.9}$$

Were the agents' conjectures given exogenously (like it was assumed in Bulavsky and Kalashnikov [37, 38]), we would allow all the influence coefficients v_i to be functions of q_i and p. However, we use the approach from the papers Bulavski [35, 36], where the (justified, or consistent) conjectures are determined simultaneously with the equilibrium price p and output values q_i by a special verification procedure. In the latter case, the influence coefficients are the scalar parameters determined only at the equilibrium. In what follows, such equilibrium is referred to as *interior* one and is described by the set of variables and parameters $(p, q_0, q_1, \ldots, q_n, v_0, v_1, \ldots, v_n)$.

7.1.3 Exterior Equilibrium

Before we introduce the verification procedure, we need an initial notion of equilibrium called *exterior* (cf. Bulavski [36]) with the parameters (influence coefficients) $v_i, i = 0, 1, \ldots, n$ given exogenously.

Definition 7.1 The collection $(p, q_0, q_1, \ldots, q_n)$ is called *exterior* equilibrium for given influence coefficients $(v_0, v_1, \ldots, v_n)$, if the market is balanced, i.e., condition (7.1) is satisfied, and for each $i, i = 0, 1, \ldots, n$, the maximum conditions (7.7) and (7.9) are valid. □

In what follows, we are going to consider only the case when the list of really producing/supplying participants is fixed (i.e., it does not depend upon the values of the model's parameters). In order to guarantee this property, we make the following additional assumption.

A3. For the price value $p_0 := \max_{1 \le j \le n} b_j$, the following (strict) inequality holds:

$$\sum_{i=0}^{n} \frac{p_0 - b_i}{a_i} < G(p_0). \tag{7.10}$$

Remark 7.1 The latter assumption, together with assumptions **A1** and **A2**, guarantees that for all nonnegative values of $v_i, i = 1, \ldots, n$, and for $v_0 \in [0, \overline{v}_0)$, where

$$0 < \overline{v}_0 = \begin{cases} a_0 \left[\dfrac{G(p_0) - \dfrac{p_0 - b_0}{a_0}}{\sum_{i=1}^{n} \dfrac{p_i - b_i}{a_i}} - 1 \right], & \text{if } \displaystyle\sum_{i=1}^{n} \frac{p_i - b_i}{a_i} > 0; \\ +\infty, & \text{otherwise,} \end{cases} \tag{7.11}$$

there always exists a unique solution of the optimality conditions (7.7) and (7.9) satisfying the balance equality (7.1), i.e., the exterior equilibrium. Moreover, conditions (7.1), (7.7) and (7.9) can hold simultaneously if, and only if $p > p_0$, that is, if and only if all outputs q_i are strictly positive, $i = 0, 1, \ldots, n$. Indeed, if $p > p_0$ then it is evident that neither inequalities $p \le b_i, i = 1, \ldots, n$, from (7.7), nor $p \le -v_0 \sum_{i=1}^{n} q_i + b_0$ from ((7.9) are possible, which means that none of q_i, $i = 0, 1, \ldots, n$, satisfying (7.7) and (7.9) can be zero.

Conversely, if all q_i, satisfying (7.7) and (7.9) are positive ($q_i > 0, i = 0, 1, \ldots, n$), then it is straightforward from conditions (7.7) that

$$p = v_i q_i + b_i + a_i q_i > b_i, \quad i = 1, \ldots, n;$$

hence $p > \max_{1 \le i \le n} = p_0$. □

The following theorem is the main result of this subsection and a tool for the introduction of the concept of interior equilibrium in the next subsection.

Theorem 7.1 *Under assumptions* **A1**, **A2** *and* **A3**, *for any* $D \ge 0$, $v_i \ge 0$, $i = 1, \ldots, n$, *and* $v_0 \in [0, v_0)$, *there exists a unique exterior equilibrium state* $(p, q_0, q_1, \ldots, q_n)$, *which depends continuously on the parameters* $(D, v_0, v_1, \ldots, v_n)$. *The equilibrium price* $p = p(D, v_0, v_1, \ldots, v_n)$ *as a function of these parameters is differentiable with respect to both D and* v_i, $i = 0, 1, \ldots, n$. *Moreover,* $p = p(D, v_0, v_1, \ldots, v_n) > p_0$, *and*

$$\frac{\partial p}{\partial D} = \frac{1}{\dfrac{v_0 + a_0}{a_0} \sum_{i=0}^{n} \dfrac{1}{v_i + a_i} - G'(p)}. \tag{7.12}$$

Proof Due to assumptions **A1–A3**, for any fixed collection of conjectures $v = (v_0, v_1, \ldots, v_n) \ge 0$, the equalities in the optimality conditions (7.7) and (7.9) determine the optimal response (to the existing clearing price) values of the producers/suppliers as continuously differentiable (with respect to p) functions $q_i = q_i(p; v_0, \ldots, v_n)$ defined over the interval $p \in [p_0, +\infty)$ by the following explicit formulas:

$$q_0 := \frac{p - b_0}{a_0} + \frac{v_0}{a_0} \sum_{i=1}^{n} \frac{p - b_i}{v_i + a_i}, \tag{7.13}$$

and

$$q_i := \frac{p - b_i}{v_i + a_i}, \quad i = 1, \ldots, n. \tag{7.14}$$

Moreover, the partial derivatives of the optimal response functions are positive:

$$\frac{\partial q_0}{\partial p} = \frac{1}{a_0} + \frac{v_0}{a_0} \sum_{i=1}^{n} \frac{1}{v_i + a_i} \ge \frac{1}{a_0} > 0, \tag{7.15}$$

and

$$\frac{\partial q_i}{\partial p} = \frac{1}{v_i + a_i} > 0, \quad i = 1, \ldots, n. \tag{7.16}$$

Therefore, the total production volume function

$$Q(p; v_0, v_1, \ldots, v_n) = \sum_{i=0}^{n} q_i(p; v_0, v_1, \ldots, v_n)$$

is continuous and strictly increasing by p. According to assumption **A3**, this function's value at the point $p = p_0$ is strictly less that $G(p_0)$. Indeed, from (7.13) and (7.14) we have:

(A) If $\sum_{i=1}^{n} \frac{p_0 - b_i}{v_i + a_i} > 0$, then

$$\begin{aligned}
Q(p_0; v_0, v_1, \ldots, v_n) &= \sum_{i=0}^{n} q_i(p_0; v_0, v_1, \ldots, v_n) \\
&= \frac{p_0 - b_0}{a_0} + \frac{v_0}{a_0} \sum_{i=1}^{n} \frac{p_0 - b_i}{v_i + a_i} + \sum_{i=1}^{n} \frac{p_0 - b_i}{v_i + a_i} \\
&= \frac{p_0 - b_0}{a_0} + \frac{v_0 + a_0}{a_0} \sum_{i=1}^{n} \frac{p_0 - b_i}{v_i + a_i} \\
&< \frac{p_0 - b_0}{a_0} + \frac{\overline{v}_0 + a_0}{a_0} \sum_{i=1}^{n} \frac{p_0 - b_i}{a_i} \\
&= \frac{p_0 - b_0}{a_0} + \left[\frac{G(p_0) - \frac{p_0 - b_0}{a_0}}{\sum_{i=1}^{n} \frac{p_0 - b_i}{a_i}} \right] \sum_{i=1}^{n} \frac{p_0 - b_i}{a_i} \\
&= \frac{p_0 - b_0}{a_0} + G(p_0) - \frac{p_0 - b_0}{a_0} = G(p_0).
\end{aligned}$$

(B) Otherwise, i.e., if $\sum_{i=1}^{n} \frac{p_0 - b_i}{v_i + a_i} = 0$, one has:

$$\begin{aligned}
Q(p_0; v_0, v_1, \ldots, v_n) &= \sum_{i=0}^{n} q_i(p_0; v_0, v_1, \ldots, v_n) \\
&= \frac{p_0 - b_0}{a_0} + \frac{v_0}{a_0} \sum_{i=1}^{n} \frac{p_0 - b_i}{v_i + a_i} + \sum_{i=1}^{n} \frac{p_0 - b_i}{v_i + a_i} \\
&= \frac{p_0 - b_0}{a_0} < G(p_0)
\end{aligned}$$

for any $v_i \geq 0, i = 1, \ldots, n$, and $v_0 \in [0, \overline{v}_0)$.

On the other hand, the total output supply $Q = Q(p; v_0, v_1, \ldots, v_n)$ clearly tends to $+\infty$ when $p \to +\infty$. Now define

$$p_* := \sup\{p : Q(p; v_0, v_1, \ldots, v_n) \le G(p) + D\}. \tag{7.17}$$

Since both functions $Q(p; v_0, v_1, \ldots, v_n)$ and $G(p)+D$ are continuous with respect to p, the former increases unboundedly and the latter, vice versa, is non-increasing by p over the whole ray $[p_0, +\infty)$, then, first, the value of p_* is finite ($p_* < +\infty$), and second, by definition (7.17) and the continuity of both functions,

$$Q(p_*; v_0, v_1, \ldots, v_n) \le G(p_*) + D.$$

Now we demonstrate that the strict inequality $Q(p_*; v_0, v_1, \ldots, v_n) < G(p_*) + D$ cannot happen. Indeed, suppose on the contrary that the latter strict inequality holds. Then the continuity of the involved functions implies that for some values $p > p_*$ sufficiently close to p_*, the same relationship is true: $Q(p; v_0, v_1, \ldots, v_n) < G(p) + D$, which contradicts definition (7.17). Therefore, the exact equality holds

$$Q(p_*; v_0, v_1, \ldots, v_n) = G(p_*) + D, \tag{7.18}$$

which, in its turn, means that the values p_* and $q_i^* = q_i(p_*; v_0, v_1, \ldots, v_n)$, $i = 0, \ldots, n$, determined by formulas (7.13) and (7.14) form an exterior equilibrium state for the collection of influence coefficients $v = (v_0, v_1, \ldots, v_n)$. The uniqueness of this equilibrium follows from the fact that the function $Q = Q(p; v_0, v_1, \ldots, v_n)$ strictly increases while the demand function $G(p)+D$ is non-increasing with respect to p. Indeed, these facts combined with (7.18) yield that $Q(p; v_0, v_1, \ldots, v_n) < G(p) + D$ for all $p \in (p_0, p_*)$, whereas $Q(p; v_0, v_1, \ldots, v_n) > G(p) + D$ when $p > p_*$. To conclude, the equilibrium price p_* and hence, the equilibrium outputs $q_i^* = q_i(p_*; v_0, v_1, \ldots, v_n)$, $i = 0, \ldots, n$, calculated by formulas (7.13) and (7.14), are determined uniquely.

Now we establish the continuous dependence of the equilibrium price (and hence, the equilibrium output volumes, too) upon the parameters $(D, v_0, \ldots, v_n)$. To do that, we substitute expressions (7.13) and (7.14) for $q_i = (p; v_0, \ldots, v_n)$ into the balance equality (7.1) and come to the following relationship:

$$\begin{aligned} q_0 + \sum_{i=1}^{n} q_i - G(p) - D &= \left(\frac{p - b_0}{a_0} + \frac{v_0}{a_0} \sum_{i=1}^{n} \frac{p - b_i}{v_i + a_i} \right) \\ &\quad + \sum_{i=1}^{n} \frac{p - b_i}{v_i + a_i} - G(p) - D \\ &= p \left(\frac{1}{a_0} + \frac{v_0 + a_0}{a_0} \sum_{i=1}^{n} \frac{1}{v_i + a_i} \right) - \frac{b_0}{a_0} \\ &\quad - \frac{v_0 + a_0}{a_0} \sum_{i=1}^{n} \frac{b_i}{v_i + a_i} - G(p) - D = 0. \end{aligned} \tag{7.19}$$

Introduce the function

$$\Gamma(p; v_0, v_1, \ldots, v_n, D) = p\left(\frac{1}{a_0} + \frac{v_0 + a_0}{a_0}\sum_{i=1}^{n}\frac{1}{v_i + a_i}\right) - \frac{b_0}{a_0} - \frac{v_0 + a_0}{a_0}\sum_{i=1}^{n}\frac{b_i}{v_i + a_i} - G(p) - D$$

and rewrite the last equality in (7.19) as the functional equation

$$\Gamma(p; v_0, v_1, \ldots, v_n, D) = 0. \tag{7.20}$$

As the partial derivative of the latter function with respect to p is (always) positive:

$$\frac{\partial \Gamma}{\partial p} = \frac{1}{a_0} + \frac{v_0 + a_0}{a_0}\sum_{i=1}^{n}\frac{1}{v_i + a_i} - G'(p) \ge \frac{1}{a_0} > 0,$$

one can apply Implicit Function Theorem and conclude that the equilibrium (clearing) price p treated as an explicit function $p = p(v_0, v_1, \ldots, v_n, D)$ is continuous and, in addition, differentiable with respect to all the parameters $v_0, v_1, \ldots, v_n, D$. Moreover, the partial derivative of the equilibrium price p with respect to D can be calculated from the full derivative equality

$$\frac{\partial \Gamma}{\partial p} \cdot \frac{\partial p}{\partial D} + \frac{\partial \Gamma}{\partial D} = 0,$$

finally yielding the desired formula (7.12)

$$\frac{\partial p}{\partial D} = \frac{1}{\frac{v_0 + a_0}{a_0}\sum_{i=0}^{n}\frac{1}{v_i + a_i} - G'(p)},$$

and thus completing the proof. □

7.1.4 Interior Equilibrium

Now we are ready to define the concept of interior equilibrium. To do that, we first describe the procedure of verification of the influence coefficients v_i as it was given in Bulavski [36]. Assume that we have an exterior equilibrium state $(p, q_0, q_1, \ldots, q_n)$ that occurs for some feasible $v = (v_0, v_1, \ldots, v_n)$ and $D \ge 0$. One of the producers, say k, $0 \le k \le n$, temporarily changes his/her behavior by *abstaining* from maximization of the conjectured profit (or domestic social surplus, as is in case $k = 0$) and making small fluctuations (variations) around his/her equilibrium output volume q_k. In mathematical terms, the latter is tantamount to restricting the model agents to the subset $I_{-k} := \{0 \le i \le n : i \ne k\}$ with the active demand reduced to $D_{-k} := D - q_k$.

A variation δq_k of the production output by agent k is then equivalent to the active demand variation in form $\delta D_{-k} := -\delta q_k$. If we consider these variations being infinitesimal, we assume that by observing the corresponding variations of the equilibrium price, agent k can evaluate the derivative of the equilibrium price with respect to the active demand in the reduced market, which clearly coincides with his/her influence coefficient.

When applying formula (7.12) from Theorem 8.1 to evaluate the player k conjecture (influence coefficient) v_k, one has to remember that agent k is temporarily absent from the equilibrium model, hence one has to *exclude* from all the sums the term with number $i = k$. Keeping that in mind, we come to the following *consistency criterion.*

7.1.4.1 Consistency Criterion

At an exterior equilibrium $(p, q_0, q_1, \ldots, q_n)$, the influence coefficients v_k, $k = 0, 1, \ldots, n$, are referred to as *consistent* if the following equalities hold:

$$v_0 = \frac{1}{\sum_{i=1}^{n} \dfrac{1}{v_i + a_i} - G'(p)}, \tag{7.21}$$

and

$$v_i = \frac{1}{\dfrac{v_0 + a_0}{a_0} \sum_{j=0, j\ne i}^{n} \dfrac{1}{v_j + a_j} - G'(p)}, \quad i = 1, \ldots, n. \tag{7.22}$$

Now we are in a position to define the concept of an interior equilibrium.

Definition 7.2 The collection $(p, q_0, \ldots, q_n, v_0, \ldots, v_n)$ where $v_i \ge 0$, $i = 0, 1, \ldots, n$, is referred to as the *interior equilibrium* if, for the coefficients $(v_0, v_1, \ldots, v_n)$ the collection $(p, q_0, \ldots, q_n)$ is an exterior equilibrium state, and the consistency criterion is satisfied for all $k = 0, 1, \ldots, n$. □

Remark 7.2 If all the agents are profit-maximizing private companies, then formulas (7.21)–(7.22) reduce to the uniform ones obtained independently in Bulavski [36] and Lui et al. [206]:

$$v_i = \frac{1}{\sum_{j\in I\setminus\{i\}} \dfrac{1}{v_j + a_j} - G'(p)}, \quad i \in I, \tag{7.23}$$

where I is an arbitrary (finite) list of the participants of the model. □

The following theorem is an extension of Theorem 2 in Bulavski [36] to the case of a mixed oligopoly.

Theorem 7.2 *Under assumptions* **A1**, **A2**, *and* **A3**, *there exists the interior equilibrium.*

Proof We are going to show that there exist $v_0 \in [0, \overline{v}_0)$; $v_i \geq 0$, $i = 1, \ldots, n$; $q_i \geq 0, i = 0, 1, \ldots, n$, and $p > p_0$ such that the vector $(p; q_0, \ldots, q_n; v_0, \ldots, v_n)$ provides for the interior equilibrium. In other words, the vector $(p, q_0, \ldots, q_n)$ is an exterior equilibrium state, and in addition, equalities (7.21)–(7.22) hold. For a technical purpose, let us introduce a parameter α so that $G'(p) := \frac{\alpha}{1+\alpha}$ for appropriate values of $\alpha \in [-1, 0]$, and then rewrite the right-hand sides of formulas (7.21)–(7.22) in the following (equivalent) form:

$$F_0\left(\alpha; v_0, \ldots, v_n\right) := \frac{1+\alpha}{(1+\alpha)\sum_{i=1}^{n} \frac{1}{v_i + a_i} - \alpha}, \tag{7.24}$$

and

$$F_i\left(\alpha; v_0, \ldots, v_n\right) := \frac{1+\alpha}{(1+\alpha)\frac{v_0 + a_0}{a_0} \sum_{j=0, j\neq i}^{n} \frac{1}{v_j + a_j} - \alpha}, \quad i = 1, \ldots, n. \tag{7.25}$$

Since $v_i \geq 0$, $a_i > 0$, $i = 0, 1, \ldots, n$, and $\alpha \in [-1, 0]$, the functions F_i, $i = 0, 1, \ldots, n$, are well-defined and continuous with respect to their arguments over the corresponding domains. Now let us introduce an auxiliary function $\Phi : [-1, 0] \times R_+^{n+1}$ as follows. For arbitrary $\alpha \in [-1, 0]$ and $(v_0, v_1, \ldots, v_n) \in [0, \overline{v}_0) \times R_+^n$, find the (uniquely determined, according to Theorem 8.1) exterior equilibrium vector $(p, q_0, q_1, \ldots, q_n)$ and calculate the derivative $G'(p)$ at the equilibrium point p. Then define the value of the function Φ as below:

$$\Phi\left(\alpha; v_0, v_1, \ldots, v_n\right) := \hat{\alpha} = \frac{G'(p)}{1 - G'(p)} \in [-1, 0]. \tag{7.26}$$

When introducing this auxiliary function Φ, we do not indicate explicitly its dependence upon D, because we are not going to vary D while proving the theorem. As the derivative $G'(p)$ is continuous by p (see assumption **A1**), and the equilibrium price $p = p\left(v_0, v_1, \ldots, v_n\right)$, in its turn, is a continuous function (cf. Theorem 8.1), then the function Φ is continuous as a superposition of continuous functions. (We also notice that its dependence upon α is fictitious.) To finish the proof, let us compose a mapping $H := (\Phi, F_0, F_1, \ldots, F_n) : [-1, 0] \times R_+^{n+1} \to [-1, 0] \times R_+^{n+1}$ and select a convex compact that is mapped by H into itself. The compact is constructed as follows: First, set $s := \max\{\overline{v}_0, a_0, a_1, \ldots, a_n\}$. Second, formulas (7.24)–(7.25) yield the following relationships: If $\alpha = -1$, then

$$F_0(-1, v_0, v_1, \ldots, v_n) = 0, \tag{7.27}$$

$$F_i(-1, v_0, v_1, \ldots, v_n) = 0, \quad i = 1, \ldots, n, \tag{7.28}$$

whereas for $\alpha \in (-1, 0]$ one has

$$\begin{aligned} 0 \le F_0(\alpha, v_0, v_1, \ldots, v_n) &= \frac{1+\alpha}{(1+\alpha)\sum_{i=1}^{n}\frac{1}{v_i + a_i} - \alpha} \\ &\le \frac{1+\alpha}{(1+\alpha)\sum_{i=1}^{n}\frac{1}{v_i + a_i}} = \frac{1}{\sum_{i=1}^{n}\frac{1}{v_i + a_i}} \le \frac{1}{\sum_{i=1}^{n}\frac{1}{v_i + s}}; \end{aligned} \tag{7.29}$$

and

$$\begin{aligned} 0 \le F_i(\alpha, v_0, v_1, \ldots, v_n) &= \frac{1+\alpha}{(1+\alpha)\frac{v_0 + a_0}{a_0}\sum_{j=0, j\neq i}^{n}\frac{1}{v_j + a_j} - \alpha} \\ &\le \frac{1}{\frac{v_0 + a_0}{a_0}\sum_{j=0, j\neq i}^{n}\frac{1}{v_j + a_j}} \le \frac{1}{\sum_{j=0, j\neq i}^{n}\frac{1}{v_j + a_j}} \\ &\le \frac{1}{\sum_{j=0, j\neq i}^{n}\frac{1}{v_j + s}}, \quad i = 1, \ldots, n. \end{aligned} \tag{7.30}$$

Relationships (7.27)–(7.30) clearly imply that for any $\alpha \in [-1, 0]$, if $0 \le v_j \le \frac{s}{n-1}$, $j = 0, 1, \ldots, n$, then the values of $F_j(\alpha, v_0, \ldots, v_n)$, $j = 0, \ldots, n$, drop within the same (closed) interval $\left[0, \frac{s}{n-1}\right]$. Therefore, we have just established that the continuous mapping $H := (\Phi, F_0, F_1, \ldots, F_n)$ maps the compact convex subset $\Omega := [-1, 0] \times \left[0, \frac{s}{n-1}\right]^{n+1}$ into itself. Thus, by Brouwer Fixed Point Theorem, the mapping H has a fixed point $(\alpha, v_0, \ldots, v_n)$, that is,

$$\begin{cases} \Phi(\alpha, v_0, v_1, \ldots, v_n) = \alpha, \\ F_0(\alpha, v_0, v_1, \ldots, v_n) = v_0, \\ F_1(\alpha, v_0, v_1, \ldots, v_n) = v_1, \\ \vdots \\ F_n(\alpha, v_0, v_1, \ldots, v_n) = v_n. \end{cases} \tag{7.31}$$

Now, for the thus obtained influence coefficients $v = (v_0, v_1, \ldots, v_n) \in [0, \overline{v}_0) \times R^n_+$, there exists (uniquely, by Theorem 8.1) the exterior equilibrium $(p, q_0, q_1, \ldots, q_n)$. Hence, we can immediately conclude (from (8.51) and the definition of function Φ) that $G'(p) = \frac{\alpha}{1+\alpha}$, and therefore, the influence coefficients satisfy conditions (7.21)–(7.22). So, according to Definition 7.2, the just constructed vector $(p; q_0, \ldots, q_n; v_0, \ldots, v_n)$ is the desired interior equilibrium. The proof is complete. □

7.1.4.2 Properties of Influence Coefficients

In our future research, we are going to extend the obtained results to the case of non-differentiable demand functions. However, some of the necessary technique can be developed now, in the differentiable case. To do that, we denote the value of the demand function's derivative by $\tau := G'(p)$ and rewrite the consistency Eqs. (7.21)–(7.22) in the following form:

$$v_0 = \frac{1}{\sum_{i=1}^{n} \frac{1}{v_i + a_i} - \tau}, \tag{7.32}$$

and

$$v_i = \frac{1}{\frac{v_0 + a_0}{a_0} \sum_{j=0, j\neq i}^{n} \frac{1}{v_j + a_j} - \tau}, \quad i = 1, \ldots, n, \tag{7.33}$$

where $\tau \in (-\infty, 0]$. When $\tau \to -\infty$ then system (7.32)–(7.33) has the unique limit solution $v_j = 0$, $j = 0, 1, \ldots, n$. For all the finite values of τ, we establish the following result.

Theorem 7.3 *For any $\tau \in (-\infty, 0]$, there exists a unique solution of equations (7.32)–(7.33) denoted by $v_k = v_k(\tau)$, $k = 0, 1, \ldots, n$, continuously depending upon τ. Furthermore, $v_k(\tau) \to 0$ when $\tau \to -\infty$, $k = 0, \ldots, n$, and $v_0 = v_0(\tau)$ strictly increases until $v_0(0)$ if τ grows up to zero.*

Proof Similar to the proof of Theorem 7.2, we introduce the auxiliary functions

$$F_0(\tau; v_0, \ldots, v_n) := \frac{1}{\sum_{i=1}^{n} \frac{1}{v_i + a_i} - \tau} = v_0, \tag{7.34}$$

and

$$F_i(\tau; v_0, \ldots, v_n) := \frac{1}{\frac{v_0 + a_0}{a_0} \sum_{j=0, j\neq i}^{n} \frac{1}{v_j + a_j} - \tau} = v_i, \quad i = 1, \ldots, n, \tag{7.35}$$

and set $s := \max\{a_0, a_1, \ldots, a_n\}$. It is easy to check that for any fixed value of $\tau \in (-\infty, 0]$, the vector-function $\mathrm{d} := (F_0, F_1, \ldots, F_n)$ maps the n-dimensional cube $M := \left[0, \frac{s}{n-1}\right]^n$ into itself. Now we show that subsystem (7.35) has a unique solution $v(v_0, \tau) = (v_1(v_0, \tau), \ldots, v_n(v_0, \tau))$ for any fixed $\tau \in (-\infty, 0]$ and $v_0 \geq 0$; moreover, the vector-function $v = v(v_0, \tau)$ is continuously differentiable with respect to both variables v_0 and τ. The Jacobi matrix of the mapping $\mathrm{d} = (F_0, F_1, \ldots, F_n)$, that is, the matrix $J := \left(\frac{\partial F_i}{\partial v_j}\right)_{i=1,j=1}^{n,n}$ has the following entries:

$$\frac{\partial F_i}{\partial v_j} = \begin{cases} 0, & \text{for } j = i; \\ \frac{v_0 + a_0}{a_0} \cdot \frac{F_i^2}{(v_j + a_j)^2}, & \text{for } j \neq i. \end{cases} \tag{7.36}$$

Thus, matrix J is nonnegative and non-decomposable. Now let us estimate the sums of the matrix entries in each row $i = 1, 2, \ldots, n$:

$$\sum_{k=1}^{n} \frac{\partial F_i}{\partial v_k} = \frac{v_0 + a_0}{a_0} F_i^2 \cdot \sum_{k=1,k\neq i}^{n} \frac{1}{(v_k + a_k)^2} \leq \frac{\frac{v_0 + a_0}{a_0} \sum_{k=1,k\neq i}^{n} \frac{1}{(v_k + a_k)^2}}{\left(\frac{v_0 + a_0}{a_0} \sum_{k=1,k\neq i}^{n} \frac{1}{v_k + a_k}\right)^2}$$

$$= \frac{a_0}{v_0 + a_0} \cdot \frac{\sum_{k=1,k\neq i}^{n} \frac{1}{(v_k + a_k)^2}}{\left(\sum_{k=1,k\neq i}^{n} \frac{1}{v_k + a_k}\right)^2} = R_i(v_1, \ldots, v_n; v_0) < 1. \tag{7.37}$$

For any fixed value $v_0 \geq 0$ (treated as a parameter), the above-mentioned functions $R_i(v_1, \ldots, v_n; v_0)$, $i = 1, \ldots, n$, are defined on the cube M, continuously depend upon the variables $v_1, \ldots, v_n$, and take only positive values strictly less than 1. Therefore, their maximum values achieved on the compact cube M are also strictly lower than 1, which implies that the matrix $(I - J)$ is invertible (here, I is the n-dimensional unit matrix), and the mapping $\mathrm{d} := (F_1, \ldots, F_n)$ defined on M is a strictly contracting mapping in the cubic norm (i.e., $\|\cdot\|_\infty$-norm). The latter allows to conclude that for any fixed values of $\tau \in (-\infty, 0]$ and $v_0 \geq 0$, the equation subsystem (7.35) has a unique solution $v(v_0, \tau) = (v_1(v_0, \tau), \ldots, v_n(v_0, \tau))$. Since $\det(I - J) \neq 0$ for any $\tau \in (-\infty, 0]$, Implicit Function Theorem also guarantees that $v(v_0, \tau)$ is continuously differentiable by both variables.

In order to establish the monotone increasing dependence of the solution $v(v_0, \tau)$ of subsystem (7.35) upon τ for any fixed value $v_0 \geq 0$, let us differentiate (7.35) with respect to τ to yield

$$\frac{\partial v_i}{\partial \tau} = F_i^2 \left[1 + \frac{v_0 + a_0}{a_0} \sum_{k=1, k \neq i}^{n} \frac{\frac{\partial v_k}{\partial \tau}}{(v_k + a_k)^2} \right], \quad i = 1, \ldots, n. \tag{7.38}$$

Rewriting (7.38) in the vector form, we come to

$$v'_\tau = J v'_\tau + F^2, \tag{7.39}$$

where

$$v'_\tau := \left(\frac{\partial v_1}{\partial \tau}, \ldots, \frac{\partial v_n}{\partial \tau} \right)^T \quad \text{and} \quad F^2 := \left(F_1^2, \ldots, F_n^2 \right)^T > 0. \tag{7.40}$$

Since all entries of the inverse of matrix $(I - J)$ are nonnegative (the latter is due to the matrix $(I - J)$ being an M-matrix, cf. e.g. Berman and Plemmons [21]) and the inverse matrix $(I - J)^{-1}$ has no zero rows, then (7.39)–(7.40) imply

$$v'_\tau = (I - J)^{-1} F^2 > 0, \tag{7.41}$$

that is, each component of the solution vector $v\,(v_0, \tau)$ of subsystem (7.35) is a strictly increasing function of τ for each fixed value of $v_0 \geq 0$. Moreover, the straightforward estimates

$$v_i\,(v_0, \tau) \leq -\frac{1}{\tau}, \quad i = 1, \ldots, n, \tag{7.42}$$

bring about the limit relationships shown below:

$$v_i\,(v_0, \tau) \to 0 \;\text{ as }\; \tau \to -\infty, \quad i = 1, \ldots, n, \quad \text{for any fixed} \;\; v_0 \geq 0. \tag{7.43}$$

To order to establish the monotone (decrease) dependence of the solution $v\,(v_0, \tau)$ of subsystem (7.35) upon $v_0 \geq 0$ for each fixed value of $\tau \in (-\infty, 0]$, we differentiate (7.35) with respect to v_0 to get:

$$\begin{aligned} \frac{\partial v_i}{\partial v_0} &= -F_i^2 \left[\frac{1}{a_0} \sum_{k=1, k \neq i}^{n} \frac{1}{v_k + a_k} - \frac{v_0 + a_0}{a_0} \sum_{k=1, k \neq i}^{n} \frac{\frac{\partial v_k}{\partial v_0}}{(v_k + a_k)^2} \right] \\ &= F_i^2 \frac{v_0 + a_0}{a_0} \sum_{k=1, k \neq i}^{n} \frac{\frac{\partial v_k}{\partial v_0}}{(v_k + a_k)^2} - \frac{F_i^2}{a_0} \sum_{k=1, k \neq i}^{n} \frac{1}{v_k + a_k}, \quad i = 1, \ldots, n. \end{aligned} \tag{7.44}$$

Again, rearrange these equalities into a system of equations

$$v'_{v_0} = J v'_{v_0} - Q, \tag{7.45}$$

where

$$v'_{v_0} := \left(\frac{\partial v_1}{\partial v_0}, \ldots, \frac{\partial v_n}{\partial v_0} \right)^T, \tag{7.46}$$

while $Q \in R^n$ is the vector with the components

$$Q_i := \frac{F_i^2}{a_0} \sum_{k=1, k \neq i}^{n} \frac{1}{v_k + a_k} > 0, \quad i = 1, \ldots, n. \tag{7.47}$$

Solving (7.45) for v'_{v_0} and making use of (7.47), one comes to the relationship

$$v'_{v_0} = -(I - J)^{-1} Q < 0, \tag{7.48}$$

which means that each component of $v\,(v_0, \tau)$ is a strictly decreasing function of $v_0 \geq 0$ for each fixed value of $\tau \in (-\infty, 0]$.

Now we are in a position to demonstrate the existence and smoothness of the unique solution $v(\tau) = (v_0(\tau), v_1(\tau), \ldots, v_n(\tau))$ of the complete system (7.34)–(7.35) for every fixed value of $\tau \in (-\infty, 0]$. To do that, we plug in the above-mentioned (uniquely defined for each fixed $\tau \in (-\infty, 0]$ and $v_0 \geq 0$) solution of subsystem (7.35) into (7.34) and arrive to the functional equation:

$$v_0 = \frac{1}{\sum_{i=1}^{n} \frac{1}{v_i\,(v_0, \tau) + a_i} - \tau}. \tag{7.49}$$

With the aim to prove the unique solvability of the latter equation, we fix an arbitrary $\tau \in (-\infty, 0]$ and introduce the function

$$\Psi\,(v_0) := \frac{1}{\sum_{i=1}^{n} \frac{1}{v_i\,(v_0, \tau) + a_i} - \tau}. \tag{7.50}$$

Since we know that

$$0 \leq v_i\,(v_0, \tau) \leq \frac{s}{n-1}, \quad n = 1, \ldots, n, \quad \text{where } s = \max\{a_0, a_1, \ldots, a_n\}, \tag{7.51}$$

it brings us to the chain of relationships

$$\Psi\left(v_0\right) \leq \frac{1}{\sum_{i=1}^{n} \frac{1}{v_i\left(v_0, \tau\right)+a_i}} \leq \frac{1}{\sum_{i=1}^{n} \frac{1}{v_i\left(v_0, \tau\right)+s}}$$
$$\leq \frac{1}{\sum_{i=1}^{n} \frac{1}{\frac{s}{n-1}+s}}=\frac{1}{\sum_{i=1}^{n} \frac{n-1}{ns}}=\frac{s}{n-1}, \tag{7.52}$$

which allows one to conclude that (for any fixed $\tau \in(-\infty, 0]$) the continuous function $\Psi=\Psi\left(v_0\right)$ maps the closed interval $\left[0, \frac{s}{n-1}\right]$ into itself. Therefore, according to Brouwer Fixed Point Theorem, there exists a fixed point $v_0=\Psi\left(v_0\right)$ within this interval.

To finish the proof of the theorem, it is sufficient to establish that the above-determined fixed point is unique for each fixed $\tau \in(-\infty, 0]$ and, in addition, is monotone increasing with respect to τ. First, (7.48) implies that (for every fixed $\tau \in(-\infty, 0]$), the functions $v_i\left(v_0, \tau\right)$, $i=1, \ldots, n$, are strictly decreasing by $v_0 \geq 0$; hence, each ratio $\frac{1}{v_i\left(v_0, \tau\right)+a_i}$, $i=1, \ldots, n$, strictly increases with respect to $v_0 \geq 0$. Now we deduce from (7.53) below that the function $\Psi=\Psi\left(v_0\right)$, in its turn, strictly decreases with respect to $v_0 \geq 0$, which means that the fixed point $v_0=\Psi\left(v_0\right)$ exists uniquely.

Differentiability of the thus determined well-defined function $v_0=v_0(\tau)$ with respect to follows from Implicit Function Theorem, because

$$\frac{\partial \Psi}{\partial v_0}=\Psi^2 \sum_{i=1}^{n} \frac{\frac{\partial v_i}{\partial v_0}}{\left(v_i+a_i\right)^2}<0, \quad \text{for any } \tau \in(-\infty, 0]. \tag{7.53}$$

It is easy to see that the vector-function

$$v(\tau):=\left[v_0(\tau), v_1\left(v_0(\tau), \tau\right), \ldots, v_n\left(v_0(\tau), \tau\right)\right]^T$$

obtained by substituting the newly constructed function $v_0=v_0(\tau)$ into the previously described solution of subsystem (7.35) represents the uniquely determined and continuously differentiable solution of the complete system (7.34)–(7.35).

In order to demonstrate the monotony of the above-described soluction's first component $v_0=v_0(\tau)$ by τ, we differentiate equation (7.49) by the chain rule and make use of (7.50) to yield

$$\frac{d v_0}{d \tau}=\left[\Psi^2 \sum_{i=1}^{n} \frac{\frac{\partial v_i}{\partial v_0}}{\left(v_i+a_i\right)^2}\right] \cdot \frac{d v_0}{d \tau}+\Psi^2 \sum_{i=1}^{n} \frac{\frac{\partial v_i}{\partial \tau}}{\left(v_i+a_i\right)^2}+\Psi^2. \tag{7.54}$$

Now solving (7.54) for the derivative $\frac{dv_0}{d\tau}$, one obtains:

$$\frac{dv_0}{d\tau} = \frac{B}{A}, \tag{7.55}$$

where, owing to (7.53),

$$A = 1 - \Psi^2 \sum_{i=1}^{n} \frac{\frac{\partial v_i}{\partial v_0}}{(v_i + a_i)^2} > 0 \tag{7.56}$$

while

$$B = \Psi^2 \left[\sum_{i=1}^{n} \frac{\frac{\partial v_i}{\partial \tau}}{(v_i + a_i)^2} + 1 \right] > 0, \tag{7.57}$$

according to (7.41). Combining (7.55)–(7.57), we conclude that $\frac{dv_0}{d\tau} = \frac{B}{A} > 0$; hence, the function $v_0 = v_0(\tau)$ strictly increases by τ. Moreover, by the evident estimate

$$v_0(\tau) \leq -\frac{1}{\tau},$$

which follows from (7.49), we deduce that $v_0 = v_0(\tau)$ vanishes as $\tau \to -\infty$. Similarly, (7.43) implies that the same is valid for all the remaining influence coefficients, i.e., $v_i(\tau) \to 0$, $i = 1, \ldots, n$, as $\tau \to -\infty$. The proof of the theorem is complete. □

7.1.5 Numerical Results

In order to illustrate the difference between the mixed and classical oligopoly cases related to the conjectural variations equilibrium with consistent conjectures (influence coefficients), we apply formulas (7.21)–(7.22) to the simple example of an oligopoly in the electricity market from Liu et al. [206]. The only difference in our modified example from the instance of Liu et al. [206] is in the following: in their case, all six agents (suppliers) are private companies producing electricity and maximizing their net profits, whereas in our case, we assume that Supplier 0 (Supplier 5 in some instances) is a public enterprise seeking to maximize domestic social surplus defined by (7.4). All the other parameters involved in the inverse demand function $p = p(G)$ and the producers' cost functions, are exactly the same.

So, following Liu et al. [206], we select the IEEE 6-generator 30-bus system to illustrate the above analysis. The inverse demand function in the electricity market is given in the form:

Table 7.1 Cost functions' parameters

Agent i	b_i	a_i
0	2.00000	0.02000
1	1.75000	0.01750
2	3.00000	0.02500
3	3.00000	0.02500
4	1.00000	0.06250
5	3.25000	0.00834

$$p(G, D) = 50 - 0.02(G + D) = 50 - 0.02 \sum_{i=0}^{n} q_i . \tag{7.58}$$

The cost functions parameters of suppliers (generators) are listed in Table 7.1. Here, agents $0, 1, \ldots, 5$ will be combined in different ways in the examples listed below. In particular, Oligopoly 1 is the classical oligopoly where each agent 0–5 maximizes its net profit; Oligopoly 2 will involve agent 0 (public one, who maximizes domestic social surplus) and $1, \ldots, 5$ (private), whereas Oligopoly 3 comprises agents 5 (public) and $0, 1, \ldots, 4$ (private).

To find the consistent influence coefficients in their classical oligopoly market (Oligopoly 1), Liu et al. [206] use formulas (7.23) for all six suppliers, whereas for our mixed oligopoly model (Oligopoly 2), we exploit formula (7.21) for Supplier 0 and formulas (7.22) for Suppliers 1 through 5. With the thus obtained influence coefficients, the (unique) equilibrium is found for each of Oligopolies 1 and 2. The equilibrium results (influence coefficients, production outputs in MWh, equilibrium price, and the objective functions' optimal values in \$ per hour) are presented in Table 7.2 through 7.6. To make our conjectures v_i comparable to those used in Kalashnykova et al. [182], Kalashnikov et al. [165], and Liu et al. [206], we divide them by $[-p'(G)] = 0.02$ and thus come to $w_i := -v_i/p'(G) = 50v_i$, $i = 0, 1, \ldots, n$, shown in all the tables.

As it could be expected, the equilibrium price in Oligopoly 1 (classical oligopoly) turns out to be equal to $p_1 = 10.4304$, whereas in Oligopoly 2 (mixed oligopoly), it

Table 7.2 Computation results in consistent CVE: w_i, generation, profits

Agent i	w_i		q_i (MWh)		Profits (\$/h)	
	Oligopoly 1	Oligopoly 2	Oligopoly 1	Oligopoly 2	Oligopoly 1	Oligopoly 2
0	0.19275	0.18779	353.405	626.006	1727.4	595.77
1	0.19635	0.16674	405.120	358.138	2076.6	1550.04
2	0.18759	0.15887	258.463	220.451	1082.9	761.90
3	0.18759	0.15887	258.463	220.451	1082.9	761.90
4	0.17472	0.14761	142.898	125.462	707.5	538.37
5	0.22391	0.19270	560.180	488.905	2709.8	1917.98

drops down to $p_2 = 9.2118$. On the contrary, the total electricity power generation is higher: $G_2 = 2039.412$ MWh—in the second case (mixed oligopoly), than in Oligopoly 1, which is $G_1 = 1978.475$ MWh. Both results are more attractive for consumers. Simultaneously, the private producers' net profit values are observed to be lower in the mixed oligopoly (Oligopoly 2) than those in the classical oligopoly (Oligopoly 1.) In Oligopoly 2, profit is minimal in the cell of Agent 0, because its real objective function is *not* the net profit but *domestic social surplus* defined by (7.4); in this instance, it happens to reach $S = \$42,187.80/\text{h}$.

It is also interesting to compare the results in CVE with consistent conjectures (Oligopolies 1 and 2) against the production volumes and profits obtained for the same cases in the classical Cournot equilibrium (i.e., with all $w_i = 1, i = 0, 1, \ldots, 5$.) Table 7.3 illustrates the yielded results, with $p_1 = 14.760$ in the classical oligopoly (Oligopoly 1) much higher than the market equilibrium price $p_2 = 9.5349$ in the mixed oligopoly (Oligopoly 2).

Again, the total electricity power generation is higher: $G_2 = 2023.256$ MWh,— in the second case (mixed oligopoly), than in Oligopoly 1: $G_1 = 1761.9$ MWh. Both results are more propitious for consumers. Simultaneously, the private producers' net profit values are observed to be much lower in the mixed oligopoly (Oligopoly 2) than those in the classical oligopoly (Oligopoly 1). In Oligopoly 2, profit is even negative in the cell of Agent 0, as its objective function is not the profit but domestic social surplus defined by (7.4); in this example, it is equal to $S = \$35,577.50/\text{h}$. The latter data, together with the market price values, suggest that the mixed oligopoly with consistent conjectures is preferable to consumers than the Cournot model.

Table 7.3 Computation results in the Cournot models: w_i, generation, profits

Agent i	w_i		q_i (MWh)		Profits (\$/h)	
	Oligopoly 1	Oligopoly 2	Oligopoly 1	Oligopoly 2	Oligopoly 1	Oligopoly 2
0	1.00000	1.00000	319.060	1200.000	3054.0	−5358.14
1	1.00000	1.00000	347.000	207.597	3461.7	1239.02
2	1.00000	1.00000	261.390	145.220	2220.5	685.38
3	1.00000	1.00000	261.390	145.220	2220.5	685.38
4	1.00000	1.00000	166.820	103.453	1426.2	548.51
5	1.00000	1.00000	406.230	221.767	3988.5	1188.70

Table 7.4 Computation results in the perfect competition model: w_i, generation, profits

Agent i	w_i		q_i (MWh)		Profits (\$/h)	
	Oligopoly 1	Oligopoly 2	Oligopoly 1	Oligopoly 2	Oligopoly 1	Oligopoly 2
0	0.00000	0.00000	348.43	348.43	1214.00	1214.00
1	0.00000	0.00000	412.49	412.49	1488.80	1488.80
2	0.00000	0.00000	238.74	238.74	712.47	712.47
3	0.00000	0.00000	238.74	238.74	712.47	712.47
4	0.00000	0.00000	127.50	127.50	507.98	507.98
5	0.00000	0.00000	685.68	685.68	1960.50	1960.50

Of course, the perfect competition model (see, Table 7.4) with $w_i = v_i = 0, i = 0, 1, \ldots, 5$, is the best for consumers in both Oligopoly 1 and 2: with $p_1 = p_2 = 8.9685$ and the total produce $G_1 = G_2 = 2051.57$ MWh. Domestic social surplus is also higher in this case than in all the previous ones: $S = \$43,303.52/\text{h}$.

It is curious to note (cf. Tables 7.2, 7.3 and 7.4) that in the classical oligopoly (Oligopoly 1), the Cournot model demonstrates to be the most profitable for the producers, whereas it is not the case for the mixed oligopoly: here, the existence of a public enterprise with domestic social surplus as its utility function makes the consistent CVE more beneficial for the rest of suppliers than the Cournot one (except for the weakest Agent 4, for which, on the contrary, the Cournot model is most gainful).

Finally, we may compare the consistent CVEs (Table 7.5), Cournot equilibria (Table 7.6) and the perfect competition for the above-defined Oligopoly 2 (mixed oligopoly with Agent 0 being a public company) against a similar Oligopoly 3, in which not Agent 0 but the (much stronger) Agent 5 is the public supplier.

With the market price $p_3 = 7.8751$ even lower and domestic social surplus $= \$44,477.30/\text{h}$ even higher than those in the perfect competition model, this consistent CVE may serve as a good example of the strong public company realizing the implicit price regulation within an oligopoly.

A bit curious are the results reflected in Table 7.6: comparing the Cournot oligopoly in Oligopolies 1, 2, and 3, one may see that with a weaker public firm (Oligopoly 2), the private producers may incline to the Cournot model of behavior (cf. Table 7.3). However, with a stronger public supplier, as it is in Oligopoly 3,

Table 7.5 Computation results in consistent CVE: w_i, generation, profits

Agent i	w_i		q_i (MWh)		Profits (\$/h)	
	Oligopoly 2	Oligopoly 3	Oligopoly 2	Oligopoly 3	Oligopoly 2	Oligopoly 3
0	0.18779	0.13208	626.006	259.480	595.77	851.16
1	0.16674	0.13497	358.138	303.229	1550.04	1052.75
2	0.15887	0.12803	220.451	176.884	761.90	471.22
3	0.15887	0.12803	220.451	176.884	761.90	471.22
4	0.14761	0.11843	125.462	105.984	538.37	377.63
5	0.19270	0.21584	488.905	1083.785	1917.98	114.52

Table 7.6 Computation results in the Cournot models: w_i, generation, profits

Agent i	w_i		q_i (MWh)		Profits (\$/h)	
	Oligopoly 2	Oligopoly 3	Oligopoly 2	Oligopoly 3	Oligopoly 2	Oligopoly 3
0	1.00000	1.00000	1200.000	122.612	−5358.14	451.01
1	1.00000	1.00000	207.597	137.452	1239.02	543.18
2	1.00000	1.00000	145.220	86.766	685.38	244.67
3	1.00000	1.00000	145.220	86.766	685.38	244.67
4	1.00000	1.00000	103.453	71.569	548.51	262.51
5	1.00000	1.00000	221.767	1649.612	1188.70	−5319.04

private companies would rather select the consistent CVE: in the Cournot model, the strong public company subdues all the rivals to the minimum levels of production and profits. Nevertheless, the Cournot model with stronger public firm provides for the very low market price: $p_3 = 6.9045$, even though at the cost of a somewhat lower domestic social surplus: $S = \$41,111.59/\text{h}$.

As it could be expected, in the perfect competition model, both Oligopolies 2 and 3 give exactly the same results, albeit different domestic social surplus values: $S = \$43,303.52/\text{h}$ in Oligopoly 2 against a bit higher $S = \$44,050.04/\text{h}$ in Oligopoly 3 with the stronger public company.

In this section, we have studied a model of mixed oligopoly with conjectural variations equilibrium (CVE). The agents' conjectures concern the price variations depending upon their production output's increase or decrease. We establish existence and uniqueness results for the conjectured variations equilibrium (called an exterior equilibrium) for any set of feasible conjectures. To introduce the notion of an interior equilibrium, we develop a consistency criterion for the conjectures (referred to as influence coefficients) and prove the existence theorem for the interior equilibrium (understood as a CVE with consistent conjectures). To prepare the base for the extension of our results to the case of non-differentiable demand functions, we also investigate the behavior of the consistent conjectures in dependence upon a parameter representing the demand function's derivative with respect to the market price.

An interesting methodological question also arises: can the mixed oligopoly be related to collaborative game theory? Formally speaking, the mixed oligopoly is rather a cooperative than competitive game, as the public company's and the private firms' interests are "neither completely opposed nor completely coincident" (Nash [248]). At first glance, collaboration can be a worthwhile strategy in a cooperative game. However, according to Zagal et al. [327], "because the underlying game model is still designed to identify a sole winner, cooperative games can encourage anti-collaborative practices in the participants. Behaving competitively in a collaborative scenario is exactly what should not happen in a collaborative game".

7.2 Toll Assignment Problems

One of the most important problems concerning the toll roads is the setting of an appropriate cost for traveling through private arcs of a transportation network. In the section this problem is considered by stating it as a bilevel optimization (BLP) model. At the upper level, one has a public regulator or a private company that manages the toll roads seeking to increase its profits. At the lower level, several companies-users try to satisfy the existing demand for transportation of goods and/or passengers, and simultaneously, to select the routes so as to minimize their travel costs. In other words, what is sought is a kind of a balance of costs that bring the highest profit to the regulating company (the upper level) and are still attractive enough to the users (the lower level).

With the aim of providing a solution to the bilevel optimization problem in question, a direct algorithm based on sensitivity analysis is proposed in this section.

In order to make it easier to move (if necessary) from a local maximum of the upper level objective function to another, the well-known "filled function" method is used. Most results in this section are taken from Kalashnikov et al. [167].

7.2.1 Introduction

During the early years of industrial development, the production facilities were established near the consumers because the transportation was expensive, time-consuming, and risky. When transportation systems appeared, they allowed the producer to compete in distant markets, promoting economies of scale by increasing sales volume.

Due to the complexity of products and globalization, supply and distribution chains have grown enormously, therefore, logistics costs have "rocketed up" sharply. According to the data from the IMF (International Monetary Fund), logistics costs represent 12 % of gross national product, and they range from 4 to 30 % of the sales at the enterprise level.

Because of this growth, many countries have attached great importance to the development and modernization of the infrastructure to achieve greater participation in the global economy. There are organizations that deal with the development of communications and transportation infrastructure, building technologies to increase the coverage, quality and competitiveness of the infrastructure. In Mexico, administration of new (private) roads is commonly conceded to private companies, state governments, or financial institutions (banks, holdings, etc.), who set transportation tolls in order to retrieve money from the road users.

It has been recently noticed that under the concession model, there is less traffic flow using these tolled highways. One of the strategies taken to increase the use of toll roads is the regulation of tolls (pass rates). However, what are the appropriate criteria to assign these toll rates?

The problem here is how to assign optimal tolls to the arcs of a multicommodity transportation network. The toll optimization problem (TOP) can be formulated as a bilevel mathematical program where the upper level is managed by a firm (or a public regulator) that raises revenues from tolls set on (some) arcs of the network, and the lower level is represented by a group of users traveling along the cheapest paths with respect to a generalized travel cost. The problem can be interpreted as finding equilibrium among tolls generating high revenues and at the same time being attractive for the users. Other possible aims of the upper level decision maker can be found in Heilporn et al. [141], Didi-Biha et al. [88], Labbé et al. [196].

The problem in question has been extensively studied. In what follows, a literature review related to the TOP is made. Almost thirty years ago, Magnanti and Wong [212] presented a very complete theoretical basis for the uses and limitations of network design based on integer optimization with several models and algorithms. This provided a unification of network design models, as well as a general framework for deriving network design algorithms. They noticed that researchers had been motivated to develop a variety of solution techniques such as linear approximation methods and the search of vertices adjacent to the lowest cost flow problem threated as a *network design problem* (NDP).

The network design issues were mentioned several years later by Yang and Bell [322], who also provided a brilliant survey of the existing literature in this area. They introduced the elasticity concept in travel demand and the reserve capacity notion of the network in the NDP, which allowed them to obtain a network design problem easier to solve when trying to maximize an appropriate objective function. Moreover, they proposed an approach to NDP involving mixed elections, i.e., simultaneously adding links and improving the capabilities. The latter approach allowed the use of formulas based upon the maximization of consumer surplus as the objective function of the NDP. The authors mentioned that the challenge remained to develop a global search algorithm that could guarantee the optimality of a solution derived with the computationally efficient manner mentioned in [212].

Bell and Iida [17] sought a unification of the theoretical analysis of transportation networks, focused primarily on the assignment of stochastic user equilibrium (SUE), estimating trip tables and network's reliability. They saw the network design as an extension of the analysis of the transportation network, where the control of traffic signals is made in terms of an NDP. The latter is considered a difficult task because of its nonconvex nature and the complexity of the networks. They mentioned that the NDP can be posed as a bilevel optimization problem, where the upper level focuses on the network design to maximize certain goals, whereas the lower level determines how users react to changes in the network. In their monograph, they presented two different approaches to solve the problem of network design, one is the iterative method of design-assignment, which is relatively simple in its application and appears to converge quickly. The other is an iterative algorithm based on sensitivity analysis, which usually consumes more computational time to converge. The authors conclude that both methods provide different local optima, with slight differences in the objective function but significantly distinct in the design structure. Finally, they mentioned that in order to have a more satisfactory approach, it is necessary to combine bilevel optimization tools (to find a local solution) and a probabilistic search method (for comparing local solutions using simulated annealing), to come to a global solution.

Marcotte [218] mentioned that the NDP mainly deals with the optimal balance either of the transportation, investment, or maintenance costs of the networks subject to congestion. The network's users behave according to Wardrop's first principle of traffic equilibrium. He also suggested that the NDP can be modeled as a multilevel mathematical optimization problem.

Mahler et al. [213] dealt with the problem of congestion in road networks represented by two problems, namely, estimation of the trip matrix and optimization of traffic signals. Both problems were formulated as bilevel programs with allocation of stochastic user equilibrium (SUE) as the problem of the lower level. The authors presented an algorithm that gives a solution to the bilevel problem of estimation of the trip matrix and optimization of signals, making use of the "logit-based" model of assignment SUE at the lower level. The algorithm used applies standard routines to estimate the matrix at every iteration, and SUE assignment to find the search direction. The authors stated that it had not been possible to demonstrate the convergence results in general; however, in case that the optimal solution can be found by direct search, they demonstrated that the algorithm is able to give a good approximation of the optimal solution.

Lawphongpanich and Hearn [199] also examined the problem of traffic congestion as a problem of fixing the toll through a formulation with static demands. They mentioned that this problem can be classified into two types: (i) the problem of the first best solution, in which all the arcs of the network are tolled, and (ii) the problem of the second best solution, where it is assumed that some roads may have tolls and others not, which does not permit them to get the maximum benefit. The authors noticed that the latter problem can be posed as a bilevel optimization model, or as a mathematical program with equilibrium constraints (MPEC). They used the results achieved for the MPEC to develop a formulation equivalent to the nonlinear optimization problem for the second best solution. The latter is done in order to establish the properties of the second-best solution, which are of a particular interest to transport economists, and in its turn, help develop another algorithm to solve the problem in the nonlinear optimization formulation.

The pricing of road systems has a long history in the literature of transportation economics, as mentioned by Morrison [246], who worked with a theoretical framework developed through the empirical evidence of viability in pricing and policy. One can also find this concept in the engineering and road planning literature, as described in Cropper and Oates [47], who talked about the implementation of environmental economics in environmental policy design road systems; they focused on reducing traffic congestion on the roads through pricing to reduce negative aspects such as pollution. Other authors who treated the problem of traffic congestion were Arnott et al. [4]; they mentioned that the allocation of a uniform toll significantly reduces this problem by taking into account parameters of time (i.e., alternating departure times for users).

Hearn and Ramana [140] worked over the definition and optimization of different objectives under a given set of tolls that promote optimal traffic systems. Shifting a focus, one finds Viton's paper [309], which makes a comparison between the viability of private toll roads and highways free to users. The concept of maximizing profits through an optimal toll system is examined by Beckmann in [13] and by Verhoef in [303].

As mentioned before, bilevel optimization offers a convenient framework for modeling the optimal toll problems, as it allows one to take into account the user's behavior explicitly. Unlike the aforementioned investigations, Labbé et al. [197] considered the TOP as a sequential game between the owners of road systems (the leaders) and road users (the followers), which fits the scheme of a bilevel optimization problem. This approach is also implemented by Brotcorne [30] on the problem of fixing tariffs on cargo transportation. In the latter case, the leader is formed by a group of competing companies, and their earnings are determined by the total profits from the rates, while the follower is a carrier who seeks to reduce travel costs, taking into account the tolls set by the leader.

One of the simplest instances was analyzed by Kalashnikov et al. [173], where a TOP defined over a congestion-free, multicommodity transportation network was considered. In this setting, a highway authority (the leader) sets tolls on a subset of arcs of the network, while the users (the follower) assign themselves to the shortest (in terms of a generalized time) paths linking their respective origin and destination nodes. The goal of the leader in this context is to maximize the toll revenue. Hence, it is not in its interest to set very high tolls, because in this case the users would be discouraged from using the tolled sub-network. The problem can be stated as a combinatorial program that subsumes NP-hard problems, such as the Traveling Salesman Problem (see, Labbé et al. [196], for a reduction method). Following the initial NP-hardness proof in [196], computational complexity and approximation results were obtained by Marcotte et al. [220].

On the other hand, Dempe et al. [58] studied this problem designing a "fuzzy" algorithm for the TOP. Next, Lohse and Dempe [69] based their studies on the analysis of an optimization problem in some sense reverse to the TOP. In addition, Didi-Biha et al. [88] developed an algorithm based on the calculation of lower and upper bounds to determine the maximum gain from the tolls on a subset of arcs of a network transporting various commodities.

Studies have been conducted with roads without congestion and capacity limits, where it is assumed that congestion is affected by the introduction of tolls. This radically changes the mathematical nature of the model, and algorithms use a different approach. Such a model was presented by Yan and Lam in [321], but these authors were limited only to a simple model with two arcs. A more extensive work on the assumption of limited capacity arcs is presented by Kalashnikov et al. in [166], which studied four different algorithms to solve this problem.

The group of authors Brotcorne et al. [34] started investigating a bilevel model for toll optimization on a multicommodity transportation network as long ago as 2001. Recently, Brotcorne et al. analyzed this problem in [32] with the difference in that they allowed subsidies in the network; that is, they considered the tolls without constraints. The authors designed an algorithm that generated paths and then formed columns for determining the optimal toll values for the current path (the lower bound). Thereafter, they adjusted the revenue upper bound and finally applied a diversification phase. Also they validated their algorithm by conducting numerical experimentation and concluded that the proposed algorithm efficiently works in networks with few toll arcs. The same authors continued their work on the same problem in [31]. In the latter paper, they designed and implemented a tabu search algorithm, and concluded that their heuristics had obtained better results than other combinatorial methods. Dempe and Zemkoho [81] also studied the TOP and proposed a reformulation based on the optimal value function. This restatement has advantage over the KKT reformulation because it keeps on the information about the congestion in the network. They obtained optimality conditions for this restatement and established some theoretical properties for it.

The aim of the present section is to propose an algorithm based on the *allowable ranges to stay optimal* (ARSO) resulting from sensitivity analysis after solving the lower level problem. With this powerful tool, one can analyze possible changes in the coefficients of some variables in the objective function which do not affect the optimal solution (cf. the region of stability in Sect. 3.6.2.1). It also permits one to examine the effects on the optimal solution when the parameters take new values beyond the ARSO. This work is inspired by the previous research undertaken by Roch et al. [271].

In addition to dealing with the allowable ranges, the proposed technique also uses the concept of a "filled function" (see Renpu [267], Wan et al. [312], Wu et al. [318]), which is applied under the assumption that a local maximum (in our case) has been found. Then the "filled functions" technique helps one either to find another local maximum, better than the previous ones, or to determine that we have found (approximately) a best feasible or an optimal solution, according to certain parameters of tolerance.

The validity and reliability of this technique are illustrated by the results of numerical experiments with test examples used to compare the proposed approach with the other ones. Finally, the numerical results also confirmed the robustness of the presented algorithm.

To resume, in this section we propose and test two versions of a heuristic algorithm to solve the Toll Optimization Problem (TOP) based upon sensitivity analysis for linear optimization problems. The algorithm makes use of a sensitivity analysis procedure for the linear optimization problem at the lower level, as well as of the "filled functions" technicalities in order to reach a global optimum when "jammed" at some local optimum. The two versions of the method differ only in the way of selecting a new toll vector, namely, by changing only one toll value at a time, or by varying several toll values applying the well-known 100 % rule of sensitivity analysis.

The proposed heuristics aim at filling in a gap in a series of numerical approaches to the solution of TOP problem listed in the Introduction. To our knowledge, no systematic attempts to apply the sensitivity analysis tools to the toll assigned problem have been made. Moreover, the combination of these powerful tools with the "filled functions" techniques brings forward some new global optimization ideas.

Numerical experiments with a series of small and medium-dimension test problems show the proposed algorithm's robustness and reasonable convergence characteristics. In particular, while ceding in efficiency to other algorithms when solving small problems, the proposed method wins in the case of medium (higher dimensional) test models.

The rest of the section is organized as follows. Section 7.2.2 contains the model statement and the definition of parameters involved. Section 7.2.3 describes the algorithm to solve the toll optimization problem, with Sect. 7.2.3.1 presenting the algorithm's structure, Sect. 7.2.3.2 justifying the reduction of the lower level equilibrium problem to a standard linear program, and Sect. 7.2.3.3 recalling the "filled functions" technique. Section 7.2.4 lists the results of numerical experiments obtained for several test problems. Supplementary material (Sect. 7.2.5) describes the data of all the test problems tested in the numerical experiments.

7.2.2 TOP as a Bilevel Optimization Model

The methodology developed to solve this problem takes the model proposed by Labbé et al. [196] as a basis. They proved that the TOP can be analyzed as a leader-follower game that takes place on a multicommodity network $G = (K, N, A)$ defined by a set of origin-destination couples K, a node set N and an arc set A. The latter is partitioned into the subset A_1 of toll arcs and the complementary subset A_2 of toll-free arcs. We endow each arc $a \in A$ with a fixed travel delay c_a representing the minimal unit travel cost. Each toll arc $a \in A_1$ also involves a toll component t_a, to be determined. The latter is also expressed in time units, for the sake of consistency. The toll vector $t = \{t_a : a \in A_1\}$ is restricted by the vector $t^{max} = \{t_a^{max} : a \in A_1\}$ from above and by zero from below.

The demand side is represented by numbers n^k denoting the demand for transportation between the origin node $o(k)$ and the destination node $d(k)$ associated with commodity $k \in K$, $|K| = r$. A demand vector b^k is associated with each commodity. Its components are defined for every node i of the network as follows:

$$b_i^k = \begin{cases} n^k, & \text{if } i = d(k); \\ -n^k, & \text{if } i = o(k); \\ 0, & \text{otherwise.} \end{cases} \tag{7.59}$$

Let $x = \{x_a^k\}_{a\in A}$ denote the set of commodity flows along the arcs $a \in A$, and $\{i^+\} \subset A$ the set of arcs having i as their head (destination) node, while $\{i^-\} \subset A$ is the set of arcs having i as their tail (origin) node, for any $i \in N$. Based on the notation introduced above, the *toll optimization problem* (TOP) can be stated as the bilevel program (7.60)–(7.63):

$$\max_{t,x} F(t,x) = \sum_{k\in K} \sum_{a\in A_1} t_a x_a^k, \tag{7.60}$$

subject to

$$0 \le t_a \le t_a^{max}, \tag{7.61}$$

$$\forall k \in K \begin{cases} \varphi_k(t) = \min\limits_{x^k} \left[\sum_{a\in A_1} (c_a + t_a)\, x_a^k + \sum_{a\in A_2} c_a x_a^k\right], \\ \textit{subject to} \\ \sum\limits_{a\in\{i^+\}} x_a^k - \sum\limits_{a\in\{i^-\}} x_a^k = b_a^k, \ \forall i \in N, k \in K, \\ x_a^k \ge 0, \ \forall a \in A, \ k \in K, \end{cases} \tag{7.62}$$

$$\sum_{k\in K} x_a^k \le q_a, \quad \forall a \in A. \tag{7.63}$$

In this (optimistic) formulation, both the toll and flow variables are controlled by the leader (the toll variables directly, the flow variables implicitly). On the other hand, the lower level constraints reflect the followers' intention to minimize their total "transportation costs", in terms of "time delay units" multiplied by the corresponding flow values, under current toll levels, and subject to the supply-demand requirements.

In order to prevent the occurrence of trivial situations, the following conditions are assumed in the same manner as in [88]:

1. For a certain amount of goods, demand from one node to another can be sent by arcs or paths that may be toll-free, depend on tolls, or combinations of both.
2. There is a transportation cost associated with each arc that is expressed as a cost at the lower level.
3. There is no profitable vector that induces a negative cost cycle in the network. This condition is satisfied if, for example, all delays c_a are nonnegative.
4. For each commodity, there exists at least one path composed solely of toll-free arcs.

7.2.3 The Algorithm

To find a solution of the TOP, we develop an algorithm dealing with the bilevel mathematical optimization model (7.60)–(7.63) starting from initial values t_a of tolls. With any toll vector fixed, we may treat the lower level problem as a linear program. After solving the latter by the simplex method, we perform sensitivity analysis for the lower level objective function. In the TOP analyzed here, the sum of the objective functions of all followers can be selected as the objective function in the lower level problem, see, Kalashnikov et al. [166]. If the analysis tells us that the current solution is a local maximum point for the upper level problem (this is so if sensitivity analysis does not allow to increase the coefficients of the basic flows along the toll arcs), we use the "filled functions" technique (described in Sect. 7.2.3.2; cf. e.g. Wu et al. [318, 319]) for the objective function of the leader. This allows us to make a "jump" to a neighborhood of another possible local maximum point, if the latter exists.

Once we have a new set of tolls, we proceed to solve the problem of the followers again and perform sensitivity analysis. If that does not allow more increases, we use the "filled functions" method again.

This procedure allows one to get an increase in the toll if the next local maximum is better; otherwise, after several fruitless attempts in a row, we stop with the last solution as approximately optimal.

7.2.3.1 Description of the Heuristic Algorithm

In this algorithm, we are going to combine the main structure of the method described by Kalashnikov et al. [173] and a new idea consisting in the following: A direct procedure may be represented as determination of the "fastest increase" direction for the upper level objective function in terms of the toll variables variations. The "formal gradient" of this objective function F from (7.60) can be determined by the current total flows along the toll arcs:

$$\frac{\partial F}{\partial t_a}(t, x) = \sum_{k \in K} x_a^k, \quad \forall a \in A_1. \tag{7.64}$$

We call it the "formal gradient" because the followers' optimal response is not taken into account in (7.64). However, as the fastest infinitesimal improvement direction, this vector can be used in our heuristic method. The possibility of solving a linear optimization problem at the lower level instead of the Nash equilibrium problem (7.60)–(7.63) has been justified in the papers [166, 173] by Kalashnikov et al.

In what follows, we present a description of the heuristic method proposed first by Kalashnikov et al. in [173] for solving the *congestion-free* case for the bilevel TOP, i.e., $q_a = +\infty, \forall a \in A$. However, the same algorithm can be also applied to solve the bilevel TOP problem with restricted capacities. This is justified by the following theoretical result.

7.2.3.2 A Simple Method to Solve a Special Generalized Nash Equilibrium Problem with Separable Payoffs

Consider a mapping $\Phi : X \to R^N$, where $X = X_1 \times X_2 \times \cdots \times X_N$ is a direct product of m subsets of Euclidean spaces: namely, $X_i \subset R^{n_i}, i = 1, \ldots, N$. Assume that the mapping Φ is *separable* in the sense that each of its components is restricted to its own domain, i.e., $\Phi_i : X_i \to R, i = 1, \ldots, N$. In other words, no two components of the mapping Φ share common variables. Many applied problems boast the latter property: cf. for example, the lower level of the Toll Optimization Problem, namely, the (generalized) Nash equilibrium problem (7.62)–(7.63).

Let us also consider two other mappings $G : X \to R^n$ and $H : X \to R^m$, which are *not necessarily* separable like the mapping Φ. Finally, let Ω be a subset of X defined as follows:

$$\Omega = \{x \in X : \ G(x) \leq 0, \ H(x) = 0\}. \tag{7.65}$$

Now assume that we search for a *generalized Nash equilibrium* (GNE): Find a vector $x^* = \left(x_1^*, \ldots, x_i^*, \ldots, x_N^*\right) \in \Omega$ such that for every player $i = 1, \ldots, N$, the corresponding sub-vector $x_i^* \in X_i$ provides a point of a (global) *maximum* of its utility function (payoff) Φ_i over the subset $\Omega_i\left(x_{-i}^*\right) \subset X_i$ defined as follows:

$$\Omega_i\left(x_{-i}^*\right) = \left\{x_i \in X_i \text{ such that } \left(x_1^*, \ldots, x_{i-1}^*, x_i, x_{i+1}^*, \ldots, x_N^*\right) \in \Omega\right\}. \tag{7.66}$$

Here, $x_{-i} = (x_1, \ldots, x_{i-1}, x_{i+1}, \ldots, x_N) \in X_1 \times \cdots \times X_{i-1} \times X_{i+1} \times \cdots \times X_N$ is the complement to the vector $x_i \in X_i$ in the direct product X. In mathematical terms, what we seek is the following:

$$\Phi_i\left(x_i^*\right) = \max\left\{\Phi_i(x_i) : \ x_i \in \Omega_i\left(x_{-i}^*\right)\right\}, \quad \text{for all } \ i = 1, \ldots, N. \tag{7.67}$$

In what follows, we always suppose that

$$\Omega_i\left(x_{-i}\right) \neq \emptyset, \quad i = 1, \ldots, N, \tag{7.68}$$

for any $x \in \Omega$, i.e., each feasible solution of our GNE problem (7.65)–(7.67).

Now consider the following (scalar) mathematical optimization (MP) problem:

$$\varphi(x) \equiv \sum_{i=1}^{N} \Phi_i\left(x_i\right) \longrightarrow \max_{x \in \Omega}. \tag{7.69}$$

We are now in a position to state and prove the main result of this subsubsection:

Lemma 7.1 *Any solution of MP problem* (7.69) *is a generalized Nash equilibrium (GNE), i.e., a solution of problem* (7.65)–(7.67).

Proof Assume that a vector $x^* = \left(x_1^*, \ldots, x_i^*, \ldots, x_N^*\right) \in \Omega$ solves problem (7.69). On the contrary, suppose that it is *not* an equilibrium for model (7.65)–(7.67). The latter means that for at least one player $i \in \{1, \ldots, N\}$, there exists another sub-vector $\bar{x}_i \in \Omega_i\left(x_{-i}^*\right)$ such that

$$\Phi_i\left(\bar{x}_i\right) > \Phi_i\left(x_i^*\right). \tag{7.70}$$

Now the mapping Φ being separable immediately implies the relationships

$$\begin{aligned} \varphi\left(\bar{x}\right) &= \sum_{j \neq i} \Phi_j\left(x_j^*\right) + \Phi_i\left(\bar{x}_i\right) > \sum_{j \neq i} \Phi_j\left(x_j^*\right) + \Phi_i\left(x_i^*\right) \\ &= \sum_{i=1}^{N} \Phi_i\left(x_i^*\right) = \varphi\left(x^*\right), \end{aligned} \tag{7.71}$$

where $\bar{x} = \left(x_1^*, \ldots, x_{i-1}^*, \bar{x}_i, x_{i+1}^*, \ldots, x_N^*\right) \in \Omega$. However, (7.71) means that $\varphi\left(\bar{x}\right) > \varphi\left(x^*\right)$, which contradicts the assumption that the above vector $x^* = \left(x_1^*, \ldots, x_i^*, \ldots, x_N^*\right) \in \Omega$ solves problem (7.69) and thus completes the proof. □

Remark 7.3 The result of Lemma 7.1 was obtained by Kalashnikov et al. in [166] in a bit more particular setting.

Now we return to the heuristic algorithm's description. Lemma 7.1 proved above permits one to justify Step 1 of the algorithm in question.

Algorithm: **`Step 1.`** `Set` $i = 0$. `Select` $t_a^{(i)} = t_a^{min} = 0$ `and minimize the aggregate lower level objective function`

$$h_{sum}(x) = \sum_{k \in K} \left[\sum_{a \in A_1} \left(c_a + t_a^{(i)}\right) x_a^k + \sum_{a \in A_2} c_a x_a^k \right], \tag{7.72}$$

subject to the flow conservation constraints and nonnegativity restrictions listed in (7.62) as well as the capacity constraints (7.63) in order to obtain the lower level's optimal response $x\left(t^{(i)}\right)$. Compute the leader's objective function value

$$F\left(t^{(i)}, x\left(\left(t^{(i)}\right)\right)\right) = \sum_{k\in K}\sum_{a\in A_1} t_a^{(i)} x_a^k. \tag{7.73}$$

If $i \geq 1$ then compare the upper level objective function value (7.73) with the same for the previous value of i, and if $F\left(t^{(i)}, x\left(\left(t^{(i)}\right)\right)\right) > F\left(t^{(i-1)}, x\left(\left(t^{(i-1)}\right)\right)\right)$ go to Step 2. Otherwise, go to Step 4. If this return to Step 4 from Step 1 occurs several times in a row (7 to 10), go to Step 5.

Step 2. Considering the *allowable ranges to stay optimal* (ARSO) given by the sensitivity analysis table obtained upon having solved the problem presented in Step 1, select the maximum increase parameters $\Delta_a^{k,+}$ for the (toll-arc) variables $x_a^k\left(t_a^{(i)}\right)$, $a \in A_1$. Denote

$$A_1^+ = \left\{a \in A_1 : \sum_{k\in K} x_a^k\left(t_a^{(i)}\right) > 0\right\}, \tag{7.74}$$

that is, the toll arcs with a positive current flow. According to (7.64) , these positive values are (nonzero) components of the "formal gradient" vector of the leader's objective function. If $A_1^+ = \emptyset$, then go to Step 4; otherwise, go to Step 3.

Step 3. The toll increment procedure can be implemented in two different ways. The first (more precautious) one consists in increasing the current toll value by the maximum allowable increment $\Delta_a^{k,+}$, $a \in A_1^+$, but not exceeding the corresponding component of the "formal gradient". More precisely, we set

$$t_a^{(i+1)} = \begin{cases} \min\left\{t_a^{max}, t_a^{(i)} + \max_{k\in K}\min\left\{\sum_{m\in K} x_a^m\left(t_a^{(i)}\right), \Delta_a^{k,+}\right\}\right\}, & \text{if } a \in A_1^+; \\ t_a^{(i)}, & \text{otherwise.} \end{cases} \tag{7.75}$$

The second mode of computing the toll increment is determined by the combination of the allowable increase values:

$$t_a^{(i+1)} = \begin{cases} \min\left\{ t_a^{max}, t_a^{(i)} + \sum\limits_{k\in K} \beta_k \min\left\{ \sum\limits_{m\in K} x_a^m\left(t_a^{(i)}\right), \Delta_a^{k,+}\right\}\right\}, & \text{if } a \in A_1^+; \\ t_a^{(i)}, & \text{otherwise.} \end{cases} \quad (7.76)$$

Here, the nonnegative coefficients $\beta_k \geq 0$, $k \in K$, and such that $\sum\limits_{k\in K} \beta_k = 1$, can be selected by the well-known 100-percent rule of sensitivity analysis. Next, if $t_a^{(i+1)} > t_a^{(i)}$ for at least one $a \in A_1^+$, then update $i := i + 1$ and close the loop by returning to Step 1 to minimize the lower level aggregate objective function with the updated toll values. Otherwise, i.e., if no toll has been increased, go to Step 4.

Step 4. The current set of tolls $\left\{t_a^{(i)}\right\}_{a\in A_1}$ apparently provides for a local maximum of the leader's objective function. In order to try to "jump" to some other possible local maximum solution, apply the "filled functions" technique described briefly in the next subsection. Then return to Step 1 and minimize the lower level aggregate objective function with the updated toll values.

Step 5. If, after a number of Steps 4 repeated (in our numerical experiments, we accepted this number as 7 to 10), one cannot improve the leader's objective function value, stop the algorithm, report the current vectors $\left\{t_a^{(i)}\right\}_{a\in A_1}$ and $x\left(t^{(i)}\right)$ as an approximation of a global optimum solution.

7.2.3.3 Application of the "Filled Functions" Technique

Our heuristic algorithm based upon sensitivity analysis also involves application of the "filled function" technique first proposed by Renpu [267]. This method works, according to the studies in [267], under the assumption that a local minimum of a function, which is continuous and differentiable in R^n, has been found. So the aim is to find another (better than the current) local minimum or determine that this is the global minimum of the function within the closed (polyhedral) constraint set $T \subset R^n$. Renpu [267] and Wu et al. [318, 319] defined "filled functions" for a minimization problem. Here, we adapt several versions of the "filled function" definitions and properties to deal with a maximization problem. For simplicity we assume that any local maximum point of the objective function has a positive value. Of course, the procedure is easily extended to the case where the value of the objective function can be negative, too.

Definition 7.3 Let $\bar{t}_0 \in T$ satisfy $\bar{t}_0 \neq t^*$ and $f\left(\bar{t}_0\right) \geq \frac{3}{4} f\left(t^*\right)$. A continuously differentiable function $P_{t^*}(x)$ is said to be a "filled function" for the maximization problem $\max_{t\in T} f(t)$ at a point t^* with $f\left(t^*\right) > 0$, if

1. t^* is a strict local minimizer of $P_{t^*}(t)$ on T;
2. any local maximizer $\bar{t}$ of $P_{t^*}(t)$ on T satisfies $f\left(\bar{t}\right) > \frac{3}{2} f(t^*)$ or $\bar{t}$ is a vertex of T;
3. any local maximizer $\hat{t}$ of the optimization problem $\max_{t\in T} f(t)$ with $f\left(\hat{t}\right) \geq \frac{7}{4} f(t^*)$ is a local maximizer of $P_{t^*}(t)$ on T;
4. any $\tilde{t} \in T$ with $\nabla P_{t^*}\left(\tilde{t}\right) = 0$ implies $f\left(\tilde{t}\right) > \frac{3}{2} f(t^*)$.

Now, define two auxiliary functions as follows: For any $d = f\left(t_k^*\right) > 0$, and $w = f(t)$, let

$$g_d(w) = \begin{cases} 1, & \text{if } w \geq \frac{1}{2}d; \\ 5 - \frac{48}{d}w + \frac{144}{d^2}w^2 + \frac{128}{d^3}w^3, & \text{if } \frac{1}{4}d \leq w < \frac{1}{2}d; \\ 0, & \text{if } w < \frac{1}{4}d, \end{cases} \tag{7.77}$$

and

$$h_d = \begin{cases} w - \frac{1}{4}d, & \text{if } w \leq \frac{1}{4}d; \\ \left(\frac{16}{d^2} - \frac{128}{d^3}\right)w^3 + \left(\frac{144}{d^2} - \frac{20}{d}\right)w^2 + \left(8 - \frac{48}{d}\right)w + 5 - d, & \text{if } \frac{1}{4}d < w \leq \frac{1}{2}d; \\ 1, & \text{if } \frac{1}{2}d < w \leq \frac{3}{2}d; \\ -\frac{128}{d^3}w^3 + \frac{624}{d^2}w^2 - \frac{1008}{d}w + 541, & \text{if } \frac{3}{2}d < w \leq \frac{7}{2}d; \\ 2, & \text{if } w > \frac{7}{4}d. \end{cases} \tag{7.78}$$

Given a $t^* \in T$ such that $f\left(t^*\right) > 0$, we define the following "filled function":

$$G_{q,t^*}(t) = -\exp\left(-\|t - t^*\|^2\right) g_{\frac{f(t^*)}{4}}\left(f(t)\right) - q h_{\frac{f(t^*)}{4}}\left(f(t)\right), \tag{7.79}$$

where $q > 0$ is a parameter. This "filled function" will be used in our algorithm.

First, based on Wu et al. [318] we have the following result:

Theorem 7.4 *Assume that the function $f : R^n \to R$ is continuously differentiable and there exists a polyhedron $T \subset R^n$ with $t_0 \in T$ such that $f(t) \leq \frac{1}{2} f\left(t_0\right)$ for all $t \in R^n \backslash int\ T$. Let $\bar{t}_0 \neq t^*$ be a point such that $f\left(t^*\right) - \left(\bar{t}_0\right) \leq \frac{1}{4} f\left(t^*\right)$. Then:*

1. *there exists a $q_{t^*}^1 \geq 0$ such that when $q > q_{t^*}^1$, then any local maximizer $\bar{t}$ of the mathematical program $\max_{t\in T} G_{q,t_0^*}(t)$ obtained via the search starting from $\bar{t}_0$ satisfies $\bar{t} \in int\ T$;*
2. *there exists a $q_{t^*}^2 > 0$ such that if $0 < q \leq q_{t^*}^2$, then any stationary point $\tilde{t} \in T$ with $\tilde{t} \neq t^*$ of the function $G_{q,t_0^*}(t)$ satisfies $f\left(\tilde{t}\right) > \frac{3}{2} f\left(t^*\right)$.*

Proof The proof is almost identical to that of Theorem 2.2 in Wu et al. [318]. □

Making use of the auxiliary function (7.79) we can detail the "jump" to a neighborhood of another local maximum point of the upper level objective function F.

Algorithm: Step 1. Let our current toll iteration $t^{(i)}$ be such that formulas (7.75) and (7.76) provide no increase in the toll values. It can be shown that maximization of the auxiliary function (7.79) instead of the original upper level function F is equivalent to a (moderate) increase of the toll parameters $t^{(i)}$ (one or several of them, depending on the mode applied: (7.75) or (7.76)).

Step 2. If the new optimal response $x\left(t^{(i+1)}\right)$ is related to new ARSO upper bounds distinct from zero, return to Step 1 of the algorithm and continue increasing the toll parameters according to formulas (7.75) or (7.76).

Step 3. Otherwise, i.e., if the new ARSO upper bounds are all zero, double the increment of the toll parameters and return to Step 2. If this happens several times without success (i.e., the ARSO upper bounds continue to be zero), go to Step 5 and finish the computational algorithm.

After having defined the above procedures, we are going to illustrate the steps of the combined proposed sensitivity analysis-"filled function" algorithm to solve the TOP.

In Fig. 7.1, we begin by assigning initial values of zero toll cost. After solving the linear optimization problem of the follower to determine the flow in the arcs and obtaining a value for the leader's objective function, sensitivity analysis of the follower is performed, taking into account only toll-arc variables of the current solution. Then having listed the possible increases in the coefficients of the objective function of the follower derived from the sensitivity analysis data, and based upon the formal gradient vector of the upper level objective function F, we update the present toll vector $\left\{t_a^{(i)}\right\}_{a\in A_1}$. When positive increments of t cannot be obtained anymore based on sensitivity analysis and the formal gradient of the function F, apply the "filled function" procedure. A new function is created based on the leader's objective function and a new toll vector is probed. Once there is a new toll vector, go to Step 1 and close the loop. The algorithms stop if neither sensitivity analysis nor the "filled function" method provide a better value for the leader's objective function after several (say,

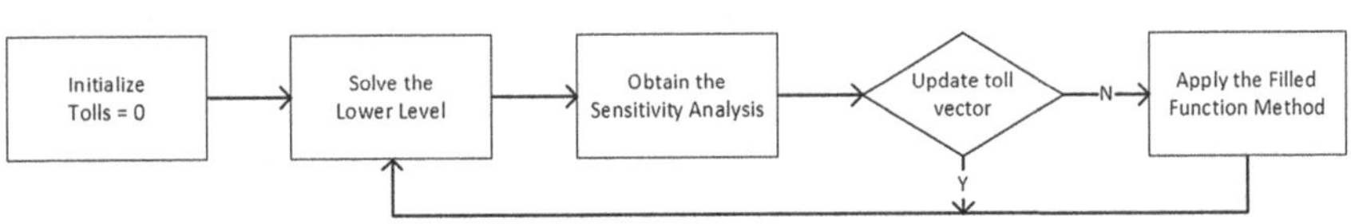

Fig. 7.1 Diagram of the combined method

7–10) attempts in a row, which can mean that an approximate global optimum has been reached, and the algorithm stops. The multicommodity flows corresponding to the final toll values give approximate optimal solutions for the follower, too.

7.2.4 Results of Numerical Experiments

In order to verify the performance of the algorithm, we conducted experiments on two different graphs, each with five different instances. In order to make valid comparisons of the efficiency and computational time of the proposed algorithm we emulated the experimentation conducted by Kalashnikov et al. [166] with their four different proposed algorithms. The following paragraphs describe the environment under which the experimentation was carried out and then describe the methodology used for the application of the algorithm.

In order to check the proposed heuristic sensitivity analysis algorithm combined with the method of "filled functions" (FF), a personal computer was used. The characteristics of the computer equipment used for the development and implementation of the algorithm were: Intel (R) Atom (TM) CPU N455 with a speed 2.00 GHz and 1.67 GB of RAM memory. The coding algorithm was written in the Matlab mathematical software in its version MatLab R2010a. This software was used due to its linear optimization tools in the "Optimization Toolbox". One of the functions used was `linprog` because the lower level of the TOP can be replaced by a corresponding linear optimization problem of the minimum cost flow.

The main parameters of the problems are the ones that define the size of the network: the number of nodes $|N|$, arcs $|A|$, toll arcs $|A_1|$, and commodities $|K|$. Each toll-free arc and toll arc has been assigned a fixed time-delay value c_a generated pseudo-randomly. The problems involved are of small size with two commodities. The graphs and the parameters of the tested instances can be found in the Supplementary material in Sect. 7.2.5. As mentioned above, the sizes of the networks were:

Network 1: 7 nodes, 12 arcs, of which 7 are toll arcs.
Network 2: 25 nodes, 40 arcs, of which 20 are toll arcs.

The results for each example can be seen in the Tables 7.7 to 7.18 below. The first column (called SA + FF) in each table shows the data related to the proposed algorithm, in which the increase in the tolls after sensitivity analysis is conducted in the first mode (cf. 7.75). Analogously, the second column (with the heading SA + FF 100 %) involves the results generated with the developed algorithm, updating the current tolls by formula (7.76). The next four columns show the results obtained after emulating the algorithms proposed in Kalashnikov et al. [166], that is, the Nelder-Mead (NM), Penalization (P), Quasi-Newton (QN), and Gradient (G) methods. The best obtained result is marked in bold.

Tables 7.7 and 7.8 may be a base for the assertion that the approximate solution obtained by all six methods applied to the test problems 1 and 2 are practically the same, which could mean that they are indeed the desired global maximum solutions for the leader.

The possible ways of measuring the algorithms efficiency are: to compare, first, the number of iterations required for each algorithm to reach an approximate solution for a given tolerance value, and second, the average computational cost (the number of iterations necessary on average) to decrease the error by one decimal order. This metric is calculated by the following formula:

$$Cost_{iter} = \frac{\#_{iter}}{\log_{10} \varepsilon_0 - \log_{10} \varepsilon_f}, \tag{7.80}$$

where $\#_{iter}$ denotes the number of iterations needed to reach the optimal value, ε_0 is the initial error computed as the difference between the initial leader's objective function value and the final one reached by the algorithm, this is, $\varepsilon_0 = |F_0 - F^*|$. In a similar manner, ε_f is the approximate final error calculated as the (absolute value of the) difference of the leader's objective function values evaluated at the last two approximate solutions. Tables 7.9 and 7.10 present the total number of iterations required for each algorithm, and Tables 7.11 and 7.12 shows the average cost of the number of iterations required to reduce the error by one order.

Tables 7.9 and 7.10 illustrate that the number of iterations the tested algorithms needed to reach approximately optimal solutions in both test sample problems have

Table 7.7 Leader's objective function value for Network 1

N1	SA + FF	SA + FF 100 %	NM	P	QN	G
1	161.9975	**162.9989**	162.9987	162.8215	162.9972	162.9134
2	274.9905	274.9979	**274.9996**	274.8320	274.9975	274.9321
3	57.98889	**58.9998**	58.9996	58.8719	58.9979	58.9229
4	**155.9806**	156.9980	156.9971	156.8504	156.9988	156.9057
5	136.9888	136.9984	136.9989	**136.8408**	136.9972	136.9750

Table 7.8 Leader's objective function value for Network 2

N2	SA + FF	SA + FF 100 %	NM	P	QN	G
1	1761.488	**1763.984**	1763.876	1762.887	1763.963	1762.629
2	2758.542	**2758.926**	2758.804	2758.237	2758.924	2758.484
3	2364.98	**2367.89**	2365.45	2365.98	2367.82	2365.33
4	3785.41	**3790.99**	3789.24	3790.11	3790.99	3790.18
5	610.99	**611.99**	611.91	610.91	611.97	611.43

Table 7.9 Number of iterations required to solve Network 1

N1	SA + FF	SA + FF 100 %	NM	P	QN	G
1	478	384	277	489	**4**	170
2	510	399	269	484	**8**	397
3	263	195	275	487	**18**	529
4	406	337	164	518	**13**	336
5	276	205	205	469	**10**	108

Table 7.10 Number of iterations required to solve Network 2

N2	SA + FF	SA + FF 100 %	NM	P	QN	G
1	592	**479**	1,412	787	581	745
2	636	546	1,587	685	**496**	812
3	734	411	1,464	633	**374**	596
4	586	497	1,286	549	**324**	893
5	556	418	1,698	591	**309**	650

Table 7.11 Average cost in the number of iteration to reduce the error for Network 1

N1	SA + FF	SA + FF 100 %	NM	P	QN	G
1	197.4995	131.8796	36.9995	348.9920	**6.3082**	41.9907
2	109.1328	80.7525	**38.8531**	284.6465	50.2147	122.6097
3	63.3016	55.5471	43.9853	275.8310	**18.9017**	185.9731
4	160.8014	168.6568	**26.0900**	236.0231	30.4811	108.9364
5	58.8175	42.5245	33.7262	301.8081	**29.4384**	35.7321

Table 7.12 Average cost in the number of iteration to reduce the error for Network 2

N2	SA + FF	SA + FF 100 %	NM	P	QN	G
1	440.6532	356.5420	915.1583	896.1183	285.9969	**239.4804**
2	266.6127	**228.8844**	876.7644	692.3814	277.0817	313.9064
3	465.8475	260.8492	774.8305	818.0137	**186.8402**	291.4508
4	314.3834	266.6357	464.7329	619.5010	**159.6143**	259.2267
5	326.4375	**245.4152**	944.1694	693.3167	360.4346	296.6317

the same order, with a single exception of the Nelder-Mead method. The latter is known to need more iterations in general. The Nelder-Mead method is a derivative-free algorithm, i.e., it does not use even the first derivatives of the upper level objective function.

Table 7.13 Number of objective function values evaluated to solve instances for Network 1

N1	SA + FF	SA + FF 100 %	NM	P	QN	G
1	1,563	1,174	3,249	492	**127**	4,506
2	1,633	807	3,860	487	**231**	10,347
3	544	398	3,948	**490**	507	13,713
4	1,821	1,495	2,367	521	**398**	7,993
5	565	416	2,939	472	**284**	2,142

It seems (from Tables 7.11 and 7.12) that our sensitivity analysis-based algorithms are quite competitive against the other methods when the dimension of the test problem is larger. Such robustness of the procedure may help when dealing with real-life problems, which are usually of larger dimensions.

In Tables 7.13 to 7.16, we also measured the number of values of the upper level objective function calculated during the performance of the algorithms and the average computational cost (measured in the number of objective function evaluations necessary to reduce the error by one decimal order). The evaluation formula used in Tables 7.15 and 7.16 is:

$$Cost_{obj} = \frac{\#_{obj}}{\log_{10}\varepsilon_0 - \log_{10}\varepsilon_f}, \tag{7.81}$$

where $\#_{obj}$ is the number of the leader's objective function evaluations until the algorithm stops.

Again, the proposed sensitivity analysis-based methods performed at a quite high level of efficiency compared to the best (quasi-Newton) algorithm even when the total number of objective function calculations is taken into account, but only for larger problems (see Table 7.14).

According to Tables 7.15 and 7.16, with respect to the *average cost* in the number of values of the leader's objective function calculated to reduce the order of error by 1, our sensitivity analysis-based methods performed better both for small and medium-sized test problems, which is a promising feature.

The last measure we checked in order to compare the algorithms' performance is the computational time that they needed to reach a good approximate solution. It is

Table 7.14 Number of objective function values evaluated to solve instances for Network 2

N2	SA + FF	SA + FF 100 %	NM	P	QN	G
1	4,717	**3,582**	15,224	17,257	4,545	18,311
2	4,840	**3,131**	12,996	14,037	6,183	63,752
3	5,312	**3,378**	9,873	16,779	5,797	49,937
4	4,454	**3,592**	8,486	12,534	4,644	73,227
5	4,210	**3,504**	8,094	13,736	5,292	51,781

Table 7.15 Average cost in the objective functions evaluations for the instances of Network 1

N1	SA + FF	SA + FF 100 %	NM	P	QN	G
1	396.7761	265.1142	529.5666	351.1331	**200.2873**	1113.0009
2	349.4392	**163.3266**	557.5213	286.4108	449.9496	3195.5751
3	130.9357	**112.8034**	631.4701	277.5302	532.3981	4820.8882
4	323.2002	342.1193	376.5560	**237.3900**	933.1916	2591.4543
5	120.4053	**86.2937**	483.5187	303.7387	836.0515	708.6885

Table 7.16 Average cost in the objective functions evaluations for the instances of Network 1

N2	SA + FF	SA + FF 100 %	NM	P	QN	G
1	1138.1844	**864.3155**	6969.9441	2013.1342	1813.1588	15966.7176
2	889.0858	**575.1503**	5949.9076	1403.6637	1876.7644	19564.7359
3	982.0206	**624.4852**	2820.4123	1397.8413	1774.8305	16041.0766
4	793.3171	**639.7833**	5228.1712	2474.2977	1730.7076	21256.8816
5	726.8940	**604.9967**	9441.2869	1151.3150	1810.6673	17110.1938

Table 7.17 Required computational time to solve the instances for Network 1 (in seconds)

N1	SA + FF	SA + FF 100 %	NM	P	QN	G
1	16.8807	13.4929	28.9564	23.4072	**1.2825**	136.8533
2	15.7290	12.2325	31.5437	24.6059	**5.7874**	131.3832
3	17.0548	13.5037	35.8244	22.2121	**6.9523**	111.3348
4	14.4474	11.7556	39.9163	24.4446	**5.3806**	145.5045
5	12.9696	10.2944	35.2639	23.2295	**3.6844**	124.4190

Table 7.18 Required computational time to solve the instances for Network 2 (in seconds)

N2	SA + FF	SA + FF 100 %	NM	P	QN	G
1	311.9787	**263.3140**	549.9019	297.6871	486.6073	657.8525
2	532.5229	444.5805	498.6304	306.3302	**185.2478**	627.73277
3	462.7594	**391.2001**	652.4052	588.3764	562.5629	595.5689
4	279.0327	238.9128	430.6348	255.2499	**133.1982**	751.8234
5	573.2400	**488.5338**	581.2160	575.3600	578.9995	507.3447

important to mention that we emulated the benchmark algorithms, so the required time is going to be valid because we have run all the experiments on the same computer. Tables 7.17 and 7.18 present the time (in seconds) used for each instance and each network.

The corresponding two Tables 7.17 and 7.18 again demonstrated that our algorithms ceded the leading position only to the quasi-Newton method that was proven to be extremely fast when applied to the low-dimensional problems. However, in the higher-dimensional examples, the sensitivity-analysis-based procedure didn't lag behind, even overwhelming all the other methods tested here.

7.2.5 Supplementary Material

In this supplementary material, we present the two networks considered during the experimentations described. In Figs. 7.2 and 7.3, the dotted lines denote the toll arcs, while the regular lines correspond to the toll-free arcs.

Also, we specify the parameters used in the two examples we solved in order to compare the algorithms' performance. Here, we list the travel costs c_a, the demands n^k, the commodities' origin-destination pairs $p = \{(o(k), d(k))\}_{k \in K}$, where $o(k)$ represents the origin node, and $d(k)$ denotes the destination node; $k \in K$, with $|K| = 2$. It is important to mention that in these experiments, we do not restrict the arc capacities.

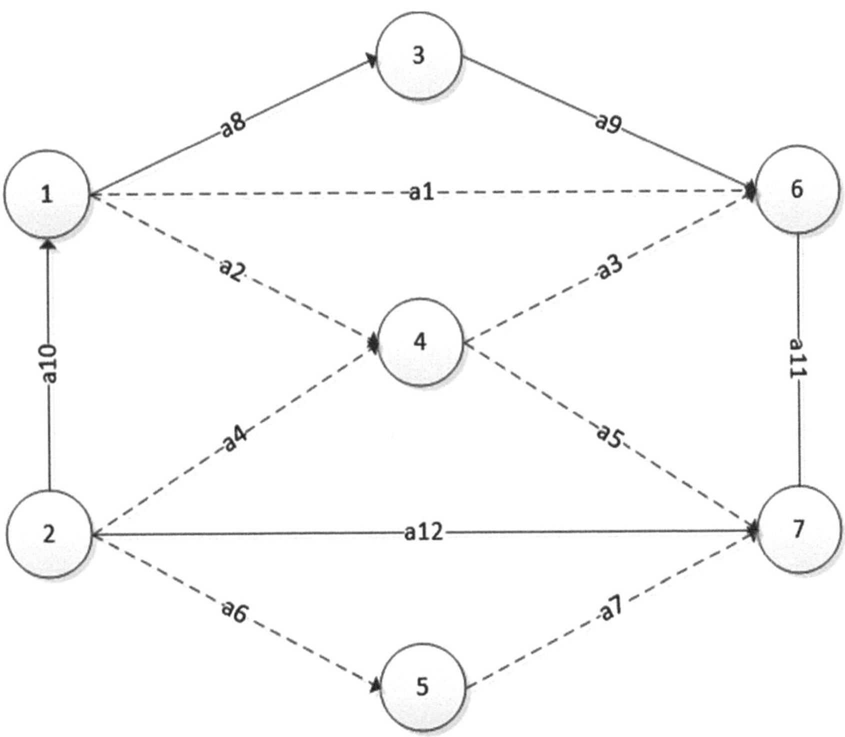

Fig. 7.2 Network 1

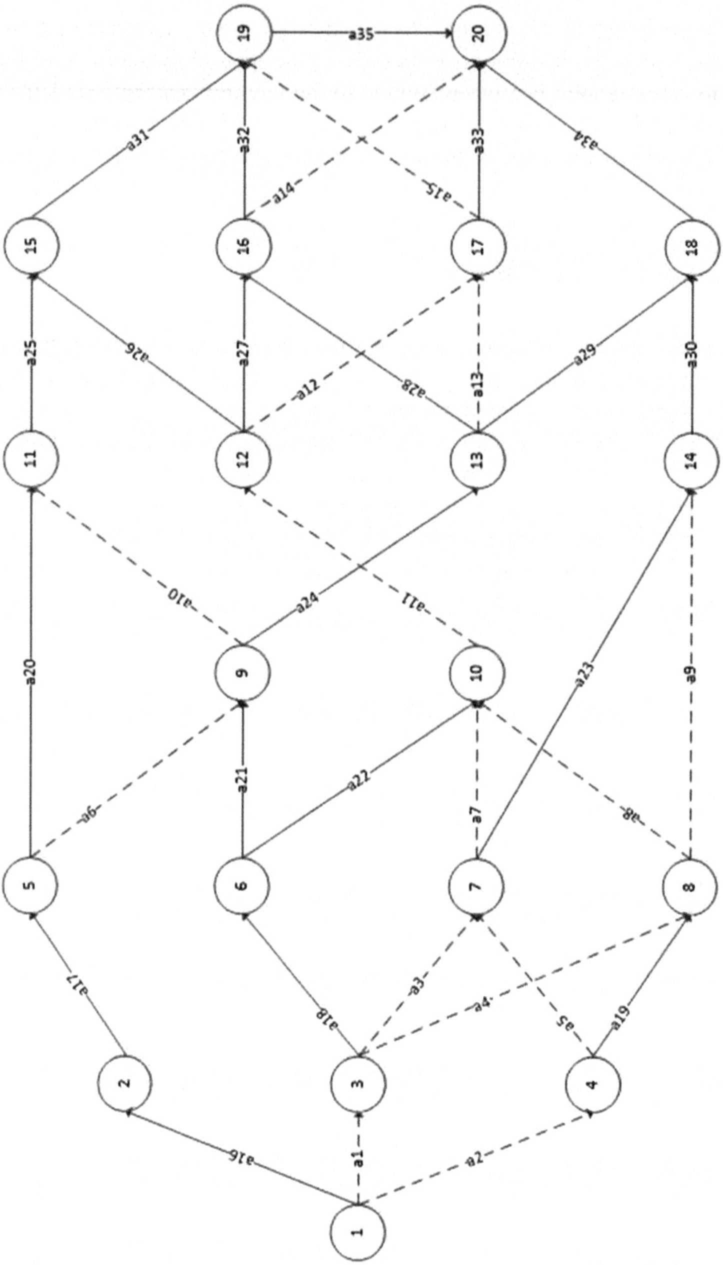

Fig. 7.3 Network 2

Table 7.19 Parameters considered in the instances for Network 1

NN of inst.	Parameters
1	$c = (1, 2, 5, 4, 3, 3, 2, 7, 4, 3, 8, 12)$; $p = \{(1, 6), (2, 7)\}$; $n^k = (10, 9)$
2	$c = (3, 4, 2, 2, 3, 3, 4, 9, 9, 5, 6, 15)$; $p = \{(1, 6), (2, 7)\}$; $n^k = (15, 5)$
3	$c = (4, 3, 2, 1, 1, 3, 2, 5, 6, 3, 1, 5)$; $p = \{(1, 6), (2, 7)\}$; $n^k = (5, 8)$
4	$c = (1, 3, 1, 2, 3, 1, 1, 5, 4, 2, 4, 13)$; $p = \{(1, 6), (2, 7)\}$; $n^k = (5, 12)$
5	$c = (3, 4, 5, 3, 3, 6, 2, 7, 7, 8, 10, 9)$; $p = \{(1, 6), (2, 7)\}$; $n^k = (10, 9)$

Table 7.20 Parameters considered in the instances for Network 2

NN of inst.	Parameters
1	$c = (1, 3, 4, 2, 1, 2, 2, 2, 2, 2, 4, 5, 1, 7, 9, 2, 4, 8, 7, 4, 4, 10, 12, 11, 11, 12, 9, 11, 4, 10, 9, 13, 16, 12, 10, 13, 12, 10, 7, 9)$; $p = \{(1, 12), (2, 19), (2, 25)\}$; $n^k = (12, 24, 30)$
2	$c = (9, 3, 7, 1, 5, 3, 4, 4, 4, 9, 1, 4, 6, 5, 6, 1, 6, 7, 7, 4, 6, 5, 2, 4, 7, 7, 8, 6, 10, 6, 5, 3, 8, 6, 11, 10, 9, 3, 5, 4)$; $p = \{(1, 12), (2, 19), (1, 25)\}$; $n^k = (31, 41, 120)$
3	$c = (4, 8, 1, 7, 3, 9, 5, 5, 2, 7, 6, 6, 4, 9, 5, 5, 9, 5, 1, 4, 9, 5, 1, 4, 9, 3, 9, 1, 8, 4, 6, 3, 9, 1, 1, 1, 2, 5, 1, 10)$; $p = \{(2, 23), (2, 19), (1, 12)\}$; $n^k = (48, 50, 31)$
4	$c = (1, 5, 2, 6, 3, 5, 2, 3, 7, 2, 5, 1, 6, 9, 3, 1, 3, 8, 1, 1, 10, 8, 9, 11, 6, 7, 10, 7, 2, 7, 7, 6, 9, 10, 6, 10, 5, 8, 5, 9)$; $p = \{(1, 25), (2, 19), (2, 25)\}$; $n^k = (84, 45, 71)$
5	$c = (4, 3, 6, 4, 4, 3, 2, 3, 3, 2, 7, 3, 4, 5, 7, 1, 6, 4, 4, 5, 7, 3, 5, 10, 10, 9, 10, 10, 10, 7, 7, 8, 11, 10, 10, 8, 8, 9)$; $p = \{(1, 25), (2, 23), (2, 25)\}$; $n^k = (10, 6, 8)$

First, we show the topology of Network 1 represented with a graph with 12 arcs and 7 nodes. For the two commodities transported within this network, we cite the parameters of the TOP problem.

The values of parameters used in the instances for Network 1 are listed in Table 7.19.

Finally, we describe Network 2, which consists in 25 nodes, 40 arcs and 3 commodities in Fig. 7.3. Here, again, the dotted lines are toll arcs, the regular lines represent toll-free highways. The values of parameters of the considered instances are collected in Table 7.20 (recall, that here $|K| = 3$).

Chapter 8
Reduction of the Dimension of the Upper Level Problem in a Bilevel Optimization Model

This section deals with a problem of reducing the dimension of the upper level problem in a bilevel optimization model, which is sometimes a crucial parameter when applying stochastic optimization tools to solving such a bilevel problem with uncertainty (cf. Sect. 6.8). Indeed, when a stochastic procedure is based upon generating scenario trees, the number of tree branches/nodes grows exponentially in dependence upon the number of upper level variables and previewed outcomes. If the number of upper level variables is large, then even when only three possible outcomes is previewed, scenario trees grow so rapidly that after 5–6 stages, the solved problems become numerically intractable.

In order to decrease the number of variables governed by the leader at the upper level, we create an artificial follower (in addition to the first follower in the original problem). The new follower is supplied with the objective function coinciding with that of the leader, and part of the originally upper level variables are passed to be controlled by the artificial follower at the lower level. Thus, the lower level problem as a whole is also transformed to become a Nash equilibrium problem confronted by both the original and the new follower. We search conditions to guarantee that the modified and the original bilevel optimization problems share at least one optimal solution.

8.1 Introduction

Bilevel optimization modeling is a new and dynamically developing area of mathematical optimization and game theory. For instance, when we study value chains, the general rule usually is: decisions are made by different parties along the chain, and these parties have often different, even opposed goals. This raises the difficulty of supply chain analysis, because regular optimization techniques (e.g., like linear optimization) cannot be readily applied, so that tweaks and reformulations are often needed (cf. Kalashnikov and Ríos-Mercado [180]).

The latter is the case with the Natural Gas Value Chain. From extraction at the wellheads to the final consumption points (households, power plants, etc.), natural

S. Dempe et al., *Bilevel Programming Problems*, Energy Systems,
DOI 10.1007/978-3-662-45827-3_8

gas goes through several processes and changes ownership many a time. Bilevel optimization is especially relevant in the case of the interaction between a Natural Gas Shipping Company (NGSC) and a Pipeline Operating Company (POC). The first one owns the gas since the moment it becomes a consumption-grade fuel (usually at wellhead/refinement complexes, from now onward called the extraction points) and sells it to Local Distributing Companies (LCD), who own small, city-size pipelines that serve final costumers. Typically, NGSCs neither engage in business with end-users, nor actually handle the natural gas physically.

Whenever the volumes extracted by the NGSCs differ from those stipulated in the contracts, we say an imbalance occurs. Since imbalances are inevitable and necessary in a healthy industry, the POC is allowed to apply control mechanisms in order to avoid and discourage abusive practices (the so called arbitrage) on part of the NGSCs. One of such tools is cash-out penalization techniques after a given operative period. Namely, if a NGSC has created imbalances in one or more pool zones, then the POC may proceed to "move" gas from positive-imbalanced zones to negative-imbalanced ones, up to the point where every pool zone has the imbalance of the same sign, i.e., either all non-negative or all non-positive thus, rebalancing the network. At this point, the POC will either charge the NGSC a higher (than the spot) price for each volume unit of natural gas withdrawn in excess from its facilities, or pay back a lower (than the sale) price, if the gas was not extracted.

Prices as a relevant factor induce us into the area of stochastic optimization instead of the deterministic approach. The formulated bilevel problem is reduced to another bilevel one but with linear constraints at both levels (cf. Kalashnikov et al. [179]). However, this reduction involves introduction of many artificial variables, on the one hand, and generation of a lot of scenarios to apply the essentially stochastic tools, on the other hand. The latter makes the dimension of the upper level problem a simply unbearable burden even for the most modern and powerful PC systems.

The aim of this section is a mathematical formalization of the task of reduction of the upper level problem's dimension without affecting (if possible!) the optimal solution of the original bilevel optimization problem. In its main part, we follow the previous papers by Kalashnikov et al. [170, 171].

8.2 An Example

We start with an example. Consider the following bilevel (linear) optimization problem (**P1**):

$$F(x, y, z) = x - 2y + z \to \min_{x,y,z}$$

subject to

$$x + y + z \geq 15,$$

$$0 \leq x, y, z \leq 10,$$

$$z \in \Psi(x, y) = \operatorname*{Argmin}_{z}\{f_2(x, y, z) = 2x - y + z : x + y - z \leq 5,\ 0 \leq z \leq 10\}.$$

It is easy to check that problem (P1) has a unique optimal solution $(x^*, y^*, z^*) = (0, 10, 5)$, with $F(x^*, y^*, z^*) = -15$. By the way, the lower level optimal value $f_2(x^*, y^*, z^*) = -5$.

Now, let us construct a modified problem (**MP1**), which is a bilevel optimization problem with two followers acting according to a Nash equilibrium between them:

$$F(x, y, z) = x - 2y + z \to \min_{x,y,z}$$

subject to

$$x + y + z \geq 15,$$

$$0 \leq x, y, z \leq 10,$$

$$(y, z) \in \Phi(x)$$

where $\Phi(x) = \{(y, z)$ solving the lower level equilibrium problem

Find a Nash equilibrium between two followers:

(1) Follower 1 has to solve the problem:

$$y \in \operatorname*{Argmin}_y\{f_1(x, y, z) = x - 2y + z : x + y - z \leq 5,\ 0 \leq y \leq 10\}$$

(2) Follower 2 has to solve the problem:

$$z \in \operatorname*{Argmin}_z\{f_2(x, y, z) = 2x - y + z : x + y - z \leq 5,\ 0 \leq z \leq 10\}\}.$$

In other words, in problem (MP1), the leader controls directly only the variable x, whereas the lower level is represented with an equilibrium problem. In the latter, there are two decision makers: the second one is the same follower from problem (P1), she/he controls the variable z, accepts the value of the leader's variable x as a parameter, and tries to reach a Nash equilibrium with the first follower, who actually also aims at finding the equilibrium with the second follower by controlling only the variable y and taking the value of the leader's variable x as a parameter. In certain sense, follower 1 is a "reflexion" of the leader, because her/his objective function is the leader's objective function's projection onto the space $\mathbb{R}^2$ of the variables (y, z) for each fixed value of the variable x.

Now it is not difficult to demonstrate that problem (MP1) is also solvable and has exactly the same solution as problem (P1): $(x^*, y^*, z^*) = (0, 10, 5)$ with optimal objective function value $F(x^*, y^*, z^*) = -15$. By the way, the lower level equilibrium problem has the optimal solution $y^* = y^*(x) = 10$, $z^* = z^*(x) = \min\{10, 5 + x\}$ for each value $0 \leq x \leq 10$ of the leader's upper level variable. Of course, the optimal value $x^* = 0$ provides for the minimum value of the upper level objective function F.

8.3 Relations Between Bilevel Problems (P1) and (MP1)

Let us consider a more general form of the bilevel optimization problem under the same name of (P1): Find a vector $(x^*, y^*, z^*) \in X \times Y \times Z \subset \mathbb{R}^{n_1} \times \mathbb{R}^{n_2} \times \mathbb{R}^{n_3}$ solving the following problem:

$$\begin{aligned} F(x, y, z) &\to \min_{x,y,z} \\ (x, y) &\in X \times Y, \\ G(x, y, z) &\leq 0, \\ z \in \Psi(x, y) &= \operatorname*{Argmin}_{z} \{f_2(x, y, z) : z \in Z,\ g(x, y, z) \leq 0.\} \end{aligned} \tag{8.1}$$

Here, $F, f_2 : \mathbb{R}^n \to \mathbb{R}$ and $G : \mathbb{R}^n \to \mathbb{R}^{m_1}$, $g : \mathbb{R}^n \to \mathbb{R}^{m_2}$ are continuous functions with $n = n_1 + n_2 + n_3$, where $n_i,\ i = 1, 2, 3$ and $m_j,\ j = 1, 2$, are fixed natural numbers. As associated to the general problem (P1), let us define the following auxiliary subset:

$$\Phi = \{(x, y) \in X \times Y : \exists z \in Z \text{ such that } g(x, y, z) \leq 0\}. \tag{8.2}$$

Now we make the following assumption:

A1. The subset $\Phi_1 \subseteq \Phi$ comprising all the pairs $(x, y) \in \Phi$ such that there exists a unique vector $z = z(x, y) \in \Psi(x, y)$ satisfying the inequality $G(x, y, z(x, y)) \leq 0$, is nonempty, convex and compact. Moreover, assume that the thus defined mapping $z : \Phi_1 \to R^{n_3}$ is continuous with respect to all variables x and y.

Next, we introduce another bilevel optimization problem: Find a vector $(x^*, y^*, z^*) \in X \times Y \times Z \subset \mathbb{R}^{n_1} \times \mathbb{R}^{n_2} \times \mathbb{R}^{n_3}$ solving the problem (**MP1**) defined as follows:

$$F(x, y, z) \to \min_{(x,y,z) \in X \times Y \times Z} \tag{8.3}$$

$$(y, z) \in \Lambda(x), \tag{8.4}$$

where $\Lambda(x)$ is a collection of generalized Nash equilibrium (GNE) points of the two person game described below. Namely, for any pair $(x, z) \in X \times Z$, player 1 selects its strategies from the set Y and minimizes its payoff function $f_1(x, y, z) \equiv F(x, y, z)$ subject to the constraints $G(x, y, z) \leq 0$ and $g(x, y, z) \leq 0$. Player 2, in its turn considering a pair $(x, y) \in X \times Y$ as fixed, uses the set of strategies Z and minimizes its payoff function $f_2(x, y, z)$ subject to the same constraints $G(x, y, z) \leq 0$ and $g(x, y, z) \leq 0$.

Remark 8.1 It is evident that if a vector $(\overline{y}, \overline{z}) \in Y \times Z$ solves the lower level equilibrium problem of (MP1) for a fixed $x \in X$, then $\overline{z} = z(x, \overline{y})$, where the mapping $z := z(x, y)$ with $z(x, y) \in \Psi(x, y)$ is defined in assumption **A1**. Conversely, if (for a fixed $x \in X$) a vector $\overline{y}$ minimizes the function $\overline{f}_1(y) \equiv f_1(x, y, z(x, y))$

over an appropriate set of vectors y, and in addition, $G(x, \overline{y}, z(x, \overline{y})) \leq 0$, then $(\overline{y}, \overline{z}) = (\overline{y}, z(x, \overline{y}))$ solves the lower level equilibrium problem in (MP1).

We are interested in establishing relationships between the solutions sets of problems (P1) and (MP1). First, we can prove the following auxiliary result.

Theorem 8.1 *Under assumption* **A1**, *there exists a nonempty convex compact subset* $D \subset X$ *such that for all* $x \in D$, *there is a generalized Nash equilibrium (GNE) solution* $(y, z) \in Y \times Z$ *of the lower level equilibrium problem of problem (MP1).*

Proof Consider a projection of the compact convex subset Φ_1 onto the set X:

$$D_1 := Pr_X \Phi_1 = \{x \in X : \exists y \text{ such that } (x, y) \in \Phi_1\}. \tag{8.5}$$

The subset D_1 is clearly nonempty, compact, and convex. Indeed, by assumption **A1**, one has $\Phi_1 \neq \emptyset$. Hence, there exists at least one pair $(x, y) \in \Phi_1$, which immediately implies that $x \in D_1$, i.e., the latter subset is not empty. Next, let $\{x_n\}_{n=1}^{\infty} \subset D_1$ be an arbitrary sequence. By definition (8.5), there are $y_n \in Y$ such that $(x_n, y_n) \in \Phi_1$, $n = 1, 2, \ldots$. Again due to assumption **A1**, the subset Φ_1 is compact, therefore, one can select a convergent subsequence $\left(x_{n_k}, y_{n_k}\right) \to (x, y) \in \Phi_1$ when $k \to \infty$. Definition (8.5) of D_1 implies that the limit point x of the sequence $\{x_n\} \subset D_1$ also belongs to D_1, which means that the latter subset is compact. Finally, we demonstrate that the subset D_1 is convex. Indeed, consider two arbitrary points $x_1, x_2 \in D_1$. By definition (8.5), there exist points $y_1, y_2 \in Y$ such that $(x_1, y_1), (x_2, y_2) \in \Phi_1$. The latter subset's convexity (assumed in **A1**) implies

$$\lambda (x_1, y_1) + (1 - \lambda)(x_2, y_2) \in \Phi_1 \quad \text{for all } \lambda \in [0, 1].$$

This inclusion can be rewritten as

$$(\lambda x_1 + (1 - \lambda)x_2, \lambda y_1 + (1 - \lambda)y_2) \in \Phi_1 \quad \text{for all } \lambda \in [0, 1],$$

which means by definition (8.5) that $\lambda x_1 + (1-\lambda)x_2 \in D_1$ for an arbitrary $\lambda \in [0, 1]$ and thus establishes the latter subset's convexity.

Now select arbitrary $x \in D_1$ and $y \in Y$ such that $(x, y) \in \Phi_1$. According to assumption **A1**, there exists the unique solution $z = z(x, y)$ of the lower level problem (LLP) of the bilevel program (P1), i.e., $z = z(x, y) \in \Psi(x, y)$. In other words, $z = z(x, y)$ is the *optimal response* of the follower to the leader's strategy (x, y):

$$f_2(x, y, z(x, y)) = \min_{z \in Z} \{f_2(x, y, z) : g(x, y, z) \leq 0\}. \tag{8.6}$$

Moreover, by assumption **A1**, this optimal response also satisfies the upper level constraints of problem (P1), i.e.

$$G(x, y, z(x, y)) \leq 0. \tag{8.7}$$

Now consider a section of Φ_1 induced by an arbitrary $x \in D_1$:

$$T(x) = \{y \in Y : (x, y) \in \Phi_1\}. \tag{8.8}$$

Again by definition (8.5) of the subset D_1, section (8.8) is clearly nonempty, i.e., $T(x) \neq \emptyset$. It is also easy to see that $T(x)$ is a compact and convex subset of Y, indeed, select an arbitrary sequence of vectors $\{y_n\}_{n=1}^{\infty} \subset Y$. Since, according to (8.8), this sequence paired with x belongs to the compact Φ_1, that is, $\{(x, y_n)\}_{n=1}^{\infty} \subset \Phi_1$, then it is possible to find a subsequence of pairs converging to an element of Φ_1, i.e., $\left(x, y_{n_k}\right) \to (x, \overline{y}) \in \Phi_1$ as $k \to \infty$. By (8.8), the latter implies $\lim_{k\to\infty} y_{n_k} = \overline{y} \in T(x)$ and thus proves the compactness of $T(x)$. It is also not difficult to verify its convexity: select two arbitrary points $y_1, y_2 \in T(x)$ and consider their convex combination $y = \lambda y_1 + (1-\lambda)y_2$. Repeating the application of definition (8.5), we note that two extended vectors (x, y_1) and (x, y_2) are elements of the convex (by assumption **A1**) subset Φ_1. Hence, their convex combination also belongs to Φ_1, that is, $(x, y) = (x, \lambda y_1 + (1-\lambda)y_2) \in \Phi_1$, which implies, according to (8.8), $y = \lambda y_1 + (1-\lambda)y_2 \in T(x)$ thus demonstrating the convexity of $T(x)$.

Next, it is evident that if there exists an equilibrium strategy $y = y(x)$ of follower 1 at the lower level GNE (general Nash equilibrium) problem in (MP1), it necessarily belongs to the subset $T(x)$. Because $T(x)$ is a nonempty compact and the mapping $z = z(x, y)$ is continuous with respect to y (according to assumption **A1**), there exists a solution $\overline{y} = \overline{y}(x)$ to the mathematical optimization problem

$$\overline{f}_1(y) \equiv f_1(x, y, z(x, y)) \to \min_{y\in T(x)}. \tag{8.9}$$

Now one can conclude (cf. Remark 8.1) that the vector

$$(\overline{y}(x), z(x, \overline{y}(x))) \in T(x) \times \Psi(x, \overline{y}(x))$$

solves the lower level problem of the modified bilevel program (MP1). Therefore, one can select the subset D_1 as the desired set D to complete the proof of the theorem. □

8.4 An Equivalence Theorem

In this subsection, we establish relationships between the optimal solution sets of problems (P1) and (MP1). We start with a rather restrictive assumption concerning problem (MP1), having in mind to relax it in Sect. 8.5.

A2. Assume that the optimal response by follower 1 (as part of the generalized Nash equilibrium (GNE) state) $y = y(x)$, the existence of which for every $x \in D$ follows from Theorem 8.1, is determined uniquely.

Remark 8.2 In assumption **A2**, it would be redundant to demand the uniqueness of the GNE state $\overline{z} = \overline{z}(x)$ since this has been already required implicitly through assumption **A1**. Indeed, if the optimal response $y = y(x)$ of follower 1 is determined uniquely, so is the follower 2's optimal response $\overline{z} = z\,(x, y(x)) = \overline{z}(x)$.

Theorem 8.2 *Under assumptions* **A1** *and* **A2***, problems (P1) and (MP1) are equivalent.*

Proof First, assumption **A1** clearly excludes the possibility for problem (P1) to have a void solution set. In other words, a solution set $\Omega_1 \subset X \times Y \times Z$ for problem (P1) is nonempty. Denote by $\Omega_2 \subset X \times Y \times Z$ the solution set for problem (MP1), then, in order to prove the theorem, it suffices to show that

$$\Omega_1 = \Omega_2. \tag{8.10}$$

We will do it in several steps.
(A) Let $(x^*, y^*, z^*) \in \Omega_1$ be an arbitrary solution of (P1). The definition of a solution to (P1), together with Theorem 8.1, imply that $x^* \in D = D_1$, $(x^*, y^*) \in \Phi_1$, and

$$F\left(x^*, y^*, z^*\right) = \min_{(x,y)\in\Phi_1} F\left(x, y, z(x, y)\right). \tag{8.11}$$

On the other hand, by assumption **A2**, there exists a unique optimal response of follower 1 $\overline{y} = y\,(x^*)$ that is part of the lower level GNE state solving the lower level GNE problem in (MP1). In other words, we have, first,

$$G\left(x^*, \overline{y}\left(x^*\right), z\left(x^*, \overline{y}\left(x^*\right)\right)\right) \le 0,$$

and second,

$$\begin{aligned} F\left(x^*, \overline{y}\left(x^*\right), z\left(x^*, \overline{y}\left(x^*\right)\right)\right) &= f_1\left(x^*, \overline{y}\left(x^*\right), z\left(x^*, \overline{y}\left(x^*\right)\right)\right) \\ &= \min_{y\in T(x^*)} f_1\left(x^*, y, z\left(x^*, y\right)\right). \end{aligned} \tag{8.12}$$

Since $(x^*, \overline{y}\,(x^*)) \in \Phi_1$, the previous relationships (8.11) and (8.12) evidently yield the inequality

$$F\left(x^*, y^*, z^*\right) = F\left(x^*, y^*, z\left(x^*, y^*\right)\right) \le F\left(x^*, \overline{y}\left(x^*\right), z\left(x^*, \overline{y}\left(x^*\right)\right)\right). \tag{8.13}$$

However, the strict inequality $F\,(x^*, y^*, z\,(x^*, y^*)) < F\,(x^*, \overline{y}\,(x^*), z\,(x^*, \overline{y}\,(x^*)))$ is impossible as contradicting the definition of GNE (8.12) because $f_1 \equiv F$ according to the description of the bilevel problem (MP1). Therefore, (8.13) actually is a series of equations

$$F\left(x^*, y^*, z^*\right) = F\left(x^*, y^*, z\left(x^*, y^*\right)\right) = F\left(x^*, \overline{y}\left(x^*\right), z\left(x^*, \overline{y}\left(x^*\right)\right)\right),$$

which implies that the vector $y^* \in T(x^*)$ is also a GNE solution of the lower level equilibrium problem in (MP1). However, by assumption **A2** we have the identity $\overline{y} = y(x^*) = y^*$, that is, the collection (x^*, y^*, z^*) solves the bilevel problem (MP1), i.e., $(x^*, y^*, z^*) \in \Omega_2$. The latter proves the inclusion

$$\Omega_1 \subseteq \Omega_2. \tag{8.14}$$

(B) Conversely, let (x^*, y^*, z^*) be an arbitrary solution of the modified bilevel optimization problem (MP1), in other words, $(x^*, y^*, z^*)\,\Omega_2$. By definition (8.3)–(8.4), first, $(y^*, z^*) \in \Lambda(x^*)$, and second, according to assumptions **A1** and **A2**, one has

$$\begin{aligned} F\left(x^*, y^*, z^*\right) &= \min_{x \in D} F\left(x, \overline{y}(x), z\left(x, \overline{y}(x)\right)\right) \\ &= F\left(x^*, \overline{y}\left(x^*\right), z\left(x^*, \overline{y}\left(x^*\right)\right)\right). \end{aligned} \tag{8.15}$$

Here, $D \subset X$ is the convex compact set whose existence is established in Theorem 8.1. Next, by assumption **A1**, $(x^*, \overline{y}(x^*)) \in \Phi_1$. Let us now assume that there exists a pair $(\tilde{x}, \tilde{y}) \in \Phi_1$ such that

$$\begin{aligned} F\left(\tilde{x}, \tilde{y}, z\left(\tilde{x}, \tilde{y}\right)\right) &= \min_{(x,y) \in \Phi_1} F\left(x, y, z(x, y)\right) \\ &< F\left(x^*, \overline{y}\left(x^*\right), z\left(x^*, \overline{y}\left(x^*\right)\right)\right). \end{aligned} \tag{8.16}$$

Due to assumption **A2**, there exists uniquely a vector $\overline{y}(\tilde{x})$ as part of the solution of the lower level equilibrium problem in (MP1), that is,

$$\begin{aligned} F\left(\tilde{x}, \overline{y}(\tilde{x}), z\left(\tilde{x}, \overline{y}(\tilde{x})\right)\right) &= \min_{y \in T(\tilde{x})} F\left(\tilde{x}, y, z\left(\tilde{x}, y\right)\right) \\ &= F\left(\tilde{x}, \tilde{y}, z\left(\tilde{x}, \tilde{y}\right)\right) \\ &< F\left(x^*, \overline{y}\left(x^*\right), z\left(x^*, \overline{y}\left(x^*\right)\right)\right). \end{aligned} \tag{8.17}$$

Comparing the first and the last terms in the chain of relationships (8.17), we come to a contradiction with the initial assumption about $(x^*, y^*, z^*) \in \Omega_2$. It means that the strict inequality in (8.16) is impossible, which, owing to (8.15), yields

$$\min_{(x,y) \in \Phi_1} F\left(x, y, z(x, y)\right) = F\left(x^*, \overline{y}\left(x^*\right), z\left(x^*, \overline{y}\left(x^*\right)\right)\right) = F\left(x^*, y^*, z^*\right).$$

The last equations immediately establish that (x^*, y^*, z^*) belongs to Ω_1, which provides the reverse inclusion $\Omega_1 \supseteq \Omega_2$ and, combined with (8.14), leads to the desired equality (8.10) thus, completing the proof of the theorem. □

8.5 Examples and Extensions

In this section, we examine linear bilevel programs and find out when assumptions **A1** and **A2** hold in this particular case. Moreover, we try to relax some of the restrictive conditions in these assumptions.

8.5.1 The Nonlinear Case

In order to investigate the conditions under which assumption **A1** holds, we first prove the following auxiliary result.

Lemma 8.1 *Consider a continuous function $F : X \times Z \to \mathbb{R}$, where $X \subseteq \mathbb{R}^n$ and $Z \subseteq \mathbb{R}^m$ are closed convex subsets. Assume that F is (strictly) monotone increasing with respect to each of the variables $z \in Z$. In other words, for each component z_k, $k = 1, \ldots, m$, and any fixed $\overline{x} \in X$ and $\overline{z}_{-k} := (\overline{z}_1, \ldots, \overline{z}_{k-1}, \overline{z}_{k+1}, \ldots, \overline{z}_m)$, one has*

$$F\left(\overline{x}, z_k^1, \overline{z}_{-k}\right) \leq F\left(\overline{x}, z_k^2, \overline{z}_{-k}\right) \text{ if and only if } z_k^1 \leq z_k^2. \tag{8.18}$$

Moreover, for each fixed $x \in X$, let the function $\varphi(z) := F(x, z)$ attain its global minimum over the (closed convex) subset Z at a unique point $z = z(x)$. Then each component $z_k = z_k(x)$, $k = 1, \ldots, m$, of the minimum point $z = z(x)$ is a convex function with respect to $x \in X$.

Proof Select an arbitrary $k \in \{1, \ldots, m\}$ and consider the corresponding component of the minimum-point-mapping $z_k = z_k(x)$ defined over X. Let x^1 and x^2 be two arbitrary points of the latter subset $X \subseteq \mathbb{R}^n$ and $\lambda \in [0, 1]$ an arbitrary parameter value. Then, we are to establish that

$$z_k\left(\lambda x^1 + (1-\lambda)x^2\right) \leq \lambda z_k\left(x^1\right) + (1-\lambda) z_k\left(x^2\right). \tag{8.19}$$

In order to do that, we recall that according to its definition, the vector $z\left(\lambda x^1 + (1-\lambda)x^2\right)$ solves the minimization problem

$$F\left(\lambda x^1 + (1-\lambda)x^2, z\left(\lambda x^1 + (1-\lambda)x^2\right)\right) = \min_{z \in Z} F\left(\lambda x^1 + (1-\lambda)x^2, z\right). \tag{8.20}$$

The convexity of the subset Z implies that $\lambda z\left(x^1\right) + (1-\lambda) z\left(x^2\right) \in Z$. Therefore, (8.20) yields the inequality

$$\begin{aligned} &F\left(\lambda x^1 + (1-\lambda)x^2, z\left(\lambda x^1 + (1-\lambda)x^2\right)\right) \\ &\leq F\left(\lambda x^1 + (1-\lambda)x^2, \lambda z_k\left(x^1\right) + (1-\lambda) z_k\left(x^2\right), z_{-k}\left(\lambda x^1 + (1-\lambda)x^2\right)\right). \end{aligned}$$

Due to the mapping F being monotone non-decreasing with respect to the variable z_k, inequality (8.18), together with the latter, produce the desired inequality (8.19) and thus complete the proof. □

Remark 8.3 The just obtained convexity of the components $z_k = z_k(x)$ of the minimum-point mapping $z = z(x)$, according to the classical result in Rockafellar [272], immediately guarantees that the said mapping $z = z(x)$ is continuous over the (relative) interior of the subset X.

It is not difficult to verify that for a solvable problem (P1) with a compact subset Φ, assumption **A1** always holds if all the components of the mappings G and g are convex continuous functions, and in addition, the upper level constraints mapping G is monotone non-decreasing, and the lower level objective function $f_2 = f_2(x, y, z)$ (strictly) monotone increasing with respect to all components of vector z for each fixed pair of values of (x, y).

Lemma 8.2 *Consider compact subsets X, Y, Z of the corresponding finite-dimensional Euclidean spaces and assume that the subset Φ defined in* (8.2) *is also compact. If the constraints of the solvable bilevel problem (P1) are defined with the mappings G and g having convex continuous components and, in addition, the upper level constraints mapping G is monotone non-decreasing, and the lower level objective function $f_2 = f_2(x, y, z)$ strictly monotone increasing with respect to all components of vector z for each fixed pair of values of (x, y), then assumption* **A1** *holds.*

Proof If problem (P1) is solvable, then the subset Φ defined by (8.2) is convex. Indeed, for any vectors $\left(x^1, y^1\right)$ and $\left(x^2, y^2\right)$ from Φ, and the corresponding z^1, $z^2 \in Z$ such that

$$g\left(x^1, y^1, z^1\right) \leq 0 \quad \text{and} \quad g\left(x^2, y^2, z^2\right) \leq 0, \tag{8.21}$$

and an arbitrary $\lambda \in [0, 1]$ consider their convex combinations $x_\lambda := \lambda x^1 + (1-\lambda)x^2$, $y_\lambda := \lambda y^1 + (1-\lambda)y^2$, and $z_\lambda := \lambda z^1 + (1-\lambda)z^2$. Now the convexity of all components of the mapping g and the subset Z, as well as relationships (8.21) allow one to evaluate

$$g\left(x_\lambda, y_\lambda, z_\lambda\right) \leq \lambda g\left(x^1, y^1, z^1\right) + (1-\lambda) g\left(x^2, y^2, z^2\right) \leq 0$$

thus having established the convexity of the subset Φ.

Hence, the subset

$$\Gamma(x, y) := \{z \in Z : \ g(x, y, z) \leq 0\} \tag{8.22}$$

is a nonempty, compact and convex subset. Because of that, for every fixed pair $(x, y) \in \Phi$, the continuous and strictly monotone increasing function $\varphi(z) := f_2(x, y, z)$ attains its (global) minimum over the subset $\Gamma(x, y)$ at a unique point

$z = z(x, y)$ depending continuously upon the variables $(x, y) \in X \times Y$. Next, the subset Φ_1 introduced in assumption **A1** as the collection of pairs $(x, y) \in \Phi$ such that the triple $(x, y, z(x, y))$ is feasible for the upper level problem, i.e., $G(x, y, z(x, y)) \le 0$, has the following properties. First, it is non-empty (otherwise, problem (P1) would be unsolvable), second, it is compact due to the continuity of the function G and the compactness of the set Φ, and third, it is convex. The latter property is established similarly to the convexity of the subset Φ. Indeed, to show that, it is enough to demonstrate that the mapping $\tilde{G}(x, y) := G(x, y, z(x, y))$ defined over $(x, y) \in \Phi$ has all its components convex as functions $\tilde{G}_i : \Phi \to \mathbb{R}, i = 1, \ldots, m_1$. Select again two arbitrary vectors (x^1, y^1) and (x^2, y^2) from Φ and the corresponding $z^1 = z(x^1, y^1), z^2 = z(x^2, y^2) \in Pr_Z\Phi$ defined (uniquely) as providing the minimum value to the functions $\varphi_1(z) := f_2(x^1, y^1, z)$ and $\varphi_2(z) := f_2(x^2, y^2, z)$, respectively. Again, take an arbitrary $\lambda \in [0, 1]$ to form their convex combinations $x_\lambda := \lambda x^1 + (1 - \lambda)x^2$, $y_\lambda := \lambda y^1 + (1 - \lambda)y^2$, and $z_\lambda := \lambda z^1 + (1 - \lambda)z^2$. By Lemma 8.1, each component $z_k = z_k(x, y)$ of the function $z = z(x, y)$ is convex over the subset Φ, $k = 1, \ldots, n_3$. Therefore, one can deduce the following chain of inequalities based upon the monotony with respect to z and convexity by x, y, z of the mapping G:

$$\begin{aligned}\tilde{G}(x_\lambda, y_\lambda) &\equiv G(x_\lambda, y_\lambda, z(x_\lambda, y_\lambda)) \le G(x_\lambda, y_\lambda, z_\lambda) \le \lambda_1 G\left(x^1, y^1, z^1\right) \\ &\quad + (1 - \lambda)G\left(x^2, y^2, z^2\right) \\ &= \lambda\tilde{G}\left(x^1, y^1\right) + (1 - \lambda)\tilde{G}\left(x^2, y^2\right). \end{aligned} \tag{8.23}$$

Now since the subset Φ_1 is in fact the intersection of the (convex) set Φ and the convex level set $G(x, y, z(x, y)) \equiv \tilde{G}(x, y) \le 0$, hence the subset Φ_1 is convex, too.

Finally, the convexity of the components of the function $z = z(x, y)$ implies their continuity over the subset Φ_1 (possibly, except only its boundary points) and thus completes the proof of the Lemma. □

Assumption **A2** is much more restrictive than **A1**: the uniqueness of a generalized Nash equilibrium (GNE) is indeed a quite rare case. In order to deal with assumption **A2**, we have to suppose additionally that the upper and lower level objective functions are (continuously) differentiable, and moreover, the combined gradient mapping $(\nabla_y F, \nabla_z f_2) : \mathbb{R}^{n_2+n_3} \to \mathbb{R}^{n_2+n_3}$ is strictly monotone (for each fixed vector $x \in X$). In mathematical terms, the latter means that

$$\begin{aligned}&\left\langle\left(\nabla_y F\left(x, y^1, z^1\right), \nabla_z f_2\left(x, y^1, z^1\right)\right) - \left(\nabla_y F\left(x, y^2, z^2\right), \nabla_z f_2\left(x, y^2, z^2\right)\right),\right. \\ &\left.\left(y^1, z^1\right) - \left(y^2, z^2\right)\right\rangle > 0\end{aligned} \tag{8.24}$$

for all $(y^1, z^1) \neq (y^2, z^2)$ from the (convex) set $\Xi = \Xi(x)$ defined below:

$$\Xi = \Xi(x) = \{(y, z) \in Y \times Z : \ G(x, y, z) \le 0 \ \text{and} \ g(x, y, z) \le 0\}, \tag{8.25}$$

which is assumed to be nonempty for some subset $K \subseteq X$. Then it is well-known (cf. Kinderlehrer and Stampachhia [187]) that for every $x \in K$, there exists a unique generalized Nash equilibrium (GNE) $(y(x), z(x))$ of the lower level problem (LLP) in (MP1), which can be found as a (unique) solution of the corresponding variational inequality problem: Find a vector $(y(x), z(x)) \in \Xi(x)$ such that

$$(y - y(x))^\top \nabla_y F(x, y(x), z(x)) + (z - z(x))^\top \nabla_z f_2(x, y(x), z(x)) \geq 0 \quad (8.26)$$

for all $(y, z) \in \Xi(x)$.

8.5.2 The Linear Case

In the linear case, when all the objective functions and the components of constraints are linear functions, the situation with providing that assumptions **A1** and **A2** hold, is a bit different. For assumption **A1** to be valid, again, it is enough to impose conditions guaranteeing the existence of a unique solution of the lower level LP problem $z = z(x, y)$ on a certain compact subset of Z. For instance, the classical conditions will do (cf. Mangasarian [214]).

As for assumption **A2**, here in the linear case, the problem is much more complicated. Indeed, the uniqueness of a generalized Nash equilibrium (GNE) at the lower level of (MP1) is a much too restrictive demand. As was shown by Rosen [275], the uniqueness of a so-called *normalized* GNE is a rather more realistic assumption. This idea was further developed later by many authors (cf. Nishimura et al. [254]).

Before we consider the general case, we examine an interesting example (a slightly modified example from Saharidis and Ierapetritou [277]), in which one of the upper level variables accepts only integer values. In other words, the problem studied in this example, is a *mixed-integer bi-level linear optimization problem* (MIBLP).

Let the mixed-integer bilevel linear optimization problem have the following form:

$$F(x, y, z) = -60x - 10y - 7z \to \min_{x,y,z} \quad (8.27)$$

subject to

$$x \in X = \{0; 1\}, \quad y \in [0, 100], \quad z \in [0, 100], \quad (8.28)$$

and

$$f_2(x, y, z) = -60y - 8z \to \min_{z} \quad (8.29)$$

subject to

$$g(x, y, z) := \begin{bmatrix} 10 & 2 & 3 \\ 5 & 3 & 0 \\ 5 & 0 & 1 \end{bmatrix} \begin{bmatrix} x \\ y \\ z \end{bmatrix} - \begin{bmatrix} 225 \\ 230 \\ 85 \end{bmatrix} \leq \begin{bmatrix} 0 \\ 0 \\ 0 \end{bmatrix}. \quad (8.30)$$

We select the mixed-integer bi-level linear program (MIBLP) (8.27)–(8.30) as problem (P1). Its modification in comparison to the original example in Saharidis and Ierapetritou [277] consists in moving the lower level variable y (in the original example) up to the upper level in our example. (However, it is curious to notice that the optimal solution of the original example coincides with that of the modified one: $(x^*, y^*, z^*) = \left(1; 75; 21\frac{2}{3}\right)$ in both cases).

It is easy to examine that assumption **A1** holds in this problem: indeed, the lower level problem (8.29)–(8.30) has a unique solution

$$z = z(x, y) = \min\left\{85 - 5x, 75 - \frac{10}{3}x - \frac{2}{3}y\right\}$$

for any pair of feasible values $(x, y) \in \Phi = \{(x, y) : x \in \{0; 1\}, 0 \le y \le 100\}$, which goes in line with the predictions by Mangasarian [214]. However, not all triples $(x, y, z(x, y))$ satisfy the lower level constraints $g(x, y, z(x, y)) \le 0$ thus, the feasible subset $\Phi_1 \subset \Phi$ described in assumption **A1** here becomes

$$\Phi_1 = \left\{(0, y) : 0 \le y \le 76\tfrac{2}{3}\right\} \cup \{(1, y) : 0 \le y \le 55\}, \tag{8.31}$$

with the optimal reaction function

$$z = z(x, y) := \begin{cases} 75 - \frac{2}{3}y, & \text{if } x = 0, \\ 71\frac{2}{3} - \frac{2}{3}y, & \text{if } x = 1. \end{cases} \tag{8.32}$$

Therefore, assumption **A1** would hold completely if the variable x were continuous. However, here the subset $\Phi_1 \subset \Phi$ is nonempty, composed of two compact and convex parts, and the function $z = z(x, y)$ is continuous with respect to the continuous variable y over each of the connected parts of Φ_1. Next, comparing the optimal values of the upper level objective function F over both connected parts of the feasible set Φ_1, we come to the conclusion that the triple $(x^*, y^*, z^*) = \left(1; 75; 21\frac{2}{3}\right)$ is the optimal solution of problem (P1). Indeed, $F\left(1, y_1^*, z_1^*\right) = F\left(1; 75; 21\frac{2}{3}\right) = -1,011\frac{2}{3}$ is strictly less than $F\left(0, y_0^* z_0^*\right) = F\left(0; 7\frac{2}{3}; 23\frac{8}{9}\right) = -933\frac{8}{9}$.

Now consider the modified problem:

$$F(x, y, z) = -60x - 10y - 7z \to \min_{x,y,z} \tag{8.33}$$

subject to

$$x \in X = \{0; 1\}, \quad y \in [0, 100], \quad z \in [0, 100], \tag{8.34}$$

and

$$\begin{cases} f_1(x,y,z) = -60x - 10y - 7z \to \min\limits_{0\le y\le 100}, \qquad f_2(x,y,z) = -60y - 8z \to \min\limits_{0\le z\le 100}, \\ \text{subject to} \\ G(x,y,z) := \begin{bmatrix} 10 & 2 & 3 \\ 5 & 3 & 0 \\ 5 & 0 & 1 \end{bmatrix} \begin{bmatrix} x \\ y \\ z \end{bmatrix} - \begin{bmatrix} 225 \\ 230 \\ 85 \end{bmatrix} \le \begin{bmatrix} 0 \\ 0 \\ 0 \end{bmatrix}. \end{cases} \tag{8.35}$$

We call problem (8.33)–(8.35) the *modified* problem (MP1). It is easy to see that for each value of x, either $x = 0$ or $x = 1$, the lower level problem has a continuous set of generalized Nash equilibria (GNE). Namely, if $x = 0$, then all the GNE points $(y, z) = (y(0), z(0))$ belong to the straight-line interval described by the equation:

$$2y + 3z = 225 \quad \text{with} \quad 0 \le y \le 76\tfrac{2}{3}. \tag{8.36}$$

In a similar manner, another straight-line interval of GNE vectors for $x = 1$, that is $(y, z) = (y(1), z(1))$, can be represented by the linear equation

$$2y + 3z = 215 \quad \text{with} \quad 0 \le y \le 75. \tag{8.37}$$

As it could be expected, the linear upper level objective function F attains its minimum value at the extreme points of the above intervals (8.36) and (8.37), corresponding to the greater value of the variable y:

$$\begin{aligned} F_0^* &= F\left(x_0^*, y_0^*, z_0^*\right) = F\left(0; 76\tfrac{2}{3}; 23\tfrac{8}{9}\right) = -933\tfrac{8}{9}, \\ F_1^* &= F\left(x_1^*, y_1^*, z_1^*\right) = F\left(1; 75; 21\tfrac{2}{3}\right) \ = -1{,}011\tfrac{2}{3}. \end{aligned}$$

Since $F_1^* < F_0^*$, the global optimal solution of problem (MP1) coincides with that of the original problem (P1): $(x^*, y^*, z^*) = \left(1; 75; 21\frac{2}{3}\right)$, although assumption **A2** is clearly not valid in this example. □

8.5.3 Normalized Generalized Nash Equilibrium

Following the line proposed in Rosen [275], we consider the concept of a normalized generalized Nash equilibrium (NGNE) defined below. First of all, we have to make our assumptions more detailed:

A3. We assume that all components $G_j = G_j(x, y, z)$, $j = 1, \ldots, m_1$, and $g_k = g_k(x, y, z)$, $k = 1, \ldots, m_2$, of the mappings G and g, respectively, are convex functions with respect to the variables (y, z). Moreover, for each fixed and feasible $x \in X$, there exists a vector $\left(y^0, z^0\right) = \left(y^0(x), z^0(x)\right) \in Y \times Z$ such that

$$G_j\left(x, y^0(x), z^0(x)\right) < 0, \quad \text{and} \quad g_k\left(x, y^0(x), z^0(x)\right) < 0, \tag{8.38}$$

for every *nonlinear* constraint $G_j(x, y, z) \le 0$ and $g_k(x, y, z) \le 0$, respectively.

Remark 8.4 Inequalities (8.38) in assumption **A3** describe a sufficient (Slater) condition for the Karush-Kuhn-Tucker (KKT) constraint qualification.

We wish to use the differential form of the necessary and sufficient KKT conditions for a constrained minimum. Therefore, we make the following additional assumption:

A4. All components $G_j = G_j(x, y, z)$, $j = 1, \ldots, m_1$, and $g_k = g_k(x, y, z)$, $k = 1, \ldots, m_2$, of the mappings G and g, respectively, possess continuous first derivatives with respect to both y and z for all feasible $(x, y, z) \in X \times Y \times Z$. We also assume that for all feasible points, the payoff function $f_i(x, y, z)$ for the i-th player, $i = 1, 2$, has continuous first derivatives with respect to the corresponding variables controlled by that player.

For our two scalar lower level objective functions in (MP1), namely, $f_i(x, y, z)$, $i = 1, 2$, we denote by $\nabla_y f_1(x, y, z)$ and $\nabla_z f_2(x, y, z)$, respectively, their gradients with respect to players' control variables.

The KKT conditions equivalent to (8.4) can now be stated as follows: First,

$$G(x, y, z) \le 0, \quad \text{and} \quad g(x, y, z) \le 0, \tag{8.39}$$

and there exist $u = (u_1, u_2) \in \mathbb{R}_+^{m_1} \times \mathbb{R}_+^{m_1}$, and $v = (v_1, v_2) \in \mathbb{R}_+^{m_2} \times \mathbb{R}_+^{m_2}$ such that

$$u_i^T G(x, y, z) = 0, \quad \text{and} \quad v_i^T g(x, y, z) = 0, \; i = 1, 2 \tag{8.40}$$

and finally,

$$\begin{cases} f_1(x, y, z) \le f_1(x, w, z) + u_1^T G(x, w, z) + v_1^T g(x, w, z), & \forall w \in Y, \\ f_2(x, y, z) \le f_2(x, y, s) + u_2^T G(x, y, s) + v_2^T g(x, y, s), & \forall s \in Z. \end{cases} \tag{8.41}$$

Since $f_i, i = 1, 2$, and the components of the mappings G and g are convex and differentiable by assumptions **A3** and **A4**, inequalities (8.41) are equivalent to

$$\begin{cases} \nabla_y f_1(x, y, z) + u_1^\top \nabla_y G(x, y, z) + v_1^\top \nabla_y g(x, y, z) = 0, \\ \nabla_z f_2(x, y, z) + u_2^\top \nabla_z G(x, y, z) + v_2^\top \nabla_z g(x, y, z) = 0. \end{cases} \tag{8.42}$$

We will also make use of the following relationships, which hold due to the convexity of the components of G and g. More exactly, for every $\left(y^0, z^0\right), \left(y^1, z^1\right) \in Y \times Z$ and each fixed $x \in X$, we have

$$\begin{cases} G_j\left(x, y^1, z^1\right) - G_j\left(x, y^0, z^0\right) \geq \left(y^1 - y^0, z^1 - z^0\right)^\top \nabla_{(y,z)} G_j\left(x, y^0, z^0\right) \\ = \left(y^1 - y^0\right)^\top \nabla_y G_j\left(x, y^0, z^0\right) + \left(z^1 - z^0\right)^\top \nabla_z G_j\left(x, y^0, z^0\right), j = 1, \ldots, m_1, \\ g_k\left(x, y^1, z^1\right) - g_k\left(x, y^0, z^0\right) \geq \left(y^1 - y^0, z^1 - z^0\right)^\top \nabla_{(y,z)} g_k\left(x, y^0, z^0\right) \\ = \left(y^1 - y^0\right)^\top \nabla_y g_k\left(x, y^0, z^0\right) + \left(z^1 - z^0\right)^\top \nabla_z g_k\left(x, y^0, z^0\right), k = 1, \ldots, m_2. \end{cases} \tag{8.43}$$

A weighted nonnegative sum of the functions $f_i, i = 1, 2$, is defined as follows:

$$\sigma(y, z; x, r) := r_1 f_1(x, y, z) + r_2 f_2(x, y, z), \quad r_i \geq 0, \tag{8.44}$$

for each nonnegative vector $r \in \mathbb{R}^2$. For every fixed $r \in \mathbb{R}^2_+$ and $x \in X$, a related mapping $p = p(y, z; x, r)$ from $\mathbb{R}^{n_2+n_3}$ into itself is defined in terms of the gradients of the functions $f_i, i = 1, 2$, by

$$p(y, z; x, r) = \begin{bmatrix} r_1 \nabla_y f_1(x, y, z) \\ r_2 \nabla_z f_2(x, y, z) \end{bmatrix}. \tag{8.45}$$

Following Rosen [275], we shall call $p(y, z; x, r)$ the *pseudo-gradient* of $\sigma(y, z; x, r)$. An important property of the latter function is given in the following

Definition 8.1 The function $\sigma(y, z; x, r)$ is called *uniformly diagonally strictly convex* for $(y, z) \in Y \times Z$ and fixed $r \geq 0$, if for every fixed $x \in X$ and for any $\left(y^0, z^0\right), \left(y^1, z^1\right) \in Y \times Z$, one has

$$\left(y^1 - y^0\right)^\top \left[p\left(y^1, z^1; x, r\right) - p\left(y^0, z^0; x, r\right)\right] > 0. \tag{8.46}$$

Repeating the arguments similar to those in Rosen [275], we will show later that a sufficient condition for $\sigma(y, z; x, r)$ to be uniformly diagonally strictly convex is that the symmetric matrix $\left[P(y, z; x, r) + P(y, z; x, r)^\top\right]$ is (uniformly with respect to x from X) positive definite for $(y, z) \in Y \times Z$, where $P(y, z; x, r)$ is the Jacobi matrix with respect to (y, z) of the mapping $p = p(y, z; x, r)$.

Again following Rosen [275], we consider a special kind of equilibrium points such that each of the nonnegative multipliers involved in the KKT conditions (8.40)–(8.41) is given by

$$\begin{cases} u_1 = u^0 / r_1 \quad \text{and} \quad v_1 = v^0 / r_1, \\ u_2 = u^0 / r_2 \quad \text{and} \quad v_2 = v^0 / r_2 \end{cases} \tag{8.47}$$

for some $r = (r_1, r_2) > 0$ and $u^0 \geq 0$, $v^0 \geq 0$. Like Rosen in [275], we call this a *normalized generalized Nash equilibrium* (NGNE) point. Now, by slightly modifying the proofs of Theorems 3 and 4 in Rosen [275], we establish the existence and uniqueness results for the NGNE points involved in the modified bilevel optimization problem (MP1).

Theorem 8.3 *Under assumptions* **A3** *and* **A4**, *there exists a normalized generalized Nash equilibrium point for the lower level equilibrium problem* (8.4) *in* (*MP1*) *for every specified* $r = (r_1, r_2) > 0$.

Proof For a fixed vector $r = \overline{r} > 0$, let

$$\rho(x, y, z; w, s; \overline{r}) := \overline{r}_1 f_1(x, w, z) + \overline{r}_2 f_2(x, y, s). \tag{8.48}$$

Consider the feasible set of the equilibrium problem (8.4) for each (fixed) $x \in X$:

$$\Theta(x) := \{(y, z) \in Y \times Z \ \text{ such that } \ G(x, y, z) \le 0 \ \text{ and } \ g(x, y, z) \le 0\} \tag{8.49}$$

and the point-to-set mapping $\Gamma : \Theta(x) \rightrightarrows \Theta(x)$ given by

$$\Gamma(y, z) := \left\{(w, s) \in \Theta(x) : \ \rho(x, y, z; w, s; \overline{r}) = \min_{(q,t)\in\Theta(x)} \rho(x, y, z; q, t; \overline{r})\right\}. \tag{8.50}$$

It follows (by assumptions **A3** and **A4**) from the continuity of the function $\rho = \rho(x, y, z; q, t; \overline{r})$ and its convexity with respect in (q, t) for a fixed (x, y, z), that Γ is an upper semi-continuous mapping that associates each point of the convex, compact set $\Theta(x)$ with a closed compact subset of the same $\Theta(x)$. Then by the Kakutani fixed point theorem, there exists a point $\left(y^0, z^0\right) \in \Theta(x)$ such that $\left(y^0, z^0\right) \in \Gamma\left(y^0, z^0\right)$, or, which is the same,

$$\rho\left(x, y^0, z^0; y^0, z^0; \overline{r}\right) = \min_{(w,s)\in\Theta(x)} \rho\left(x, y^0, z^0; w, s; \overline{r}\right). \tag{8.51}$$

The fixed point $\left(y^0, z^0\right) \in \Theta(x)$ is an equilibrium point solving problem (8.4). Indeed, suppose on the contrary that this point is not an equilibrium. Then, for example, for player 1, there would exist a vector y^1 such that $\left(y^1, z^0\right) \in \Theta(x)$ and $f_1\left(x, y^1, z^0\right) < f_1\left(x, y^0, z^0\right)$. However, in this case, one has $\rho\left(x, y^0, z^0; y^1, z^0; \overline{r}\right) < \rho\left(x, y^0, z^0; y^0, z^0; \overline{r}\right)$, which contradicts (8.51).

Now by the necessity of the KKT conditions, (8.51) implies the existence of $u^0 \in \mathbb{R}_+^{m_1}$ and $v^0 \in \mathbb{R}_+^{m_2}$ such that

$$\left(u^0\right)^T G(x, y, z) = 0, \quad \left(v^0\right)^T g(x, y, z) = 0, \tag{8.52}$$

and

$$\begin{cases} \overline{r}_1 \nabla_y f_1(x, y, z) + \left(u^0\right)^\top \nabla_y G(x, y, z) + \left(v^0\right)^\top \nabla_y g(x, y, z) = 0, \\ \overline{r}_2 \nabla_z f_2(x, y, z) + \left(u^0\right)^\top \nabla_z G(x, y, z) + \left(v^0\right)^\top \nabla_z g(x, y, z) = 0. \end{cases} \tag{8.53}$$

But these are just conditions (8.40) and (8.42) with

$$\begin{cases} u_1 = u^0/r_1 \text{ and } v_1 = v^0/r_1, \\ u_2 = u^0/r_2 \text{ and } v_2 = v^0/r_2, \end{cases}$$

which, together with (8.39), are sufficient to ensure that $\left(y^0, z^0\right) \in \Theta(x)$ satisfies (8.4). Therefore, $\left(y^0, z^0\right)$ is a normalized generalized Nash equilibrium (NGNE) point for the specified value of $r = \overline{r}$. □

Theorem 8.4 *Let assumptions* **A3** *and* **A4** *be valid, and* $\sigma(y, z; x, r)$ *be (uniformly with respect to* $x \in X$*) diagonally strictly convex for every* $r \in Q$*, where* Q *is a convex subset of the positive orthant* $\mathbb{R}^2_+$*. Then for each* $r \in Q$ *there exists a unique normalized generalized Nash equilibrium (NGNE) point.*

Proof The existence of at least one NGNE for any $r \in Q$ is a consequence of Theorem 8.3. Assume that for some $r = \overline{r} \in Q$ there are two distinct NGNE points $\left(y^0, z^0\right) \neq \left(y^1, z^1\right)$ both belonging to $\Theta(x)$. Then we have for $t = 0, 1$, first,

$$G\left(x, y^t, z^t\right) \leq 0 \text{ and } g\left(x, y^t, z^t\right) \leq 0 \tag{8.54}$$

and second, there are vectors $u^t \in \mathbb{R}^{m_1}_+$, $v^t \in \mathbb{R}^{m_2}_+$, such that

$$\left(u^t\right)^T G\left(x, y^t, z^t\right) = 0, \quad \left(v^t\right)^T g\left(x, y^t, z^t\right) = 0, \tag{8.55}$$

and

$$\begin{cases} \overline{r}_1 \nabla_y f_1\left(x, y^t, z^t\right) + \left(u^t\right)^\top \nabla_y G\left(x, y^t, z^t\right) + \left(v^t\right)^\top \nabla_y g\left(x, y^t, z^t\right) = 0, \\ \overline{r}_2 \nabla_z f_2\left(x, y^t, z^t\right) + \left(u^t\right)^\top \nabla_z G\left(x, y^t, z^t\right) + \left(v^t\right)^\top \nabla_z g\left(x, y^t, z^t\right) = 0. \end{cases} \tag{8.56}$$

Now multiply the first row in (8.56) by $\left(y^0 - y^1\right)$ for $t = 0$ and by $\left(y^1 - y^0\right)$ for $t = 1$. In a similar manner, multiply the second row in (8.56) by $\left(z^0 - z^1\right)$ for $t = 0$ and by $\left(z^1 - z^0\right)$ for $t = 1$. Finally, sum all these four terms. That yields the expression $\beta + \gamma = 0$, where

$$\beta = \begin{pmatrix} y^1 - y^0 \\ z^1 - z^0 \end{pmatrix}^\top \left[p\left(y^1, z^1; x, r\right) - p\left(y^0, z^0; x, r\right)\right] \tag{8.57}$$

and

$$\begin{aligned}
\gamma &= \left(u^0\right)^\top \nabla_y G\left(x, y^0, z^0\right)\left(y^0 - y^1\right) + \left(u^1\right)^\top \nabla_y G\left(x, y^1, z^1\right)\left(y^1 - y^0\right) \\
&\quad + \left(v^0\right)^\top \nabla_y g\left(x, y^0, z^0\right)\left(y^0 - y^1\right) + \left(v^1\right)^\top \nabla_y g\left(x, y^1, z^1\right)\left(y^1 - y^0\right) \\
&\quad + \left(u^0\right)^\top \nabla_z G\left(x, y^0, z^0\right)\left(z^0 - z^1\right) + \left(u^1\right)^\top \nabla_z G\left(x, y^1, z^1\right)\left(z^1 - z^0\right) \\
&\quad + \left(v^0\right)^\top \nabla_z g\left(x, y^0, z^0\right)\left(z^0 - z^1\right) + \left(v^1\right)^\top \nabla_z g\left(x, y^1, z^1\right)\left(z^1 - z^0\right) \\
&\geq \left(u^0\right)^\top \left[G\left(x, y^0, z^0\right) - G\left(x, y^1, z^1\right)\right] + \left(u^1\right)^\top \left[G\left(x, y^1, z^1\right) - G\left(x, y^0, z^0\right)\right] \\
&\quad + \left(v^0\right)^\top \left[g\left(x, y^0, z^0\right) - g\left(x, y^1, z^1\right)\right] + \left(v^1\right)^\top \left[g\left(x, y^1, z^1\right) - g\left(x, y^0, z^0\right)\right] \\
&= -\left(u^0\right)^T G\left(x, y^1, z^1\right) - \left(u^1\right)^T G\left(x, y^0, z^0\right) - \left(v^0\right)^T g\left(x, y^1, z^1\right) \\
&\quad - \left(v^1\right)^T g\left(x, y^0, z^0\right) \geq 0.
\end{aligned} \tag{8.58}$$

Next, since $\sigma(y, z; x, r)$ is (uniformly with respect to $x \in X$) diagonally strictly convex we have $\beta > 0$ by (8.46), which, together with (8.58) contradicts the equality $\beta + \gamma = 0$ and thus proves the theorem. □

We complete this section by giving (similarly to Rosen in [275]) a sufficient condition on the functions f_i, $i = 1, 2$, that insures that $\sigma(y, z; x, r)$ is (uniformly by x from X) diagonally strictly convex. The condition is given in terms of the $(n_2 + n_3) \times (n_2 + n_3)$-matrix $P(y, z; x, r)$, which is the Jacobi matrix of $p(y, z; x, r)$ with respect to (y, z) for a fixed $r > 0$. That is, the jth column of $P(y, z; x, r)$ is $\partial p(y, z; x, r)/\partial y_j$, if $1 \leq j \leq n_2$, and $\partial p(y, z; x, r)/\partial z_{j-n_2}$, if $n_2 + 1 \leq j \leq n_2 + n_3$, where the function $p = p(y, z; x, r)$ is defined by (8.45).

Theorem 8.5 *Under assumptions* **A3** *and* **A4**, *a sufficient condition for $\sigma = \sigma(y, z; x, r)$ to be (uniformly with respect to $x \in X$) diagonally strictly convex for $(y, z) \in \Theta(x)$ and a fixed $r = \overline{r} > 0$ is that the symmetric matrix*

$$\left[P(y, z; x, \overline{r}) + P(y, z; x, \overline{r})^\top\right]$$

is (uniformly with respect to $x \in X$) positive definite for $(y, z) \in \Theta(x)$.

Proof Let $\left(y^0, z^0\right), \left(y^1, z^1\right) \in \Theta(x)$, with $\left(y^0, z^0\right) \neq \left(y^1, z^1\right)$, be any two distinct points in $\Theta(x)$, and define $(y(\alpha), z(\alpha)) := \alpha \left(y^1, z^1\right) + (1-\alpha)\left(y^0, z^0\right)$, so that (due to the convexity of $\Theta(x)$) one has $(y(\alpha), z(\alpha)) \in \Theta(x)$ for all $0 \le \alpha \le 1$. Now, since $P(y, z; x, \overline{r})$ is the Jacobi matrix for the mapping $p(y, z; x, \overline{r})$ with respect to the variables (y, z), one has

$$\begin{aligned}\frac{d\, p(y(\alpha), z(\alpha); x, \overline{r})}{d\,\alpha} &= P(y(\alpha), z(\alpha); x, \overline{r}) \frac{d\,(y(\alpha), z(\alpha))}{d\,\alpha} \\ &= P(y(\alpha), z(\alpha); x, \overline{r}) \begin{pmatrix} y^1 - y^0 \\ z^1 - z^0 \end{pmatrix}, \end{aligned} \tag{8.59}$$

which implies

$$p\left(y^1, z^1; x, \overline{r}\right) - p\left(y^0, z^0; x, \overline{r}\right) = \int_0^1 P(y(\alpha), z(\alpha); x, \overline{r}) \begin{pmatrix} y^1 - y^0 \\ z^1 - z^0 \end{pmatrix} d\,\alpha. \tag{8.60}$$

Left-multiplying both sides by $\begin{pmatrix} y^1 - y^0 \\ z^1 - z^0 \end{pmatrix}^\top$ brings one to

$$\begin{aligned} &\begin{pmatrix} y^1 - y^0 \\ z^1 - z^0 \end{pmatrix}^\top \left[p\left(y^1, z^1; x, \overline{r}\right) - p\left(y^0, z^0; x, \overline{r}\right)\right] \\ &= \int_0^1 \begin{pmatrix} y^0 - y^1 \\ z^0 - z^1 \end{pmatrix}^\top P(y(\alpha), z(\alpha); x, \overline{r}) \begin{pmatrix} y^0 - y^1 \\ z^0 - z^1 \end{pmatrix} d\,\alpha \\ &= \frac{1}{2} \int_0^1 \begin{pmatrix} y^0 - y^1 \\ z^0 - z^1 \end{pmatrix}^\top \left[P(y(\alpha), z(\alpha); x, \overline{r}) + P(y(\alpha), z(\alpha); x, \overline{r})^\top\right] \\ &\quad \times \begin{pmatrix} y^0 - y^1 \\ z^0 - z^1 \end{pmatrix} d\,\alpha > 0, \end{aligned}$$

which proves that the relationship (8.46) is true. □

References

1. R.K. Ahuja, T.L. Magnanti, J.B. Orlin, *Network Flows-Theory, Algorithms, and Applications* (Prentice Hall, Englewood Cliffs, 1993)
2. G.B. Allende, G. Still, Solving bilevel programs with the KKT-approach. Math. Program. **138**, 309–332 (2013)
3. M. Anitescu, On using the elastic mode in nonlinear programming approaches to mathematical programs with complementarity constraints. SIAM J. Optim. **15**, 1203–1236 (2006)
4. R. Arnott, A. de Palma, R. Lindsey, Depurate time and route choice for the morning commute. Transp. Res. **24**, 209–228 (1990)
5. D. Aussel, R. Correa, M. Marechal, *Spot Electricity Markets with Transmission Losses*, Technical Report, Université de Perpignan, France (2011)
6. H. Babbahadda, N. Gadhi, Necessary optimality conditions for bilevel optimization problems using convexificators. J. Glob. Optim. **34**, 535–549 (2006)
7. T. Başar, G.J. Olsder, *Dynamic Non-cooperative Game Theory* (Academic Press, New York, 1982)
8. B. Bank, J. Guddat, D. Klatte, B. Kummer, K. Tammer, *Non-linear Parametric Optimization* (Birkhäuser Verlag, Basel, 1983)
9. B. Bank, R. Hansel, Stability of mixed-integer quadratic programming problems. Math. Program. Stud. **21**, 1–17 (1984)
10. J.F. Bard, *Practical Bilevel Optimization: Algorithms and Applications* (Kluwer Academie Publishers, Dordrecht, 1998)
11. J.F. Bard, J. Falk, An explicit solution to the multi-level programming problem. Comput. Oper. Res. **9**, 77–100 (1982)
12. M.S. Bazaraa, C.M. Shetty, *Nonlinear Programming-Theory and Algorithms* (Wiley, New York, 1979)
13. M.J. Beckmann, On optimal tolls for highway tunnels and bridges, in *Vehicular Traffic Science*, eds. by R. Herman, L. Edie, R. Rothery (Elsevier, New York, 1965), pp. 331–341
14. C.R. Bector, S. Chandra, J. Dutta, *Principles of Optimization Theory* (Alpha Science, UK, 2005)
15. K. Beer, *Lösung Großer Linearer Optimierungsaufgaben* (Deutscher Verlag der Wissenschaften, Berlin, 1977)
16. J.G. Beierlein, J.W. Dunn, J.C. McConnon-Jr, The demand for electricity and natural gas in the northeastern united states. Rev. Econ. Stat. **63**(81), 403–408 (2014)
17. M.G.H. Bell, Y. Iida, *Transportation Network Analysis* (Wiley, Chichester, 1997)
18. O. Ben-Ayed, Bilevel linear programming. Comput. Oper. Res. **20**(5), 485–501 (1993)

S. Dempe et al., *Bilevel Programming Problems*, Energy Systems,
DOI 10.1007/978-3-662-45827-3

19. K.P. Bennett, J. Hu, G. Kunapuli, J.-S. Pang, Model selection via bilevel optimization, International Joint Conference on Neural Networks (IJCNN 2006), 2006, pp. 1922–1929
20. K.P. Bennett, G. Kunapuli, J. Hu, J.-S. Pang, Bilevel optimization and machine learning, in *Computational Intelligence: Research Frontiers*, Lecture Notes in Computer Science, ed. by J.M. Zurada, G.G. Yen, J. Wang (Springer, Berlin, 2008), pp. 25–47
21. A. Berman, D. Plemmons, *Nonnegative Matrices in Mathematical Sciences* (Academic Press, New York, 1979)
22. J.H. Bigelow, N.Z. Shapiro, Implicit function theorems for mathematical programming and for systems of inequalities. Math. Program. **6**, 141–156 (1974)
23. J.F. Bonnans, A. Shapiro, *Perturbation Analysis of Optimization Problems* (Springer, New York, 2000)
24. D. Bös, *Public Enterprise Economics* (North-Holland, Amsterdam, 1986)
25. D. Bös, *Privatization: A Theoretical Treatment* (Clarendon Press, Oxford, 1991)
26. A.L. Bowleyl, *The Mathematical Groundwork of Economics* (Oxford University Press, Oxford, 1924)
27. J. Bracken, J. McGill, Mathematical programs with optimization problems in the constraints. Oper. Res. **21**, 37–44 (1973)
28. T.F. Bresnahan, Duopoly models with consistent conjectures. Am. Econ. Rev. **71**, 934–945 (1991)
29. P.J. Brockwell, R.A. Davis, *Time Series: Theory and Methods* (Springer, New York, 1991)
30. L. Brotcorne, *Approches opérationnelles et stratégiques des problémes de trafic routier*, Ph.D. thesis, Université Libre de Bruxelle (1998)
31. L. Brotcorne, E. Cirinei, P. Marcotte, G. Savard, A tabu search algorithm for the network pricing problem. Comput. Oper. Res. **39**, 2603–2611 (2012)
32. L. Brotcorne, F. Cirinei, P. Marcotte, G. Savard, An exact algorithm for the network pricing problem. Discret. Optim. **8**, 246–258 (2011)
33. L. Brotcorne, S. Hanafi, R. Mansi, A dynamic programming algorithm for the bilevel Knapsack problem. Oper. Res. Lett. **37**(3), 215–218 (2009)
34. L. Brotcorne, M. Labbé, P. Marcotte, G. Savard, A bilevel model for toll optimization on a multicommodity transportation network. Transp. Sci. **35**, 345–358 (2001)
35. V.A. Bulavsky, An imagined experiment in the framework of the generalized Cournot model. Econ. Math. Methods (Ekonomika i Matematicheskie Metody) **32**, 128–137 (1996). in Russian
36. V.A. Bulavsky, Structure of demand and equilibrium in a model of oligopoly. Econ. Math. Methods (Ekonomika i Matematicheskie Metody) **33**, 112–124 (1997). in Russian
37. V.A. Bulavsky, V.V. Kalashnikov, A one-parametric method to study equilibrium. Econ. Math. Methods (Ekonomika i Matematicheskie Metody) **30**, 123–138 (1994). in Russian
38. V.A. Bulavsky, V.V. Kalashnikov, Equilibrium in generalized Cournot and Stackelberg models. Econ. Math. Methods (Ekonomika i Matematicheskie Metody) **31**, 164–176 (1995). in Russian
39. W. Candler, R. Norton, *Multilevel Programming and Development Policy, Technical Report 258* (World Bank Staff, Washington, 1977)
40. W. Candler, R. Townsley, A linear two-level programming problem. Comput. Oper. Res. **9**, 59–76 (1982)
41. R.W. Chaney, Piecewise C^k functions in nonsmooth analysis. Nonlinear Anal. Methods Appl. **15**, 649–660 (1990)
42. F.H. Clarke, *Optimization and Nonsmooth Analysis* (Wiley, New York, 1983)
43. R.C. Cornes, T. Sandler, Easy riders, joint production, and public goods. Econ. J. **94**, 580–598 (1984)
44. R.C. Cornes, M. Sepahvand, *Cournot Vs Stackelberg Equilibria with a Public Enterprise and International Competition*, Technical Report, University of Nottingham, School of Economics, United Kingdom, Discussion Paper No. 03/12 (2003)
45. C. Cortes, V. Vapnik, Support-vector networks. Mach. Learning **20**, 273–297 (1995)
46. Z. Coulibaly, D. Orban, An ℓ_1 elastic interior-point method for mathematical programs with complementarity constraints. SIAM J. Optim. **22**(1), 187–211 (2012)
47. M.L. Cropper, W. Oates, Environmental economics: a survey. J. Econ. Lit. **30**, 675–740 (1992)

48. V. Demiguel, M.P. Friedlander, F.J. Nogales, S. Scholtes, A two-sided relaxation scheme for mathematical programs with equilibrium constraints. SIAM J. Optim. **16**(2), 587–609 (2005)
49. S. Dempe, A simple algorithm for the linear bilevel programming problem. Optimization **18**, 373–385 (1987)
50. S. Dempe, A necessary and a sufficient optimality condition for bilevel programming problems. Optimization **25**, 341–354 (1992)
51. S. Dempe, A bundle algorithm applied to bilevel programming problems with non-unique lower level solutions. Comput. Optim. Appl. **15**, 145–166 (2000)
52. S. Dempe, *Foundations of Bilevel Programming* (Kluwer Academie Publishers, Dordrecht, 2002)
53. S. Dempe, Annotated bibliography on bilevel programming and mathematical programs with equilibrium constraints. Optimization **52**, 333–359 (2003)
54. S. Dempe, N. Dinh, J. Dutta, Optimality conditions for a simple convex bilevel programming problem, in *Variational Analysis and Generalized Differentiation in Optimization and Control*, ed. by B.S. Mordukhovich (Springer, Berlin, 2010), pp. 149–161
55. S. Dempe, J. Dutta, Is bilevel programming a special case of a mathematical program with complementarity constraints? Math. Program. **131**, 37–48 (2012)
56. S. Dempe, J. Dutta, B. Mordukhovich, New necessary optimality conditions in optimistic bilevel programming. Optimization **56**, 577–604 (2007)
57. S. Dempe, V. Kalashnikov (eds.), *Optimization with Multivalued Mappings: Theory, Applications and Algorithms* (Springer, Berlin, 2006)
58. S. Dempe, D. Fanghänel, T. Starostina, Optimal toll charges: fuzzy optimization approach, in *Methods of Multicriteria Decision-Theory and Applications*, ed. by F. Heyde, A. Löhne, C. Tammer (Shaker Verlag, Aachen, 2009), pp. 29–45
59. S. Dempe, S. Franke, Bilevel programming: stationarity and stability. Pac. J. Optim. **9**, 183–199 (2013)
60. S. Dempe, S. Franke, Solution algorithm for an optimistic linear Stackelberg problem. Comput. Oper. Res. **41**, 277–281 (2014)
61. S. Dempe, N. Gadhi, Necessary optimality conditions for bilevel set optimization problems. J. Glob. Optim. **39**, 529–542 (2007)
62. S. Dempe, V.V. Kalashnikov, *Discrete Bilevel Programming with Linear Lower Level Problems*, Technical Report, TU Bergakademie Freiberg (2005). Preprint
63. S. Dempe, V.V. Kalashnikov, N.I. Kalashnykova, A.A. Franco, A new approach to solving bi-level programming problems with integer upper level variables. ICIC Express Lett. **3**, 1281–1286 (2009)
64. S. Dempe, V.V. Kalashnikov, G.A. Pérez-Valdés, N.I. Kalashnykova, Natural gas bilevel cash-out problem: convergence of a penalty function method. Eur. J. Oper. Res. **215**(3), 532–538 (2011)
65. S. Dempe, V.V. Kalashnikov, R.Z. Ríos-Mercado, Discrete bilevel programming: application to a natural gas cash-out problem. Eur. J. Oper. Res. **166**, 469–488 (2005)
66. S. Dempe, O. Khamisov, *Global Solution of Bilevel Programming Problems* (Technical Report, TU Bergakademie Freiberg, 2008)
67. S. Dempe, S. Lohse, Inverse linear programming, in *Recent Advances in Optimization*, ed, by A. Seeger. Proceedings of the 12th French-German-Spanish Conference on Optimization held in Avignon, 20–24 September, 2004, Lectures Notes in Economics and Mathematical Systems, vol. 563 (Springer, Berlin, 2006), pp. 19–28
68. S. Dempe, S. Lohse, *Dependence of Bilevel Programming on Irrelevant Data*, Technical Report, TU Bergakademie Freiberg (2011). Preprint Fakultät für Mathematik und Informatik. Available at http://www.optimization-online.org
69. S. Dempe, S. Lohse, Optimale Mautgebühren-Ein Modell und ein Optimalitätstest. at-Automatisierungstechnik **60**(4), 225–232 (2012)
70. S. Dempe, B. Luderer, Z.H. Xu, *Global Optimization of A Mixed-integer Bilevel Programming Problem* (Technical Report, TU Bergakademie Freiberg, 2014)

71. S. Dempe, B.S. Mordukhovich, A.B. Zemkoho, Sensitivity analysis for two-level value functions with applications to bilevel programming. SIAM J. Optim. **22**, 1309–1343 (2012)
72. S. Dempe, B.S. Mordukhovich, A.B. Zemkoho, Necessary optimality conditions in pessimistic bilevel programming. Optimization **63**(4), 505–533 (2014)
73. S. Dempe, D. Pallaschke, Quasidifferentiability of optimal solutions in parametric nonlinear optimization. Optimization **40**, 1–24 (1997)
74. S. Dempe, M. Pilecka, Necessary optimality conditions for optimistic bilevel programming problems using set-valued programming. J. Glob. Optim. **2014**, 1–20 (2014)
75. S. Dempe, K. Richter, Bilevel programming with knapsack constraints. Central Eur. J. Oper. Res. **8**, 93–107 (2000)
76. S. Dempe, H. Schmidt, On an algorithm solving two-level programming problems with nonunique lower level solutions. Comput. Optim. Appl. **6**, 227–249 (1996)
77. S. Dempe, H. Schreier, *Operations Research: Deterministische Modelle und Methoden* (Teubner Verlag, Wiesbaden, 2006)
78. S. Dempe, T. Unger, Generalized PC^1 functions. Optimization **46**, 311–326 (1999)
79. S. Dempe, S. Vogel, The generalized Jacobian of the optimal solution in parametric optimization. Optimization **50**, 387–405 (2001)
80. S. Dempe, A.B. Zemkoho, The generalized Mangasarian-Fromowitz constraint qualification and optimality conditions for bilevel programs. J. Optim. Theory Appl. **148**(1), 46–68 (2011)
81. S. Dempe, A.B. Zemkoho, Bilevel road pricing: theoretical analysis and optimality conditions. Ann. Oper. Res. **196**(1), 223–240 (2012)
82. S. Dempe, A.B. Zemkoho, On the Karush-Kuhn-Tucker reformulation of the bilevel optimization problem. Nonlinear Anal. Theory Methods Appl. **75**, 1202–1218 (2012)
83. V.F. Demyanov, Unconstrained optimization problems, in *Nonlinear Optimization* eds. by I.M. Bomze, V.F. Demyanov, R. Fletcher, T. Terlaky. Lecture Notes in Mathematics, vol. 1989 (Springer, Berlin, 2010), pp. 84–116
84. V.F. Dem'yanov, A.M. Rubinov, *Quasidifferential Calculus* (Optimization Software Inc., Publications Division, New York , 1986)
85. X. Deng, Complexity issues in bilevel linear programming, in *Multilevel Optimization: Algorithms and Applications* eds. by A. Migdalas, P.M. Pardalos, P. Värbrand (Kluwer Academic Publishers, Dordrecht, 1998), pp. 149–164
86. L. DeSilets, B. Golden, Q. Wang, R. Kumar, Predicting salinity in the Chesapeake Bay using backpropagation. Comput. Oper. Res. **19**, 277–285 (1992)
87. A. Dhara, J. Dutta, *Optimality Conditions in Convex Optimization, A Finite-dimensional View* (CRC Press, Boca Raton, 2012)
88. M. Didi-Biha, P. Marcotte, G. Savard, Path-based formulations of a bilevel toll setting problem, in *Optimization with Multivalued Mappings: Theory, Applications and Algorithms*, ed. by S. Dempe, V. Kalashnikov, Optimization and Its Applications, vol. 2 (Springer Science+Business Media, LLC, New York, 2006), pp. 29–50
89. P.H. Dien, Locally Lipschitzian set-valued maps and general extremal problems with inclusion constraints. Acta Math. Vietnam. **1**, 109–122 (1983)
90. P.H. Dien, On the regularity condition for the extremal problem under locally Lipschitz inclusion constraints. Appl. Math. Optim. **13**, 151–161 (1985)
91. N. Dinh, M.A. Goberna, M.A. López, T.Q. Son, New Farkas-type results with applications to convex infinite programming. ESAIM. Control Optim. Cal. Var. **13**, 580–597 (2007)
92. N. Dinh, T.T.A. Nghia, G. Vallet, A closedness condition and its applications to DC programs with convex constraints. Optimization **59**(4), 541–560 (2010)
93. M.J. Doane, D.F. Spulber, Open access and the evolution of the us spot market for natural gas. J. Law Econ. **34**, 447–517 (1994)
94. L.F. Domínguez, E.N. Pistikopoulos, Multiparametric programming based algorithms for oure integer and mixed-integer bilevel programming problems. Comput. Chem. Eng. **34**(12), 2097–2106 (2010)

95. R. Driskill, S. McCafferty, Dynamic duopoly with output adjustment costs in international markets: Taking the conjecture out of conjectural variations, in *Trade Policies for International Competitiveness*, ed. by R.E. Feenstra (University of Chicago Press, Chicago, 1989), pp. 124–144
96. J. Dutta, S. Chandra, Convexificators, generalized convexity and optimality conditions. J. Optim. Theory Appl. **113**, 41–65 (2002)
97. B.C. Eaves, On the basic theorem of complementarity. Math. Program. **1**, 68–75 (1971)
98. Energy Information Administration (2008), FERC order 636: the restructuring rule, Retrieved 21 June 2008 from http://www.eia.doe.gov/oil_gas/natural_gas/analysis_publications/ngmajorleg/ferc636.html
99. Energy Information Administration, Natural gas model description, Retrieved 9th August from http://www.eia.doe.gov/emeu/steo/pub/document/textng.html
100. Energy Information Administration (2005), Ferc policy on system ownership since 1992, Retrieved 21 June 2008 from http://www.eia.doe.gov/oil_gas/natural_gas/analysis_publications/ngmajorleg/fercpolicy.html
101. Environmental Protection Agency (2008), The impacts of ferc order 636 on coal mine gas project development, Retrieved 21 June 2008 from http://www.epa.gov/cmop/docs/pol004.pdf
102. EuroGas (2009), Eurogas long term outlook to 2030, Retrieved 22 July 2009 from http://www.eurogas.org/uploaded/Eurogas%20long%20term%20outlook%20to%202030%20-%20final.pdf
103. N.P. Faísca, V. Dua, P.M. Saraiva, B. Rustem, E.N. Pistikopoulos, Parametric global optimization for bilevel programming. J. Glob. Optim. **38**, 609–623 (2007)
104. D. Fanghänel, Optimality criteria for bilevel programing problems using the radial subdifferential, in Optimization with Multivalued Mappings: Theory, Applications and Algorithms eds. by S. Dempe, V. Kalashnikov. Optimization and its Applications, vol. 2 (Springer Science+Business Media, LLC, New York, 2006), pp. 73–95
105. D. Fanghänel, Zwei-Ebenen-Optimierung mit diskreter unterer Ebene und stetiger oberer Ebene, Ph.D. thesis, TU Bergakademie Freiberg (2006)
106. D. Fanghänel, S. Dempe, Bilevel programming with discrete lower level problems. Optimization **58**, 1029–1047 (2009)
107. C. Fershtman, The interdependence between ownership status and market structure: the case of privatization. Economica **57**, 319–328 (1990)
108. C. Fershtman, M. Kamien, Dynamic duopolistic competition with sticky prices. Econometrica **55**, 1151–1164 (1987)
109. P. Festa, M.G.C. Resende, in Grasp: An Annotated Bibliography, Essays and Surveys, in *Metaheuristics*, ed. by C.C. Ribeiro, P. Hansen (Kluwer Academic Publishers, New York, 2002), pp. 325–367
110. C. Figuières, A. Jean-Marie, N. Quérou, M. Tidball, *Theory of Conjectural Variations* (World Scientific, New Jersey, 2004)
111. M.L. Flegel, *Constraint Qualifications and Stationarity Concepts for Mathematical Programs with Equilibrium Constraints*, Ph.D. thesis (Universität Würzburg, 2005)
112. M.L. Flegel, C. Kanzow, On the Guignard constraint qualification for mathematical programs with equilibrium constraints. Optimization **54**(6), 517–534 (2005)
113. M.L. Flegel, C. Kanzow, A direct proof for M-stationarity under MPEC-GCQ for mathematical programs with equilibrium constraints, in *Optimization with Multivalued Mappings: Theory, Applications and Algorithms*, ed. by S. Dempe, V. Kalashnikov, Optimization and its Applications, vol. 2 (Springer Science+Business Media, LLC, New York, 2006), pp. 111–122
114. R. Fletcher, S. Leyffer, Solving mixed integer nonlinear programs by outer approximation. Math. Program. **66**, 327–349 (1994)
115. R. Fletcher, S. Leyffer, *Numerical Experience with Solving MPECs as NLPs, Technical Report NA/210* (University of Dundee, Department of Mathematics , 2002)
116. R. Fletcher, S. Leyffer, D. Ralph, S. Scholtes, Local convergence of SQP methods for mathematical programs with equilibrium constraints. SIAM J. Optim. **17**(1), 259–286 (2006)

117. C.A. Floudas, Z.H. Gümüş, M.G. Ierapetritou, Global optimization in design under uncertainty: feasibility test and flexibility index problem. Ind. Eng. Chem. Res. **40**, 4267–4282 (2001)
118. G. De Frajas, F. Delbono, Game theoretic models of mixed oligopoly. J. Econ. Surv. **4**, 1–17 (1990)
119. R. Frisch, Monopole, polypole-la notion de force en économie, Nationaløkonomisk Tidsskrift **71**, 241–259 (1933)
120. R. Frisch, Monopoly, polypoly: the concept of force in the economy. Int. Econ. Pap. **1**, 23–36 (1951)
121. M. Fukushima, J.-S. Pang, *Convergence of a smoothing continuation method for mathematical programs with complementarity constraints, Ill-Posed Variational Problems and Regularization Techniques, LNEMS*, vol. 477 (Springer, Berlin, 1999)
122. A. Galántai, Properties and constructions of NCP functions. Comput. Optim. Appl. **52**, 805–824 (2012)
123. GAMS, *A User's Guide* (Gams Development Corporation, Washington, 2013)
124. D.Y. Gao, Canonical duality theory and solutions to constrained nonconvex quadratic programming. J. Glob. Optim. **29**, 377–399 (2004)
125. D.Y. Gao, Solutions and optimality criteria to box constrained nonconvex minimization problems. J. Ind. Manag. Optim. **3**, 293–304 (2007)
126. M.R. Garey, D.S. Johnson, *Computers and Intractability: A Guide to the Theory of NP-Completeness* (W.H. Freeman and Co., San Francisco, 1979)
127. J. Gauvin, F. Dubeau, Differential properties of the marginal function in mathematical programming. Math. Program. Study **19**, 101–119 (1982)
128. C. Geiger, C. Kanzow, *Theorie und Numerik Restringierter Optimierungsaufgaben* (Springer, Berlin, 2002)
129. A. Gilat, *Matlab: An introduction with Applications* (Wiley, New York, 2004)
130. N. Giocoli, The escape from conjectural variations: the consistency condition in duopoly theory from bowley to fellner. J. Econ. (Cambridge) **29**, 601–618 (2005)
131. J.M. Gowdy, Industrial demand for natural gas: interindustry variation in New York state. Energy Econ. **5**, 171–177 (1983)
132. L. Grygarová, Qualitative Untersuchung des I. Optimierungsproblems in mehrparametrischer programmierung. Appl. Math. **15**, 276–295 (1970)
133. J. Guddat, F. Guerra Vasquez, H.Th. Jongen, *Parametric Optimization: Singularities, Pathfollowing and Jumps* (Wiley, Chichester, 1990)
134. Z.H. Gümüş, C.A. Floudas, Global optimization of mixed-integer-bilevel programming problems. Comput. Manag. Sci. **2**, 181–212 (2005)
135. R. Gutiérrez, A. Nafidi, R. Gutiérrez Sánchez, Forecasting total natural-gas consumption in spain by using the stochastic gompertz innovation diffusion mode. Appl. Energy **80**, 115–124 (2005)
136. P. Hansen, B. Jaumard, G. Savard, New branch-and-bound rules for linear bilevel programming. SIAM J. Sci. Stat. Comput. **13**, 1194–1217 (1992)
137. P.T. Harker, S.-C. Choi, A penalty function approach for mathematical programs with variational inequality constraints. Inf. Decis. Technol. **17**, 41–50 (1991)
138. P.T. Harker, J.-S. Pang, Finite-dimensional variational inequality and nonlinear complementarity problems: a survey of theory, algorithms and applications. Math. Program. **48**, 161–220 (1990)
139. R.G. Harris, E.G. Wiens, Government enterprise: an instrument for the internal regulation of industry. Can. J. Econ. **13**, 125–132 (1980)
140. D.W. Hearn, M.V. Ramana, Solving congestion toll pricing models, in *Equilibrium and Advanced Transportation Modeling*, eds. by P. Marcotte, N. Sang (Kluwer Academic Publishers, Dordrecht, 1998), pp. 125–151
141. G. Heilporn, M. Labbé, P. Marcotte, G. Savard, A polyhedral study of the network pricing problem with connected toll arcs. Networks **55**, 234–246 (2010)

142. S.R. Hejazi, A. Memariani, G. Jahanshaloo, M.M. Sepehri, Linear bilevel programming solution by genetic algorithm. Comput. Oper. Res. **29**, 1913–1925 (2002)
143. R. Henrion, T. Surowiec, On calmness conditions in convex bilevel programming. Appl. Anal. **90**(5–6), 951–970 (2011)
144. J. Hertz, A. Krogh, R.G. Palmer, *Introduction to the Theory of Neural Computation* (Addison-Wesley, Redwood City, 1991)
145. J.-P. Hiriart-Urruty, C. Lemarechal, *Convex Analysis and Minimization Algorithms*, vol. 1 (Springer, Berlin, 1993)
146. T. Hoheisel, C. Kanzow, A. Schwartz, Theoretical and numerical comparison of relaxation methods for mathematical programs with complementarity constraints. Math. Program. **137**(1–2), 257–288 (2013)
147. R. Horst, H. Tuy, *Global Optimization: Deterministic Approaches* (Springer, Berlin, 1990)
148. X.P. Hu, Y.X. Li, J W. Guo, L.J. Sun, A.Z. Zeng, A simulation optimization algorithms with heuristic transformation and its application to vehicle routing problems. Int. J. Innovative Comput Inf. Control **4**, 1169–1182 (2008)
149. H.G. Huntington, Federal price regulation and the supply of natural gas in a segmented field market. L. Econ. **54**, 337–347 (1978)
150. IHS Engineering (2007), Ec proposes new legislation for european energy policy, Retrieved 21 June 2008 from http://engineers.ihs.com/news/eu-en-energy-policy-9-07.html
151. G. Isac, V.A. Bulavsky, V.V. Kalashnikov, *Complementarity, Equilibrium, Efficiency and Economics* (Kluwer Academic Publishers, Dordrecht, 2002)
152. The Mathworks Inc. (2008), Statistics toolbox: linkage, Retrieved in November, 2008 from http://www.mathworks.com
153. Y. Ishizuka, E. Aiyoshi, Double penalty method for bilevel optimization problems. Ann. Oper. Res. **34**, 73–88 (1992)
154. R.H. Jan, M.S. Chern, Nonlinear integer bilevel programming. Eur. J. Oper. Res. **72**, 574–587 (1994)
155. V. Jeyakumar, D.T. Luc, Nonsmooth calculus, minimality, and monotonicity of convexificators. JOTA **101**, 599–621 (1999)
156. V. Jeyakumar, A.M. Rubinov, B.M. Glover, Y. Ishizuka, Inequality systems and global optimization. J. Math. Anal. Appl. **202**(3), 900–919 (1996)
157. Y. Jiang, X. Li, C. Huang, X. Wu, Application of particle swarm optimization based on CHKS smoothing function for nonlinear bilevel programming problem. Appl. Math. Comput. **219**, 4332–4339 (2013)
158. H.Th. Jongen, J.-J. Rückmann, V. Shikhman, MPCC: critical point theory, SIAM J. Optim. **20**, 473–484 (2009)
159. H.Th. Jongen, V. Shikhman, Bilevel optimization. Math. Program. **136**, 65–90 (2012)
160. H.Th. Jongen, V. Shikhman, S. Steffensen, Characterization of strong stability for C-stationary points in MPCC. Math. Program. **132**, 295–308 (2012)
161. H.Th. Jongen, G.-W. Weber, Nonlinear optimization: characterization of structural optimization, J. Glob. Optim. **1**, 47–64 (1991)
162. A. Juris, *The Emergence of Markets in the Natural Gas Industry* (World , Washington, 1998)
163. A. Kadrani, Nouvelles régularisations pour l'optimisation avec contraintes de complémentarité, Ph.D. thesis, Univerité de Sherbrooke (2008)
164. A. Kadrani, J.-P. Dussault, A. Benchakroun, A new regularization scheme for mathematical programs with complementarity constraints. SIAM J. Optim. **20**, 78–103 (2009)
165. V.V. Kalashnikov, V.A. Bulavsky, N.I. Kalashnykova, F.J. Castillo, Mixed oligopoly with consistent conjectures. Eur. J. Oper. Res. **210**, 729–735 (2011)
166. V.V. Kalashnikov, F. Camacho, R. Askin, N.I. Kalashnykova, Comparison of algorithms solving a bilevel toll setting problem. Int. J. Innov. Comput. Inf. Control **6**, 3529–3549 (2010)
167. V.V. Kalashnikov, J.F. Camacho-Vallejo, R.C.Herrera-Maldonado, N.I. Kalashnykova, A heuristic algorithm solving bilevel toll optimization problems. Int. J. Logist. Manag. to appear
168. V.V. Kalashnikov, E. Cordero, V.V. Kalashnikov-Jr., Cournot and Stackelberg equilibrium in mixed duopoly models. Optimization **59**, 689–706 (2010)

169. V.V. Kalashnikov, E. Cordero, V.V. Kalashnikov-Jr., Cournot and Stackelberg equilibrium in mixed duopoly models. Optimization **59**, 689–706 (2010)
170. V.V. Kalashnikov, S. Dempe, G.A. Pérez-Valdés, N.I. Kalashnykova, Reduction of dimension of the upper level problem, in a bilevel programming model. Part 1, *Smart Innovation, Systems and Technologies*, ed. by L.C. Jain, J. Watada, G. Phillips-Wren, R.J. Howlett (Springer, Berlin, 2011), pp. 255–264
171. V.V. Kalashnikov, S. Dempe, G.A. Pérez-Valdés, N.I. Kalashnykova, Reduction of dimension of the upper level problem in a bilevel programming model. Part 2, Intelligent Decision Technologies, *Smart Innovation, Systems and Technologies*, eds. by L.C. Jain J. Watada, G. Phillips-Wren, R.J. Howlett, vol. 10 (Springer, Berlin, 2011), pp. 265–272
172. V.V. Kalashnikov, N.I. Kalashnikova, Solving two-level variational inequality. J. Glob. Optim. **17**, 289–294 (1991)
173. V.V. Kalashnikov, N.I. Kalashnykova, R.C. Herrera Maldonado, Solving bilevel toll optimization problems by a direct algorithm using sensitivity analysis, in *Proceedings of the 2011 New Orleans International Academic Conference, New Orleans, USA, 21–23 March 2011*, ed. by R. Clute (The Clute Institute, 2011), pp. 1009–1018
174. V.V. Kalashnikov, C. Kemfert, Conjectural variations equilibrium in a mixed duopoly. Eur. J. Oper. Res. **192**, 717–729 (2009)
175. V.V. Kalashnikov, C. Kemfert, Conjectural variations equilibrium in a mixed duopoly. Eur. J. Oper. Res. **192**, 717–729 (2009)
176. V.V. Kalashnikov, T. Matis, G.A. Pérez-Valdés, Time series analysis applied to construct US natural gas price functions for groups of states. Energy Econ. **32**, 887–900 (2010)
177. V.V. Kalashnikov, G.A. Pérez-Valdés, N.I. Kalashnykova, A linearization approach to solve the natural gas cash-out bilevel problem. Ann. Oper. Res. **181**, 423–442 (2010)
178. V.V. Kalashnikov, G.A. Pérez-Valdés, T.I. Matis, N.I. Kalashnykova, US natural gas market classification using pooled regression. Math. Probl. Eng. **2014**, 1–9 (2014)
179. V.V. Kalashnikov, G.A. Pérez-Valdés, A. Tomasgard, N.I. Kalashnykova, Natural gas cash-out problem: bilevel stochastic optimization approach. Eur. J. Oper. Res. **206**, 18–33 (2010)
180. V.V. Kalashnikov, R.Z. Ríos-Mercado, A natural gas cash-out problem: a bi-level programming framework and a penalty function method. Optim. Eng. **7**, 403–420 (2006)
181. V.V. Kalashnikov, R.Z. Ríos-Mercado, A natural gas cash-out problem: a bi-level programming framework and a penalty function method. Optim. Eng. **7**, 403–420 (2006)
182. N.I. Kalashnykova, V.A. Bulavsky, V.V. Kalashnikov, F.J. Castillo-Pérez, Consistent conjectural variations equilibrium in a mixed duopoly. J. Adv. Comput. Intell. Intell. Inf. **15**, 425–432 (2011)
183. P. Kall, S.W. Wallace, *Stochastic Programming* (Wiley, New York, 1994)
184. S. Karamardian, An existence theorem for the complementarity problem. J. Optim. Theory Appl. **18**, 445–454 (1976)
185. P.W. Keat, P.K.Y. Young, *Managerial Economics: Economic Tools for Today's Decision Makers* (Prentice Hall, New Jersey, 2006)
186. N. Keyaerts, L. Meeus, W. D'haeseleer, *Analysis of balancing-system design and contracting behavior in the natural gas markets, in European Doctoral Seminar on Natural Gas Research* (Delft, The Netherlands, 2008)
187. D. Kinderlehrer, G. Stampacchia, *An Introduction to Variational Inequalities and Their Applications* (Academic Press, New York, 1980)
188. D. Klatte, B. Kummer, Stability properties of infima and optimal solutions of parametric optimization problems, in *Nondifferentiable Optimization: Motivations and Applications, Proceedings of the IIASA Workshop, Sopron, 1984*, ed. by V.F. Demyanov, Lecture Notes, in Economics and Mathmatical Systems, vol. 255, (Springer, Berlin, 1984), pp. 215–229
189. D. Klatte, B. Kummer, *Nonsmooth Equations in Optimization; Regularity, Calculus, Methods and Applications* (Kluwer Academic Publishers, Dordrecht, 2002)
190. B. Kohli, Optimality conditions for optimistic bilevel programming problem using convexifactors. J. Optim. Theory Appl. **152**(3), 632–651 (2012)

191. M. Kojima, Strongly stable stationary solutions in nonlinear programs, in *Analysis and Computation of Fixed Points*, ed. by S.M. Robinson (Academic Press, New York, 1980), pp. 93–138
192. M. Köppe, M. Queyranne, C.T. Ryan, Parametric integer programming algorithm for bilevel mixed integer programs. J. Optim. Theory Appl. **146**, 137–150 (2010)
193. N. Krichene, World crude oil and natural gas: a demand and supply model. Energy Econ. **24**, 557–576 (2002)
194. B. Kummer, *Newton's Method for Non-differentiable Functions*, vol. 45 (Advances in Mathematical Optimization, Mathematical Research (Akademie-Verlag, Berlin, 1988)
195. G. Kunapuli, K.P. Bennett, J. Hu, J.-S. Pang, *Bilevel Model Selection for Support Vector Machines*, Data mining and mathematical programming. Chapters of the book are based on lectures at the workshop, Montreal, Canada, 10–13 October 2006 (Providence, RI: American Mathematical Society (AMS), 2008), pp. 129–158
196. M. Labbé, P. Marcotte, G. Savard, A bilevel model of taxation and its applications to optimal highway pricing. Manag. Sci. **44**, 1608–1622 (1998)
197. M. Labbé, P. Marcotte, G. Savard, On a class of bilevel programs, in *Nonlinear Optimization and Related Topics*, eds. by G. Di Pillo, F. Giannessi (Kluwer Academic Publishers, Dordrecht, 2000), pp. 183–206
198. J. Laitner, "Rational" duopoly equilibria. Q. J. Econ. **95**, 641–662 (1980)
199. S. Lawphongpanich, D.W. Hearn, An mpec approach to second best toll pricing. Math. Program. **101B**, 33–55 (2004)
200. S. Leyffer, Integrating SQP and branch-and-bound for mixed integer nonlinear programming. Comput. Optim. Appl. **18**, 295–309 (2001)
201. S. Leyffer, G. López-Calva, J. Nocedal, Interior methods for mathematical programs with complementarity constraints. SIAM J. Optim. **17**, 52–77 (2006)
202. X.F. Li, J.Z. Zhang, Necessary optimality conditions in terms of convexificators in Lipschitz optimization. J. Optim. Theory Appl. **131**, 429–452 (2006)
203. W.T. Lin, Y.H. Chen, R. Chatov, The demand for natural gas, electricity and heating oil in the united states. Resour. Energy **9**, 233–258 (1987)
204. T. Lindhi, The inconsistency of consistent conjectures, coming back to Cournot. J. Econ. Behav. Optim. **18**, 69–90 (1992)
205. C. Liu, Y. Wang, A new evolutionary algorithm for multi-objective optimization problems. ICIC Express Lett. **1**, 93–98 (2007)
206. Y.F. Liu, Y.X. Ni, F.F. Wu, B. Cai, Existence and uniqueness of consistent conjectural variation equilibrium in electricity markets. Int. J. Electr. Power Energy Syst. **29**, 455–461 (2007)
207. R. Lucchetti, F. Mignanego, G. Pieri, Existence theorem of equilibrium points in Stackelberg games with constraints. Optimization **18**, 857–866 (1987)
208. Z.-Q. Luo, J.-S. Pang, D. Ralph, *Mathematical Programs with Equilibrium Constraints* (Cambridge University Press, Cambridge, 1996)
209. F.K. Lyness, Gas demand forecasting. J. R. Stat. Soc. Ser. D (The Statistician) **33**, 9–12 (1984)
210. C.M. Macal, A.P. Hurter, Dependence of bilevel mathematical programs on irrelevant constraints. Comput. Oper. Res. **24**, 1129–1140 (1997)
211. P.W. MacAvoy, *The Natural Gas Market: Sixty Years of Regulation and Deregulation* (Yale University Press, New Haven and London, 2000)
212. T.L. Magnanti, R.T. Wong, Network design and transportation planning: models and algorithms. Transp. Sci. **18**, 1–55 (1984)
213. M. Maher, X. Zhang, D. Van Vliet, A bi-level programming approach for trip matrix estimation and traffic control problems with stochastic user equilibrium link flows. Transp. Res. **35B**, 23–40 (2001)
214. O. Mangasarian, Uniqueness of solution in linear programming. Linear Algebra Appl. **25**, 151–162 (1979)
215. O.L. Mangasarian, *Mathematical Programming in Neural Networks*, University of Wisconsin, Technical Report, Computer Science Department (1992)
216. O.L. Mangasarian, Misclassification minimization. J. Glob. Optim. **5**, 309–323 (1994)

217. P. Marcotte, A new algorithm for solving variational inequalities with application to the traffic assignment problem. Math. Program. **33**(3), 339–351 (1985)
218. P. Marcotte, Network design problem with congestion effects: a case of bilevel programming. Math. Program. **34**, 142–162 (1986)
219. P. Marcotte, J.-P. Dussault, A sequential linear programming algorithm for solving monotone variational inequalities. SIAM J. Control Optim. **27**(6), 1260–1278 (1989)
220. P. Marcotte, G. Savard, F. Semet, A bilevel programming approach to the traveling salesman problem. Oper. Res. Lett. **32**, 240–248 (2004)
221. H.M. Markowitz, *Portfolio Selection: Efficient Diversification of Investments, Cowles Foundation for Research in Economics at Yale University, Monograph*, vol. 16 (Wiley, New York, 1959)
222. T. Matshumura, O. Kanda, Mixed oligopoly at free entry markets. J. Econ. **84**, 27–48 (2005)
223. T. Matsumura, Stackelberg mixed duopoly with a foreign competitor. Bull. Econ. Res. **55**, 275–287 (2003)
224. N. Matsushima, T. Matsumura, Mixed oligopoly with homogeneous goods. Can. J. Econ. **64**, 367–393 (1993)
225. W. Merril, N. Schneider, Government firms in oligopoly industries: a short-run analysis. Q. J. Econ. **80**, 400–412 (1966)
226. A.G. Mersha, *Solution Methods for Bilevel Programming Problems*, Ph.D. thesis, TU Bergakademie Freiberg, Germany, 2008 (Dr. Hut, Munich, 2009)
227. A.G. Mersha, S. Dempe, Linear bilevel programming with upper level constraints depending on the lower level solution. Appl. Math. Comput. **180**, 247–254 (2006)
228. A.G. Mersha, S. Dempe, Feasible direction method for bilevel programming problem. Optimization **61**(4–6), 597–616 (2012)
229. C.A. Meyer, C.A. Floudas, Convex underestimation of twice continuously differentiable functions by piecewise quadratic perturbation: spline αBB underestimators. J. Glob. Optim. **32**, 221–258 (2005)
230. K.T. Midthun, *Optimization Models for Liberalized Natural Gas Markets*, Ph.D. thesis, Norwegian University of Science and Technology, Faculty of Social Science and Technology Management, Department of Industrial Economics and Technology Management (Trondheim, Norway, 2009)
231. A. Migdalas, P.M. Pardalos, P. Värbrand (eds.), *Multilevel Optimization: Algorithms and Applications* (Kluwer Academic Publishers, Dordrecht, 1998)
232. J.A. Mirrlees, The theory of moral hazard and unobservable behaviour: part I. Rev. Econ. Stud. **66**, 3–21 (1999)
233. A. Mitsos, Global solution of nonlinear mixed-integer bilevel programs. J. Glob. Optim. **47**, 557–582 (2010)
234. A. Mitsos, P. Lemonidis, P.I. Barton, Global solution of bilevel programs with a nonconvex inner program. J. Glob. Optim. **42**(4), 475–513 (2008)
235. J. Moore, J.F. Bard, The mixed integer linear bilevel programming problem. Oper. Res. **38**, 911–921 (1990)
236. J.T. Moore, J.F. Bard, The mixed integer linear bilevel programming problem. Oper. Res. **38**, 911–921 (1990)
237. B.S. Mordukhovich, Nonsmooth analysis with nonconvex generalized differentials and conjugate mappings. Dokl. Akad. Nauk BSSR **28**, 976–979 (1984)
238. B.S. Mordukhovich, Generalized differential calculus for nonsmooth and set-valued mappings. J. Math. Anal. Appl. **183**, 250–288 (1994)
239. B.S. Mordukhovich, Lipschitzian stability of constraint systems and generalized equations. Nonlinear Anal. **22**(2), 173–206 (1994)
240. B.S. Mordukhovich, Stability theory for parametric generalized equations and variational inequalities via nonsmooth analysis. Trans. Am. Math. Soc. **343**(2), 609–657 (1994)
241. B.S. Mordukhovich, *Variational Analysis and Generalized Differentiation, Basic Theory*, vol. 1 (Springer, Berlin, 2006)

242. B.S. Mordukhovich, *Variational Analysis and Generalized Differentiation, Applications*, vol. 2 (Springer, Berlin, 2006)
243. B.S. Mordukhovich, N.M. Nam, N.D. Yen, Subgradients of marginal functions in parametric mathematical programming. Math. Program. **116**, 369–396 (2009)
244. B.S. Mordukhovich, N.M. Nam, Variational stability and marginal functions via generalized differentiation. Math. Oper. Res. **30**(4), 800–816 (2005)
245. B.S. Mordukhovich, J.V. Outrata, Coderivative analysis of quasi-variational inequalities with applications to stability and optimization. SIAM J. Optim. **18**, 389–412 (2007)
246. S.A. Morrison, A survey of road pricing. Trans. Res. **20**, 87–97 (1986)
247. Ch. Müller, A. Rehkopf, Optimale Betriebsführung eines virtuellen Kraftwerks auf Basis von gasbetriebenen Mikro-Blockheizkraftwerken. at-Automatisierungstechnik **59**(3), 180–187 (2011)
248. J. Nash, Two-person cooperative games, in *The Essential John Nash*, eds. by H.W. Kuhn, S. Nasar (Prinston University Press, Prinston, 2002), pp. 99–114
249. C. Nelder, Natural gas price forecast, the future of natural gas: It's time to invest, Retrived 22nd July 2009 from http://www.energyandcapital.com/articles/natural-gas-price-forecast/916 (2009)
250. J.A. Nelder, R. Mead, A simplex method for function minimization. Comput. J. **7**, 308–313 (1965)
251. G.L. Nemhauser, L.A. Wolsey, *Integer and Combinatorial Optimization* (Wiley, New York, 1988)
252. Y. Nesterov, A. Nemirovskii, *Interior-point Polynomial Algorithms in Convex Programming* (SIAM, Philadelphia, 1994)
253. L. Nett, Mixed oligopoly with homogeneous goods. Ann. Publ. Coop. Econ. **64**, 367–393 (1993)
254. R. Nishimura, S. Hayashi, M. Fukushima, Robust Nash equilibria in n-person non-cooperative games: uniqueness and reformulation. Pac. J. Optim. **5**, 237–259 (2005)
255. I. Nishizaki, M. Sakawa, T. Kan, Computational methods through genetic algorithms for obtaining Stackelberg solutions to two-level integer programming problems, part 3. Electron. Commun. (Japan) **86**, 1251–1257 (2003)
256. W. Novshek, On the existence of Cournot equilibrium. Rev. Econ. Stud. **52**, 85–98 (1985)
257. F. Nožička, J. Guddat, H. Hollatz, B. Bank, *Theorie der Linearen Parametrischen Optimierung* (Akademie-Verlag, Berlin, 1974)
258. W. Oeder, Ein Verfahren zur Lösung von Zwei-Ebenen-Optimierungsaufgaben in Verbindung mit der Untersuchung von chemischen Gleichgewichten, Ph.D. thesis, Technische Universität Karl-Marx-Stadt (1988)
259. J. Outrata, M. Kočvara, J. Zowe, *Nonsmooth Approach to Optimization Problems with Equilibrium Constraints* (Kluwer Academic Publishers, Dordrecht, 1998)
260. J.V. Outrata, Necessary optimality conditions for Stackelberg problems. J. Optim. Theory. Appl. **76**, 305–320 (1993)
261. J.V. Outrata, On optimization problems with variational inequality constraints. SIAM J. Optim. **4**, 340–357 (1994)
262. J.-S. Pang, M. Fukushima, Complementarity constraint qualifications and simplified B-stationarity conditions for mathematical programs with equilibrium constraints. Comput. Optim. Appl. **13**, 111–136 (1999)
263. M.K. Perry, Oligopoly and consistent conjectural variations. Bell J. Econ. **13**, 197–205 (1982)
264. M. Pilecka, *Combined Reformulation of Bilevel Programming Problems*, Master's thesis, TU Bergakademie Freiberg, Fakultät für Mathematrik und Informatik (2011)
265. D. Ralph, S. Dempe, Directional derivatives of the solution of a parametric nonlinear program. Math. Program. **70**, 159–172 (1995)
266. P. Recht, *Generalized Derivatives: An Approach to a New Gradient in Nonsmooth Optimization*, vol. 136 (Meisenheim, Mathematical Systems in Economics (Anton Hain, 1993)
267. G.E. Renpu, A filled function method for finding a global minimizer of a function of several variables. Math. Program. **46**, 191–204 (1990)

268. S.M. Robinson, Regularity and stability for convex multivalued functions. Math. Oper. Res. **1**, 130–143 (1976)
269. S.M. Robinson, Some continuity properties of polyhedral multifunctions. Math. Program. Study **14**, 206–214 (1981)
270. S.M. Robinson, Generalized equations and their solutions, part II: applications to nonlinear programming. Math. Program. Study **19**, 200–221 (1982)
271. S. Roch, G. Savard, P. Marcotte, Design and analysis of an algorithm for Stackelberg network pricing. Networks **46**, 57–67 (2005)
272. R.T. Rockafellar, *Convex Analysis* (Princeton University Press, Princeton, 1970)
273. R.T. Rockafellar, Maximal monotone relations and the second derivatives of nonsmooth functions. Ann. Inst. H. Poincare Anal. Non Lineaire **2**, 167–184 (1985)
274. R.T. Rockafellar, R.J.-B. Wets, *Variational Analysis* (Springer, Berlin, 1998)
275. J. Rosen, Existence and uniqueness of equilibrium points for concave n-person games. Econometrica **33**, 520–534 (1965)
276. R.J. Ruffin, Cournot oligopoly and competitive behavior. Rev. Econ. Stud. **38**, 493–502 (1971)
277. G.K. Saharidis, M.G. Ierapetritou, Resolution method for mixed integer bi-level linear problems based on decomposition techinque. J. Glob. Optim. **44**, 29–51 (2009)
278. K.H. Sahin, A.R. Ciric, A dual temperature simulated annealing approach for solving bilevel programming problems. Comput. Chem. Eng. **23**, 11–25 (1998)
279. NOAA Satellite and Information Service, Retrieved 11th November 2010 from ftp://ftp.ncdc.noaa.gov/pub/data/cirs/ (2010)
280. H. Scheel, S. Scholtes, Mathematical programs with equilibrium constraints: stationarity, optimality, and sensitivity. Math. Oper. Res. **25**, 1–22 (2000)
281. H. Schmidt, *Zwei-Ebenen-Optimierungsaufgaben mit mehrelementiger Lösung der unteren Ebene*, Ph.D. thesis, Fakultät für Mathematik, Technische Universität Chemnitz-Zwickau (1995)
282. S. Scholtes, Convergence properties of a regularization scheme for mathematical programs with complementarity constraints. SIAM J. Optim. **11**, 918–936 (2001)
283. S. Scholtes, *Introduction to Piecewise Differentiable Equations, SpringerBriefs in Optimization* (Springer, New York, 2012)
284. S. Scholtes, M. Stöhr, How stringent is the linear independence assumption for mathematical programs with stationarity constraints? Math. Oper. Res. **26**, 851–863 (2001)
285. A. Schrijver, *Theory of Linear and Integer Programming* (Wiley, Chichester, 1998)
286. J.W. Shavlik, R.J. Mooney, G.G. Towell, Symbolic and neural network learning algorithms: an experimental comparison. Mach. Learn. **6**, 111–143 (1991)
287. J. Shawe-Taylor, N. Christianini, *Kernel Methods for Pattern Analysis* (Cambridge University Press, Cambridge, 2004)
288. K. Shimizu, Y. Ishizuka, J.F. Bard, *Nondifferentiable and Two-level Mathematical Programming* (Kluwer Academic Publishers, Dordrecht, 1997)
289. P.K. Simpson, *Artifical Neural Systems* (Pergamon Press, New York, 1990)
290. M.J. Smith, A descent algorithm for solving monotone variational inequalities and monotone complementarity problems. J. Optim. Theory Appl. **44**, 485–496 (1984)
291. W.R. Smith, R.W. Missen, *Chemical Reaction Equilibrium Analysis: Theory and Algorithms* (Wiley, New York, 1982)
292. M.V. Solodov, An explicit descent method for bilevel convex optimization. J. Convex Anal. **14**(2), 227–237 (2007)
293. A. Soto, *FERC Order 636 & 637*, Retrieved 21 June 2008 from http://www.aga.org/Legislative/issuesummaries/FERCOrder636637.html (2008)
294. H.v. Stackelberg, *Marktform und Gleichgewicht* (Springer, Berlin, 1934), English Translated: The Theory of the Market Economy (Oxford University Press, Oxford, 1952)
295. R. Sugden, Consistent conjectures and voluntary contributions to public goods: why the conventional theory does not work. J. Publ. Econ. **27**, 117–124 (1985)
296. T. Surowiec, *Explicit Stationarity Conditions and Solution Characterization for Equilibrium Problems with Equilibrium Constraints*, Ph.D. thesis, Humboldt-Universität zu Berlin (2010)

297. E.-G. Talbi (ed.), *in Metaheuristics for Bi-level Optimization, Studies in Computational Intelligence*, vol. 482 (Springer, Heidelberg, 2013)
298. K.T. Talluri, G.J. Van Ryzin, *The Theory and Practice of Revenue Management* (Springer, New York, 2004)
299. I. Ekeland, On the variational principle. J. Math. Anal. Appl. **47**, 324–353 (1974)
300. A. Tomasgard, F. Rømo, M. Forstad, K. Midthun, Optimization models for the natural gas value chain, in *Geometric Modeling, Numerical Simulation, and Optimization: Applied Mathematics at SINTEF*, ed. by E. Quak, G. Hasle, K.-A. Lie (Springer, Berlin, 2007), pp. 521–558
301. D.M. Topkis, A.F. Veinott, On the convergence of some feasible direction algorithms for nonlinear programming. SIAM J. Control **5**, 268–279 (1967)
302. V. Vapnik, *The Nature of Statistical Learning Theory* (Springer, New York, 2000)
303. E. Verhoef, *Economic Efficiency and Social Feasibility in the Regulation of Road Transport Externalities*, Ph.D. thesis, University of Amsterdam (1996)
304. L. Vicente, G. Savard, J. Judice, Discrete linear bilevel programming problem. J. Optim. Theory Appl. **89**, 597–614 (1996)
305. L.N. Vicente, Bilevel programming: introduction, history and overview, in *Encyclopedia of Optimization*, eds. by C.A. Floudas, P.M. Pardalos (Kluwer Academic Publishers, Dordrecht, 2001)
306. L.N. Vicente, P.H. Calamai, Bilevel and multilevel programming: a bibliography review. J. Glob. Optim. **5**(3), 291–306 (1994)
307. L.N. Vicente, P.H. Calamai, Geometry and local optimality conditions for bilevel programs with quadratic strictly convex lower levels, in Minimax and Applications, eds. by D. Du, P.M. Pardalos (Kluwer Academic Publishers, Dordrecht, 1995)
308. J. Vickers, G. Yarrow, *Privatisation-An Economic Analysis* (MIT Press, Cambridge, 1988)
309. P.A. Viton, Private roads. J. Urban Econ. **37**, 260–289 (1995)
310. S. Wagner, *Mathematisches Modell zur Transportverlagerung auf den kombinierten Verkehr*, Master's thesis, TU Bergakademie Freiberg, Faculty of Business Administration (2009)
311. S.W. Wallace, Decision making under uncertainty: is sensitivity analysis of any use? Oper. Res. **48**(1), 20–25 (2000)
312. Z. Wan, L. Yuan, J. Chen, A filled function method for nonlinear systems of equalities and inequalities. Comput. Appl. Math. **31**(2), 391–405 (2012)
313. U.-P. Wen, S.-T. Hsu, Linear bi-level programming problems-a review. J. Oper. Res. Soc. **42**(2), 125–133 (1991)
314. U.P. Wen, Y.H. Yang, Algorithms for solving the mixed integer two level linear programming problem. Comput. Oper. Res. **17**, 133–142 (1990)
315. R.E. Wendel, A preview of a tolerance approach to sensitivity analysis in linear programming. Discret. Math. **38**, 121–124 (1982)
316. W. Wiesemann, A. Tsoukalas, P.-M. Kleniati, B. Rustem, Pessimistic bi-level optimisation. SIAM J. Optim. **23**, 353–380 (2013)
317. R. Winter, *Zwei-Ebenen-Optimierung mit einem stetigen Knapsack-Problem in der unteren Ebene: Optimistischer und pessimistischer Zugang* (Technical Report, TU Bergakademie Freiberg, 2010)
318. Z.Y. Wu, F.C. Bai, M. Mammadov, Y.J. Yang, A new heuristic auxiliary function method for systems of nonlinear equations. Optim. Methods Softw. (2012), accepted for publication
319. Z.Y. Wu, M. Mammadov, F.S. Bai, Y.J. Yang, A filled function method for nonlinear equations. Appl. Math. Comput. **189**, 1196–1204 (2007)
320. P. Xu, L. Wang, An exact algorithm for the bilevel mixed integer linear programming problem under three simplifying assumptions. Comput. Oper. Res. **41**, 309–318 (2014)
321. H. Yan, W.H.K. Lam, Optimal road tolls under conditions of queuing and congestion. Trans. Res. **30**, 319–332 (1996)
322. H. Yang, M.G.H. Bell, Models and algorithms for road network design: a review and some new developments. Trans. Rev. **18**, 257–278 (1998)
323. J.J. Ye, Nondifferentiable multiplier rules for optimization and bilevel optimization problems. SIAM J. Optim. **15**, 252–274 (2004)

324. J.J. Ye, D. Zhu, New necessary optimality conditions for bilevel programs by combining the MPEC and value function approaches. SIAM J. Optim. **20**(4), 1885–1905 (2010)
325. J.J. Ye, D.L. Zhu, Optimality conditions for bilevel programming problems. Optimization **33**, 9–27 (1995)
326. S.H. Yoo, H.J. Lim, S.J. Kwak, Estimating the residential demand function for natural gas in seoul with correction for sample selection bias. Appl. Energy **86**, 460–465 (2009)
327. J.P. Zagal, J. Rick, I. Hsi, Collaborative games: lessons learned from board games. Simul. Gaming **37**, 24–40 (2006)
328. A.B. Zemkoho, *Bilevel Programming: Reformulations, Regularity, and Stationarity*, Ph.D. thesis, TU Bergakademie Freiberg (2012)
329. R. Zhang, C. Wu, A decomposition-based optimization algorithm for scheduling largescalejob shops. Int. J. Innov. Comput. Inf. Control **5**, 2769–2780 (2009)

Index

S. Dempe et al., *Bilevel Programming Problems*, Energy Systems,
DOI 10.1007/978-3-662-45827-3

The manufacturer's authorised representative in the EU is Springer Nature Customer Service Centre GmbH, Europaplatz 3, 69115 Heidelberg, Germany. If you have any concerns regarding our products, please contact ProductSafety@springernature.com

Printed and bound by CPI Group (UK) Ltd, Croydon, CR0 4YY
22/09/2026
02221061-0016